T0214569

IFIP Advances in Information and Communication Technology 521

Editor-in-Chief

Kai Rannenberg, Goethe University Frankfurt, Germany

Editorial Board

TC 1 – Foundations of Computer Science
Jacques Sakarovitch, Télécom ParisTech, France

TC 2 – Software: Theory and Practice
Michael Goedicke, University of Duisburg-Essen, Germany

TC 3 – Education
Arthur Tatnall, Victoria University, Melbourne, Australia

TC 5 – Information Technology Applications
Erich J. Neuhold, University of Vienna, Austria

TC 6 – Communication Systems
Aiko Pras, University of Twente, Enschede, The Netherlands

TC 7 – System Modeling and Optimization
Fredi Tröltzsch, TU Berlin, Germany

TC 8 – Information Systems
Jan Pries-Heje, Roskilde University, Denmark

TC 9 – ICT and Society
Diane Whitehouse, The Castlegate Consultancy, Malton, UK

TC 10 – Computer Systems Technology
Ricardo Reis, Federal University of Rio Grande do Sul, Porto Alegre, Brazil

TC 11 – Security and Privacy Protection in Information Processing Systems
Steven Furnell, Plymouth University, UK

TC 12 – Artificial Intelligence
Ulrich Furbach, University of Koblenz-Landau, Germany

TC 13 – Human-Computer Interaction
Marco Winckler, University Paul Sabatier, Toulouse, France

TC 14 – Entertainment Computing
Matthias Rauterberg, Eindhoven University of Technology, The Netherlands

IFIP – The International Federation for Information Processing

IFIP was founded in 1960 under the auspices of UNESCO, following the first World Computer Congress held in Paris the previous year. A federation for societies working in information processing, IFIP's aim is two-fold: to support information processing in the countries of its members and to encourage technology transfer to developing nations. As its mission statement clearly states:

> *IFIP is the global non-profit federation of societies of ICT professionals that aims at achieving a worldwide professional and socially responsible development and application of information and communication technologies.*

IFIP is a non-profit-making organization, run almost solely by 2500 volunteers. It operates through a number of technical committees and working groups, which organize events and publications. IFIP's events range from large international open conferences to working conferences and local seminars.

The flagship event is the IFIP World Computer Congress, at which both invited and contributed papers are presented. Contributed papers are rigorously refereed and the rejection rate is high.

As with the Congress, participation in the open conferences is open to all and papers may be invited or submitted. Again, submitted papers are stringently refereed.

The working conferences are structured differently. They are usually run by a working group and attendance is generally smaller and occasionally by invitation only. Their purpose is to create an atmosphere conducive to innovation and development. Refereeing is also rigorous and papers are subjected to extensive group discussion.

Publications arising from IFIP events vary. The papers presented at the IFIP World Computer Congress and at open conferences are published as conference proceedings, while the results of the working conferences are often published as collections of selected and edited papers.

IFIP distinguishes three types of institutional membership: Country Representative Members, Members at Large, and Associate Members. The type of organization that can apply for membership is a wide variety and includes national or international societies of individual computer scientists/ICT professionals, associations or federations of such societies, government institutions/government related organizations, national or international research institutes or consortia, universities, academies of sciences, companies, national or international associations or federations of companies.

More information about this series at http://www.springer.com/series/6102

Luis M. Camarinha-Matos · Kankam O. Adu-Kankam
Mohammad Julashokri (Eds.)

Technological Innovation for Resilient Systems

9th IFIP WG 5.5/SOCOLNET
Advanced Doctoral Conference on Computing,
Electrical and Industrial Systems, DoCEIS 2018
Costa de Caparica, Portugal, May 2–4, 2018
Proceedings

 Springer

Editors
Luis M. Camarinha-Matos ⓘ
NOVA University of Lisbon
Monte da Caparica
Portugal

Mohammad Julashokri ⓘ
NOVA University of Lisbon
Monte da Caparica
Portugal

Kankam O. Adu-Kankam ⓘ
NOVA University of Lisbon
Monte da Caparica
Portugal

ISSN 1868-4238   ISSN 1868-422X (electronic)
IFIP Advances in Information and Communication Technology
ISBN 978-3-030-08735-7   ISBN 978-3-319-78574-5 (eBook)
https://doi.org/10.1007/978-3-319-78574-5

© IFIP International Federation for Information Processing 2018
Softcover re-print of the Hardcover 1st edition 2018
This work is subject to copyright. All rights are reserved by the Publisher, whether the whole or part of the
material is concerned, specifically the rights of translation, reprinting, reuse of illustrations, recitation,
broadcasting, reproduction on microfilms or in any other physical way, and transmission or information
storage and retrieval, electronic adaptation, computer software, or by similar or dissimilar methodology now
known or hereafter developed.
The use of general descriptive names, registered names, trademarks, service marks, etc. in this publication
does not imply, even in the absence of a specific statement, that such names are exempt from the relevant
protective laws and regulations and therefore free for general use.
The publisher, the authors and the editors are safe to assume that the advice and information in this book are
believed to be true and accurate at the date of publication. Neither the publisher nor the authors or the editors
give a warranty, express or implied, with respect to the material contained herein or for any errors or
omissions that may have been made. The publisher remains neutral with regard to jurisdictional claims in
published maps and institutional affiliations.

Printed on acid-free paper

This Springer imprint is published by the registered company Springer International Publishing AG
part of Springer Nature
The registered company address is: Gewerbestrasse 11, 6330 Cham, Switzerland

Preface

This proceeding book, which collects selected results produced in engineering doctoral programs, focuses on research and development on resilient systems. There is a growing need for development and integration of resilience into technological systems owing to the increased rate of disruptive events around the globe. Such events are due to a variety of causes, e.g., globalization, natural hazards and climate change, economic crisis, demographic shifts, fast technological evolution, cyber-attacks, rise of nationalisms, to name a few, and they challenge the way systems are designed, implemented, and managed. Subsequently, their adverse impact on lives and property are immeasurable. It is also evident that these occurrences are becoming widespread and highly pervasive. Furthermore, rapid advances in smart sensors, actuators, embedded intelligence, and their seamless integration into multiple systems architecture and platforms have revolutionized the technological world, and even the way we live. However, the pervasive nature and growing complexity of these technologies have also subjected our world to high levels of vulnerabilities, which pose a big threat to human existence. It is therefore a matter of necessity and undeniable urgency for the research community to take proactive and pragmatic actions by exploring new ways and measures to mitigate these emerging global problems.

The DoCEIS series of advanced Doctoral Conferences on Computing, Electrical and Industrial Systems aims at creating a space for sharing and discussing ideas and results from doctoral research in these inter-related areas of engineering, while promoting a strong multidisciplinary dialog. As such, participants were challenged to look beyond their specific research question and relate their work to the selected theme of the conference, namely, to identify in which ways their research topics can benefit from or contribute to resilient systems. Current trends in strategic research programs are confirming the fundamental role of multidisciplinary and interdisciplinary approaches in innovation. More and more funding agencies are including this element as a key requirement in their calls for proposals. In this way, the "exercise" requested by DoCEIS can be seen as a contribution to the process of acquiring such skills, which are mandatory in the profession of a PhD.

The ninth edition of DoCEIS, which was sponsored by SOCOLNET, IFIP, and IEEE IES, attracted a good number of paper submissions from a large number of PhD students and their supervisors from 21 countries. This book comprises the works selected by the international Program Committee for inclusion in the main program and covers a wide spectrum of application domains. As such, research results and on-going work are presented, illustrated, and discussed in areas such as:

- Collaborative systems
- Decision support systems
- Supervision systems
- Energy management
- Smart grids

- Sensing systems
- Electrical systems
- Simulation and analysis
- Monitoring systems
- Energy distribution systems

We expect that this book will provide readers with an inspiring set of promising ideas and new challenges, presented in a multidisciplinary context, and that by their diversity these results can trigger and motivate richer research and development directions.

We would like to thank all the authors for their contributions. We also appreciate the efforts and dedication of the DoCEIS international Program Committee members, who both helped with the selection of articles and contributed with valuable comments to improve their quality.

February 2018
Luis M. Camarinha-Matos
Kankam O. Adu-Kankam
Mohammad Julashokri

Organization

**9th IFIP/SOCOLNET Advanced Doctoral
Conference on** COMPUTING, ELECTRICAL
AND INDUSTRIAL SYSTEMS
Costa de Caparica, Portugal, May 2–4, 2018

Conference and Program Chair

Luis M. Camarinha-Matos   NOVA University of Lisbon, Portugal

Organizing Committee Co-chairs

Luis Gomes   NOVA University of Lisbon, Portugal
João Goes   NOVA University of Lisbon, Portugal
João Martins   NOVA University of Lisbon, Portugal

International Program Committee

Vanja Ambrozic, Slovenia
Amir Assadi, USA
Ezio Bartocci, Austria
Olga Battaia, France
Marko Beko, Portugal
Luis Bernardo, Portugal
Nik Bessis, UK
Andrea Bottino, Italy
Erik Bruun, Denmark
Barbora Buhnova, Czech Republic
Giuseppe Buja, Italy
Luis M. Camarinha-Matos, Portugal
Laura Carnevali, Italy
Wojciech Cellary, Poland
Noelia Correia, Portugal
Luis M. Correia, Portugal
Jose de la Rosa, Spain
Stefano Di Carlo, Italy
Dirk Dirk Lehmhus, Germany
Ruggero Donida Labati, Italy
Florin G. Filip, Romania

Maria Helena Fino, Portugal
José M. Fonseca, Portugal
Diego Gachet, Spain
Adriana Giret, Spain
João Goes, Portugal
Luis Gomes, Portugal
Antoni Grau, Spain
Paul Grefen, The Netherlands
Michael Huebner, Germany
Oleksandr Husev, Estonia
José Igreja, Portugal
Ricardo Jardim-Gonçalves, Portugal
Vladimir Katic, Serbia
Asal Kiazadeh, Portugal
Hans-Jörg Kreowski, Germany
Zbigniew Leonowicz, Poland
Marin Lujak, France
João Martins, Portugal
Rui Melicio,Portugal
Paulo Miyagi, Brazil
Renato Moraes, Brazil

Filipe Moutinho, Portugal
Horacio Neto, Portugal
Rodolfo Oliveira, Portugal
Luis Oliveira, Portugal
Eugenio Oliveira, Portugal
Angel Ortiz, Spain
Gordana Ostojic, The Netherlands
Peter Palensky, Austria
Luis Palma, Portugal
Nuno Paulino, Portugal
Pedro Pereira, Portugal
Duc Pham, UK
João Pimentão, Portugal
Paulo Pinto, Portugal
Armando Pires, Portugal
Ricardo J. Rabelo, Brazil

Rita Ribeiro, Portugal
Juan Rodriguez-Andina, Spain
Enrique Romero-Cadaval, Spain
Carlos Roncero, Spain
Thilo Sauter, Austria
Eduard Shevtshenko, Estonia
Pierluigi Siano, Italy
Thomas Strasser, Austria
Damien Trentesaux, France
Antonios Tsourdos, UK
Manuela Vieira, Portugal
Ramon Vilanova, Spain
Soufi Youcef, France
Ahmed F. Zobaa, UK
Tamus Zoltán Ádám, Hungary

Organizing Committee (PhD Students)

Kankam O. Adu-Kankam, Ghana
Andreia Artifice, Portugal
Koorosh Aslansefat, Iran
Adriana Jesus, Portugal
Mohammad Julashokri, Iran
Paulo Lourenço, Portugal
Ricardo Madeira, Portugal

Impact and Dissemination Task Force

Filipe Moutinho, Portugal
Rudolfo Oliveira, Portugal
Luis Palma, Portugal

Technical Sponsors

      Society of Collaborative Networks

      IFIP WG 5.5 COVE
Co-Operation infrastructure for Virtual Enterprises and electronic
business

IEEE−Industrial Electronics Society

Organizational Sponsors

Organized by: PhD Program on Electrical and Computer Engineering, FCT - NOVA University of Lisbon

Contents

Simulation and Analysis

Monitoring Systems

Energy Distribution Systems

Collaborative Systems

Learning Through Mass Collaboration - Issues and Challenges

Majid Zamiri$^{(\boxtimes)}$ (ID) and Luis M. Camarinha-Matos$^{(\boxtimes)}$ (ID)

Faculty of Sciences and Technology, UNINOVA - CTS,
NOVA University of Lisbon, 2829-518 Caparica, Portugal
zamiri_majid@yahoo.com, cam@uninova.pt

Abstract. A growing number of successful mass collaboration projects in various fields show profound changes in the way communities operate and act collectively. One emerging application of mass collaboration is for collective learning, in which a mass of minds jointly drives the effort of building and acquiring knowledge. Such attempt builds on a reservoir of raw knowledge that develops as each contributor shares his/her own partial experience and knowledge. A key element in this process is to ensure that such created knowledge is reliable and trustworthy. This leads to the need of effective assessment mechanisms. Furthermore, the process of learning through mass collaboration needs to be better understood. For this purpose, this work includes a summary of a systematic review of recent literature with the aim of identifying affecting factors and constituents of mass collaborative learning namely, the type of organizational structures, collaborative learning approaches, adopted technologies, and adopted methods for evaluating the quality of performance and knowledge. Based on the findings, a research strategy focused on the quality of collective learning is then proposed.

Keywords: Mass collaboration · Learning · Collaborative networks

1 Introduction

Advances in communication technology and internet created possibilities for people across the globe to increasingly join into mass collective projects and share their contributions to create value. Emerging of mass collaboration and its application to different domains enabled multitudes of humans to build powerful hubs of resources, skills and knowledge, helping to find solutions for a wide variety of problems. Indeed, it opens a wide range of opportunities to truly harness the power of groups, leveraging resources and driving profound societal changes. In comparison with other forms of collective action, it is amazing that even when each contributor pursues his/her own interests in a mass collaborative project, still a coherent product may come out of it.

There are many applications of mass collaboration. In education for example, it emerges in the light of collaborative learning where a large number of uncoordinated contributors give themselves the chance to learn, adapt, and achieve impact together. It refers to a mechanism in which learners at various levels of performance not only proactively acquire and share a wide variety of materials but also autonomously

© IFIP International Federation for Information Processing 2018
Published by Springer International Publishing AG 2018. All Rights Reserved
L. M. Camarinha-Matos et al. (Eds.): DoCEIS 2018, IFIP AICT 521, pp. 3–17, 2018.
https://doi.org/10.1007/978-3-319-78574-5_1

contribute to knowledge creation and consolidation. It is, indeed, a great shift from a formalized and centralized to an informal and self-directed form of learning [1]. Mass collaborative learning accommodates recent changes in technology and methods of learning leading to a new paradigm of education. Opposite to formal learning delivered by instructors in a systematic intentional way within an educational setting, in this case knowledge is created, revised, and shared in large scale within informal collaborative communities.

Knowledge and information can appear in a different variety of kinds (e.g., stories, interpretations, opinions, and facts) and created for various purposes (e.g., to sell, to inform, to present a viewpoint, to encourage). For each one of these diverse kinds and purposes, knowledge and information can enormously vary and differ in terms of value, reliability, nature, granularity and lifespan. It can range from high to poor quality and include every shade in between unlike traditional printed materials in newspapers, magazines, books and academic libraries which are somehow regulated for quality and accuracy [2, 3]. Therefore, along with informal learning in mass collaborative projects, there is a strong need for being able to discern the quality of knowledge or information created in whatever format on the internet and social media. Forasmuch as the degree of learners' skills and proficiency in creating and sharing right information and knowledge is varied so the accuracy and reliability of created and shared content then becomes a main concern. Even though large amounts of high quality information and knowledge are available on the internet and social media, there is also countless incorrect knowledge (mistakenly created and spread by honest people), false knowledge (deliberately generated and diffused by dishonest people), half-truths, fallacies, distortions, exaggerations, urban legends and plain old lies [4]. Hence, it is for learners indispensable that instead of easily "accepting" or "rejecting" each knowledge claim, they can adopt a skeptical attitude towards all received piece of knowledge or information, and as a critical and savvy user put it in a "this is claimed" pile, and neither accept or reject it upon receiving. Moreover, they should not be deceived by the appearance of knowledge or information that looks just as professional or reported as highly credible and reliable sources.

The rapid development of knowledge and information repositories, proliferation of web-based knowledge applications and services, and easy access to diverse sources by knowledge users and learners have augmented the awareness of, and the need for high quality knowledge sharing in communities. In this context, quality is, indeed, a buzzword which evaluation is quite complicated. Many concerns have been raised recently about it, particularly in created and shared knowledge on the internet and social networks, and the possibility of detrimental effects that emanate from unreliable knowledge or information. That is the case, for instance, of the impact of fake news on politics. Learners require being certain that acquired knowledge is up to date, reliable, accurate, relevant, objective, and the degree of its quality is high. Nowadays, quality is considered a crucial issued for education in general and for mass collaborative learning. Although it remains an open challenge, over the last years it has been increasingly the focus of attention for many researchers to meet the needs of communities trying to evaluate and promote the quality of knowledge [5–7]. The literature also shows that there has been a great deal of effort to identify and/or introduce mechanisms for evaluating the quality of knowledge and information in different domains. As result, various mechanisms start to appear today for this purpose [8–10], nevertheless, the issue is far from being solved.

This paper results from a systematic literature review aiming to *identify what kind of knowledge quality assessment mechanisms and Supportive tools could be developed to make a mass collaborative learning community more resilient against unreliable knowledge*. A summary of the survey is included to address some of identified factors and constituents namely: the type of organizational structures, collaborative learning approaches, adopted technologies, and adopted methods for evaluating the quality of information and knowledge that have influential impact on mass collaborative leaning.

It is expected that the findings of this study provide a comprehensive overview of the affecting elements on mass collaborative learning and help developing a better insight into how to evaluate the quality of created and shared knowledge in communities combined with supportive tools that can make their members more resilient in face of unreliable sources. It is also envisaged that the research findings of this study build a solid foundation for better developing the next research phase, which is generically guided by the following question:

Q: *What kind of methods and supportive tools can provide an appropriate basis to help evaluating the quality and reliability of co-created knowledge or information in mass collaborative learning projects?*

The reminder of this paper includes a short review of community resilience in Sect. 2. A synthesis of the state of the art is then presented in Sect. 3, and the research directions and plan are the topics of discussion in Sect. 4, followed by concluding remarks in Sect. 5.

2   Relationship to Resilience

Resilience has become a significant issue in many fields, reflecting the capacity of an economy, organization, city, forest or individual to deal with perturbations, and cope with all kinds or traumatic experiences. It represents the ability to successfully maintain a stable healthy level of physical or psychological functioning. In psychological terms, resilience can be employed to give us a scanty sense of hazard and let us get back to feeling normal again after any shape change [11]. Positive adaptation to development, reorganization, and renewal is also another essential aspect of resilience, but it has been less in the focus of attention [12]. Evidences in the literature show that resilience is in fact an ordinary, not uncommon function. People normally demonstrate resilience, for example, they respond to destructions caused by a storm, and make efforts to rebuild their houses that were destroyed by that stressful event. Resilience is in fact important for several reasons [13, 14], including:

- Provision of opportunities to protect people against conditions that might be overwhelming;
- Enabling people to develop mechanisms for managing extreme events during disasters;
- Helping governments, communities, and people to create more prepared and safer environment;

- Helping people to use resources and assets quickly;
- Promoting physical, psychological health and well-being, and mitigating the rate of mortality;
- Helping to decrease the stress on health care and the rate of risk-taking behaviors such as addiction, overuse of drugs, excessive smoking and drinking;
- Helping to promote studying and learning achievements, and
- Enhancing engagement in family activities and community collaboration.

A community that is resilient can harness, utilize, and develop nearly all possible resources to properly react and withstand against adverse situations to mitigate the rate of risk, and recover from emergencies. Resilience, indeed, enables community members to come together, intentionally promote their personal and collective capabilities, raise awareness of sustainability, respond effectively to turbulent changes, minimize impact of disasters, implement required plans and pay needed attention to urgencies, return to normal situations, and build development trajectory for future success [15].

Governments, organizations (specifically charitable ones), academics and communities show growing interest on programs that enable building resilience in face of turbulent changes and mishaps although the process of resilience-building is not precisely clear and there is not much understanding about what are its constituents. Nor is also there as a single approach or good model that could be used to build resilience for communities of all kinds. In addition, the literature is scant about what hinders or helps the community to be resilient in a disaster context. Evidences show that it is essential to consider what the community is used for; what is the vision of community; what are available resources and what are needed, who are members, how much understanding do members share about self-resilience, community-resilience and the risks they face, how much members are active and integrated, what factors cause community growth and decay, how much community is dynamic, and so forth. Nevertheless, it is notable that resilient communities share certain characteristics such as involving active participants, communication, cooperation, collaboration, loyalty, defined roles, diversity, sufficient resources to meet community needs, etc.

To make clear the process of building resilient communities in the scope of collaborative projects, some steps are proposed in [16, 17] which are summarized in Table 1.

It is a widely held view that the more a community can leverage disasters as an opportunity to improve, the more resilient it is. To build and bolster community resilience, and thus augment the capacity to cope with perturbations, various tools, mapping methods and guidelines are suggested that can streamline the process. For instance, creating networks, sharing knowledge, and utilizing diversity of ideas and experiences [18] by developing collaborative approaches [19] can improve the levels of community learning to tackle complicated problems [20, 21]. It is assumed that incorporating specific kinds of approaches such as, extending the size of community to mass level, shifting to collaborative networks, exploiting mass knowledge co-production, finding proper mechanisms for evaluating the quality of co-created knowledge (e.g., creating nodes of expertise, and feedback loop), and adopting quality measuring tools can be helpful in making a community more resilient.

Table 1. Proposed steps to clarify the process of building resilient communities in collaborative projects

Steps	Needed action to take
Step 1 Explore threats	(1) Identify changes, threats and hazards, (2) Identify environmental impact, (3) Identify resources and assets, (4) Identify potential members, (5) Developing objectives
Step 2 Evaluate risks	(1) Evaluate sensitivities, (2) Evaluate adaptive capacities, (3) Evaluate risk, (4) Evaluate vulnerabilities
Step 3 Assess options	(1) Identify possible and feasible options to decrease risks and vulnerabilities
Step 4 Prioritize acts	(1) Assess possible and feasible option, (2) Prioritize them according to their risk and vulnerability
Step 5 Put it into action	(1) Make needed plan, (2) Define responsibilities, (3) Monitor progress and productivity, (4) Reiterate

It is expected that by taking the advantages of fundamental properties of the collaborative communities such as, adaptability, efficiency, diversity, and cohesion we can leverage the opportunities to design a system with inherent resilience. Such resilient system creates the possibilities to reduce the risks associated with the attack of false inputs and their adverse impacts. That is, a developed sustainable system can help collaborative-networked learning groups to constantly maintain high level of preparedness against unreliable materials. Besides, it can strengthen communities and their members specifically those that are more vulnerable to withstand major threats related to wide spreading of unreliable knowledge or information in online environments.

In this context, our work focuses on the impact that untrustworthy information and unreliable knowledge can have on a community through mass collaborative learning processes. We are particularly interested in contributing to a better understanding of the mass collaborative learning concept and identifying approaches to deal with unreliability. In this way, we expect to contribute to more resilient communities.

3   State of the Art Overview

Literature shows that there is growing tendency in response to the need of communities for fostering collaborative learning in effective ways. Several contributing factors, e.g., pedagogical approaches, ICT-infrastructures, educational programs [22], learning environments, learning designs, and learning interactions [23, 24], etc., are highlighted in recent years as vehicles to better engage learners in collective learning, incite their passion for constructive social impact, and develop a foundation for next generation of learning approaches.

The main findings of the literature survey, which included reviewing about 100 papers in mass collaboration context and related areas, whose findings are succinctly presented in the following subsections.

3.1 Organizational Structures and Mass Collaboration

An organizational structure acts as an "instruction" for decision makers to more easily assign plans, strategies or decisions which are useful for their group [25]. Small or large, every community must operate with an appropriate organizational structure because, for example, it assists better identifying responsibilities and roles, utilizing and controlling resources, binding group members and pointing them common goals, facilitating decision making processes, making easier communication, etc. [26]. The type of structure indicates in which ways internal works can be carried out at all levels of the community. Basically, the goals and strategies of the community, and the type of members' or customers' needs are the main determinants for selecting a structure.

The organizational structure has profound impact on collaborative networks, and largely builds the level of autonomy and collaboration with and amongst the members. Evidences show that as the communities are more and more evolving from small and medium size to large scale collectives, and from non-computerized to a digital-based model, there is a need for structural adaptability. That is, shifting away from traditional structures (e.g., hierarchical, centralized, etc.) towards unconventional models (e.g., informal, self-directed, etc.). Although there are vast amounts of literature on organizational structures, there is very little work trying to specifically evaluate the role of organizational structures in large-size networked collaborative learning. Having reviewed several suggested models in relevant areas, from which no suitable organizational structure for mass collaborative learning could be found, the taxonomy recommended in [27] was selected as the closest fit with the nature and type of structures applied by virtual communities in collected papers on survey. As illustrated in Table 2, two main forms of collaboration in networks - collaborative networked organization and ad-hoc collaboration - root the main classes of this taxonomy.

Table 2. Taxonomy of collaborative network [27]

Collaborative network				
Collaborative networked organization				Ad-hoc collaboration
Long-term strategic network		Goal-oriented network		- Mass collaboration
VBE - Virtual organization Breeding Environment	**PVC** - Professional Virtual Community	Grasping opportunity driven network	Continuous production driven network	- Flash mob - Informal network - One-to-one informal collaboration
- Industry cluster - Industry district - Disaster rescue network - Business ecosystem - Collaborative innovation network		- Extended enterprise - Virtual enterprise - Virtual organization - Virtual team	- Supply chain - Collaborative e-government - Collaborative smart grid - Distributed manufacturing	

Considering this taxonomy, an analysis of all collected papers shows that the issue of organizational structure was addressed in 32 papers. Details on the percentage and type of applied structure in those papers are illustrated in Fig. 1.

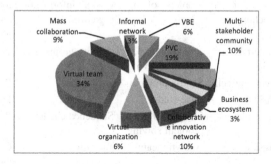

Fig. 1. Percentages of applied organizational structures in collected papers

It is noteworthy that the type of organization structure which was used in three articles in collected papers falls between VBE and PVC categories in above taxonomy (a kind of hybrid). As it can be seen in the figure, informal networks and business ecosystem structures were the least and virtual team structure was the most addressed structure in the selected papers. From the findings it can be inferred that the organizational structure should be adopted based on community purpose, type, size, needs, communication devices, and environment. Furthermore, as the level of collaboration is shifting to large scale, structures should be dynamically reconfigurable. In other words, it needs blending available resources, skills and competencies across the community to take advantage of collaboration opportunities. Finally, in mass collaborative projects the more the organizational structure is dynamic and the better members' network position fit with the organizational context, the more knowledge they can acquire.

3.2 Collaborative Learning Techniques (CoLTs) and Mass Collaboration

Collaborative learning in broad sense is a type of education approach in which learners in a group attempt to accomplish a common goal. In such group members are responsible for both their own tasks. CoLTs can make straightforward the process of discussion, and sharing knowledge, opinions and experiences for learners. Applying these techniques can provide supportive directions for development of learning and collaboration in communities of all sizes. Different areas of study over the years have benefited from using a variety of CoLTs [28]. However, their application to mass collaboration and learning has not received much attention. Therefore, there are not enough evidences in the literature showing that what types of CoLTs are exactly required for this purpose, and how these techniques can facilitate learning where the process is entirely self-directed. Despite such limitations, many proposed techniques from relevant domains were reviewed in order to pick up the ones that are most promising. To this end, the taxonomy offered by [28] was selected as it is reasonably comprehensive. It encompasses 5 major categories of general learning activities and 31 CoLTs (see details in Table 3).

Table 3. Collaborative learning techniques [28]

Collaborative learning techniques				
Techniques for discussions	Techniques for reciprocal teaching	Techniques for problem solving	Techniques using graphic information organizers	Techniques focusing on collaborative writing
Think-pair-share	Note-taking pairs	Think-aloud pair problem solving	Affinity grouping	Dialogue journals
Round Robin	Learning cell	Send-a-problem	Group grid	Round table
Buzz groups	Fishbowl	Case study	Team matrix	Dyadic essays
Talking chips	Role play	Structured/group problem solving	Sequence chains	Peer editing
Three-step interview	Jigsaw	Analytic teams	Word webs	Collaborative writing
Paired annotations	Test-taking teams	Group investigation		Team anthologies
Critical debates				Paper seminar

The analysis of collected papers reveals that CoLTs were considered in 23 papers. In this regard, techniques for discussion received the most attention and techniques for reciprocal teaching received the least attention in those papers. More details about the percentage of applied CoLTs are depicted in Fig. 2. Although this taxonomy is not specifically designed for mass collaborative learning, findings reveal that some of the techniques such as group problem solving, peer editing, and paired annotation have seemingly potential structures to guide the development of mass collaborative learning projects. Moreover, techniques like note-taking, which provide elaborated explanations and reflective feedback from partners, can enhance the chance of learning in mass collaboration.

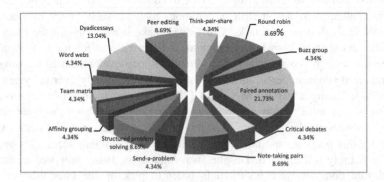

Fig. 2. Type and percentage of adopted CoLTs in collected papers

3.3   Supportive Tools and Mass Collaborative Learning

Recent interest in technology support to collaborative learning represents a confluence of trends such as the emergence of constructive approaches for learning [29], the aim to build more influential learning environment [30], and the advance of new technologies to support collaborative learning [31]. However, neither every form of collaborative learning necessarily needs the same type of technology, nor can a single tool provide all required features.

Literature shows that new supportive tools such as CSCL, social media, web-based and mobile technologies along with Internet, have equipped large number of learners around the world to comfortably communicate anytime and anywhere, and empowered them to exchange their resources, knowledge, and experiences. However, the real use case for technology in mass collaboration is still evolving, and as such, improvements in learning outcomes for self-directed learners are yet to be proven. There are also several open issues, such as: how can supportive tools efficiently process the massive load of content?; or how can needed training or information be provided for a single learner in the community who does not have enough technical knowledge? As such, a comprehensive list of specific applicable tools for this purpose is not yet proposed. Therefore, considering different related models in the literature, the Project-Based Collaborative Learning Model [32] was selected to check which of proposed tools in this model are also used in the analyzed papers. Seven distinct phases of this model and more details are exhibited in Fig. 3.

From this analysis, it was evidenced that nearly one third of collected papers (35 papers) evaluate the role of technologies in support collaboration and learning. It also shows that communicative tools are the most, and consensus building tools are the least employed tools. More detail and percentage of each applied tool are shown in Fig. 4. Therefore, it can be inferred that resource management tools can bring some opportunities for mass learners to access, evaluate, utilize, and share their resources more readily. These tools can also help transforming complex tasks into easy-achievable works. Moreover, tools such as Routing, Milestones, and Calendaring seem to not have as high chance as Wiki, Discussion board, and Blog for application in mass collaborative projects.

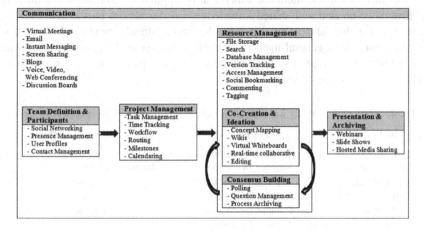

Fig. 3. Technology support for project-based collaborative learning model [32]

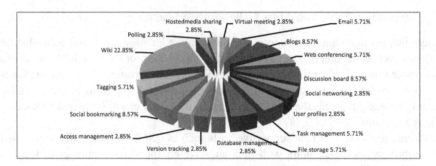

Fig. 4. Type and percentage of adopted technologies in collected papers

3.4   Evaluating the Quality of Created Knowledge in Mass Collaboration

Since knowledge is recognized as one of the most strategic assets for every organization and community, evaluating the identified, acquired, created, shared and/or retrieved knowledge influences community's prosperity. For every community it is important to know how to identify the quality and reliability of acquired/possessed knowledge. Evaluating the quality of knowledge can help in, for example, identifying strengthens and weakness, providing guidelines that could be helpful for future plans and development, improving effectiveness, and so forth [33]. The intangibility of knowledge, however, makes the process of evaluation somehow difficult in both practice and research. On the other hand, when large amounts of knowledge and information are exchanged through the Internet and social networks by known or unknown users serving different purposes, make the evaluation crucial. It becomes more complicated as online knowledge and information can be easily altered, misrepresented, built up, or plagiarized.

Despite quality of knowledge has been a topic in different fields of study and businesses, and various types of strategy, methods, and question have been proposed for this purpose, it has not been well studied yet particularly when integrated with mass collaboration. Therefore, we could not find a comprehensive mechanism or list of applicable methods in the literature which can be applied for gauging the quality and reliability of co-created and shared knowledge or information. Furthermore, no systematic research for addressing this issue has been already conducted. Hence, we collected a number of general methods, namely available in Wikipedia (Table 4) and compared them with solutions proposed in the analyzed papers in order to gain better insight about this scope of study.

Table 4.   Suggested methods for evaluating the quality of knowledge in collected papers

Suggested methods	Suggested methods	Suggested methods
- Credit assignment	- Type of contributor activity	- Positioning
- Machine learning	- Number of anonymous contributors	- Argumentation
- User feedback	- Top contributor experience	- Consensus
- Experts evaluation	- Ranking method	- Selection
- Initialization	- Content facilitation	- Reputation mechanism
- Computing user weights	- Process facilitation	- Peer review
		- Group observation

It is worth noting that for appraising the quality of articles in Wikipedia, eight major criteria are under consideration including, accuracy, comprehensiveness, stability, well written, uncontroversial, compliance with standards of Wikipedia, having appropriate style, and having appropriate images. Furthermore, in this approach some methods are commonly used for example, nominating qualified articles, reputation mechanism, peer review, and feedback, to name a few [34].

From the findings of collected papers it can be concluded that:

- User feedback and expert evaluation were the most suggested methods;
- The role of top contributors is deniable;
- Publishing the result of evaluation could be helpful for all learners;
- Both qualitative and quantitative approaches should be considered; and
- Combination of machine learning and human factors seems work better.

4 Research Direction

In this section, we propose a research approach for tracking the reliability and quality of online knowledge or information in the context of mass collaborative learning. The goal of our approach is to integrate human and computer support to reach an optimal balance between simplicity and speed on one hand, and validity of result on the other. Hence, at the current stage, we envisage a prototype comprising two main parts: human part and computer part. The human part involves an individual phase and a community phase. The community phase benefits from the contribution of both ordinary and expert members. The computer part provides supportive tools (e.g., fake news detection, website or resource detectors) which can raise red flags on unreliable and questionable contents. More details are presented in Table 5.

Table 5. Suggested issues for evaluating the quality of online knowledge/information

Human part			Computer part
Individual phase	Community phase (crowd sourcing)		Detector tools
1. Manual filtering	Ordinary members	Expert members	- Fact check
2. Completing checklist:	1. Completing checklist	1. Completing checklist	extension
- Authority	2. Evidence-based reasoning	2. Evidence-based reasoning	- Fake news detector, etc.
- Accuracy			- Other (novel) tools
- Currency	3. Formal argumentation	3. Formal argumentation	
- Accessibility			
- Coverage	4. Making decision	4. Making decision	
- Relevancy			
- Purpose			
- Bias			
- Soundness			
- Clarity			
- Safety			
- Reference			
3. Making decision			

The planned approach for the collaborative evaluation consists of the following steps:

- **Step 1**: individual phase - a community member quickly checks the knowledge or information manually to decide whether it is worthy enough to warrant further evaluation or not. After manual filtering, a defined check list will be completed for those items that are accepted to investigate in more detail. By applying cognitive skills (critical thinking and critical appraising) the member assesses by self-checking the reliability of knowledge or information based on suggested criteria (mentioned in Table 5) and gives each of them an emotional rate. The individual decision can then be made based on the given rates.
- **Step 2**: once individual decision is made, that is the time to take advantage of crowd sourcing that takes place in both levels (ordinary or expert members) of community phase. In this step, at first ordinary and expert members separately but in parallel complete the check list and give emotional rates. The reasons and evidences for given rates are then shared among contributing members. Afterwards, developing formal argumentation and collective evaluation not only enables community members to gain common sense about the findings but also helps reaching results that are beyond individual's ability. The results of decision in this phase should be visible for all.
- **Step 3**: the final decision about the quality and reliability of knowledge or information evolves from the evaluation of results in both phases.

It is worth mentioning that detector tools can be used to support community members throughout the evaluation process. Figure 5 exhibits these three steps.

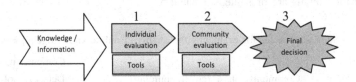

Fig. 5. Three steps for evaluation of knowledge or information in mass collaborative learning

5   Conclusions and Future Work

Emerging collaborative forms of learning in open networks and communities provide new opportunities for joint learning. With the objective of gaining understanding of current state of the art in mass collaborative learning, an extensive literature survey was conducted. As a result, various findings on the organizational structures, collaborative learning techniques, and support tools were highlighted.

However, mass collaborative learning also confronts community members with the problem of dissemination of unreliable knowledge or information through Internet and social networks. To prevent the negative side effects of such problem, an important goal is making community members more secure and resilient against vulnerabilities caused by online fraud. To this end, in this study a preliminary approach is suggested aiming at a combination of human and computer support to enable contributors taking advantage of collaborative evaluation in dealing with threats.

This work is still at a preliminary level, but it is expected that collective evaluation in different steps and along with support tools can provide learners and communities with helpful guidelines for achieving a high level of consciousness about the quality of acquired knowledge. In next phases of this work we intend to investigate what organizational structure for the suggested approach should be established? What kind of mechanism does the community need for evidence acquisition and combination? What kind of mechanism can help appropriately inferring the final evaluations and decisions?

Acknowledgement. This work was funded in part by the Center of Technology and Systems of Uninova and the Portuguese FCT-PEST program UID/EEA/00066/2013 (Impactor project).

References

1. Cress, U., Moskaliuk, J., Jeong, H. (eds.): Mass Collaboration and Education. CCLS, vol. 16. Springer, Cham (2016). https://doi.org/10.1007/978-3-319-13536-6
2. Dillenbourg, P., Baker, M., Blaye, A., O'malley, C.: The evolution of research on collaborative learning. In: Spada, E., Reiman, P. (eds.) Learning in Humans and Machine: Towards an Interdisciplinary Learning Science, pp. 189–211. Elsevier, Oxford (1996)
3. Weitzel, L., Quaresma, P., Palazzo, J.M.D.O.: Evaluating quality of health information sources. In: 26th IEEE International Conference on Advanced Information Networking and Applications (2012). https://doi.org/10.1109/aina.2012.41
4. Eppler, M.J., Muenzenmayer, P.: Measuring information quality in the web context: a survey of state-of-the-art instruments and an application methodology. In: Proceedings of the Seventh International Conference on Information Quality, ICIQ 2002 (2002)
5. Huang, K., Lee, Y., Wang, R.: Quality Information and Knowledge. Prentice Hall, Upper Saddle River (1999)
6. Lee, Y.W., Strong, D.M., Kahn, B.K., Wang, R.Y.: AIMQ: a methodology for information quality assessment. J. Info. Manage. **40**(2), 133–146 (2002)
7. Agichtein, E., Castillo, C., Donato, D., Gionis, A., Mishne, G.: Finding high-quality content in social media. In: Proceedings of the 2008 International Conference on Web Search and Data Mining, WSDM 2008, pp. 183–194 (2008). https://doi.org/10.1145/1341531.1341557
8. Knight, S., Burn, J.: Developing a framework for assessing information quality on the World Wide Web. J. Informing Sci. **8**, 159–172 (2005). https://doi.org/10.28945/493
9. Todoran, I.G., Lecornu, L., Khenchaf, A.: Information quality evaluation in fusion systems. In: Proceedings of the 16th International Conference on Information Fusion (2013)
10. Fritc, J.W., Cromwell, R.L.: Evaluating Internet resources: identity, affiliation, and cognitive authority in a networked world. J. Am. Soc. Inf. Sci. Technol. **52**(6), 499–507 (2001)
11. Windle, G.: What is resilience? A review and concept analysis. Rev. Clin. Gerontol. **21**(2), 152–169 (2011). https://doi.org/10.1017/S0959259810000420
12. Berkes, F., Colding, J., Folke, C. (eds.): Navigating Social-Ecological Systems: Building Resilience for Complexity and Change. Cambridge University Press, Cambridge (2003)
13. Romac, S.: The Importance of Community Resilience: Developing the American Red Cross International Services Department in the New Hampshire Region. SIT Graduate Institute, Brattleboro (2014)

14. Anderson, B.: What kind of thing is resilience. J. Politics, **35**(1), 60–66 (2015). https://doi.org/10.1111/1467-9256.12079

15. Magis, K.: Community resilience: an indicator of social sustainability. J. Soc. Nat. Resour. **23**(5), 401–416 (2010). https://doi.org/10.1080/08941920903305674

16. Chandra, A., Acosta, J., Stern, S., Uscher-Pines, L., Williams, M.V., Douglas Yeung, D., Garnett, J., Meredith, L.S.: Building community resilience to disasters a way forward to enhance national health security. Rand. Health Q. **1**(1), 6 (2011)

17. Thornley, L., Ball, J., Signal, L., Aho, K.L.T., Rawson, E.: Building community resilience: learning from the Canterbury earthquakes, Kōtuitui: New Zealand. J. Soc. Sci. Online **10**(1), 23–35 (2015). https://doi.org/10.1080/1177083X.2014.934846

18. Olsson, P., Folke, C., Berkes, F.: Adaptive co-management for building resilience in social-ecological systems. Environ. Manag. **34**, 75–90 (2004)

19. Olsson, P., Folke, C., Galaz, V., Hahn, T., Schultz, L.: Enhancing the fit through adaptive co-management: Creating and maintaining bridging functions for matching scales in the Kristianstads Vattenrike Biosphere Reserve Sweden. Ecol. Soc. **12**, 28 (2007)

20. Berkes, F.: Evolution of co-management: Role of knowledge generation, bridging organizations and social learning. J. Environ. Manage. **90**, 1692–1702 (2009)

21. Berkes, F.: Environmental governance for the Anthropocene? Social-ecological systems, resilience, and collaborative learning. J. Sustain. **9**(7), 1232 (2017)

22. Rippel, M., Schaefer, D., Mistree, F.: Fostering collaborative learning and mass-customization of education in a graduate engineering design course. Int. J. Eng. Educ. **25**(4), 729–744 (2009)

23. Razali, S.N., Shahbodin, F., Hussin, H., Bakar, N.: Factors that affecting the effective online collaborative learning environment. In: Pattern Analysis, Intelligent Security and the Internet of Things, pp. 215–224 (2015)

24. Whittaker, S.: Talking to strangers: an evaluation of the factors affecting electronic collaboration. In: Proceedings of the 1996 ACM Conference on Computer Supported Cooperative Work, pp. 409–418 (1996). https://doi.org/10.1145/240080.240352

25. Morten, E.: How bureaucratic structure matters: an organizational perspective. In: Peters, B. G., Pierre, J. (eds.) The SAGE Handbook of Public Administration, pp. 157–168. SAGE Publications, Los Angeles (2003)

26. Latif, K.I., Baloch, Q.B., Naser Khan, M.: Structure, corporate strategy and the overall effectiveness of the organisation. Abasyn J. Soc. Sci. **5**(2) (2012)

27. Camarinha-Matos, L.M., Afsarmanesh, H., Galeano, N., Molina, A.: Collaborative networked organizations – concepts and practice in manufacturing enterprises. J. Comput. Ind. Eng. **57**(1), 46–60 (2009)

28. Barkley, E.F., Cross, K.C., Major, C.H.: Collaborative Learning Techniques: A Handbook for College Faculty, 1st edn. (2014). ISBN-13: 978-0787955182

29. Kirschner, P.A., Martens, R.L., Strijbos, J.W.: CSCL in higher education? A framework for designing multiple collaborative environments. In: Strijbos, J.W., Kirschner, P.A., Martens, R.L. (eds.) What We Know About CSCL: And Implementing it in Higher Education, pp. 3–30. Kluwer, Boston (2004)

30. Oblinger, D.G., Oblinger, J.L. (eds.): Educating the Net Generation. Educause, Washington (2005). http://www.educause.edu/educatingthenetgen

31. Johnson, D.W., Johnson, R.T.: Cooperation and the use of technology. In: Jonassen, D. (ed.) Handbook of Research for Educational Communications and Technology, pp. 785–812. MacMillan, London (1996)

32. Deal, A.: A teaching with technology White Paper: Collaboration Tools (2009). https://www.cmu.edu/teaching/technology/whitepapers/CollaborationTools_Jan09.pdf
33. Chapman, D.D., Wiessner, C.A., Storberg-Walker, J., Hatcher, T.: New learning: a different way of approaching conference evaluation. Knowl. Manage. Res. Pract. **5**, 261–270 (2007)
34. Stvilia, B., Twidale, M.B., Smith, L.C., Gasser, L.: Information quality work organization in Wikipedia. J. Am. Soc. Inf. Sci. Technol. **59**(6), 983–1001 (2008). https://doi.org/10.1002/asi.20813

Semantic Modelling of User Interactions in Virtual Reality Environments

Jacek Sokołowski[(✉)] and Krzysztof Walczak

Poznań University of Economics and Business, Niepodległości 10, 61-875 Poznań, Poland
{sokolowski,walczak}@kti.ue.poznan.pl
http://www.kti.ue.poznan.pl

Abstract. Virtual reality (VR) enables building a new class of applications that provide the capability to immerse users in 3D virtual environments. A key element of VR applications is interactivity, which largely depends on the interaction channels provided by a particular virtual reality system used. Vast diversity of available VR system configurations forces application designers to build VR applications for specific systems. In this paper, the use of semantic techniques is proposed to enable building resilient VR applications that can be automatically adapted to a particular VR system configuration. Interaction ontologies used both at the VR system part and the VR application part enable automatic mapping of available interaction channels to specific interaction requirements of a given application, enabling deployment of the same application on different VR systems.

Keywords: Virtual reality · Immersive environments · Contextual interaction
Semantic modelling

1 Introduction

Non-trivial virtual reality (VR) environments usually require implementation of complex forms of user interaction, including spatial navigation within the whole environment as well as various forms of interaction with particular components of the virtual world. The forms of interaction available in a given environment depend on the interaction channels provided by the particular VR system used. There is a great diversity of virtual reality systems – from simple 3D-enabled desktop environments, providing interaction through a mouse and a keyboard, through different kinds of head-mounted displays, where interaction is typically implemented by head-tracking, to high-end immersive systems, featuring advanced interaction techniques, such as body tracking and gesture recognition. Often specialized interaction devices are used, but this creates a barrier to the expansion of virtual reality into new areas. Vast diversity of available VR system configurations forces application designers to build VR applications for specific systems, which results in large fragmentation of the market. Migrating an application from one environment to another is difficult and time-consuming. These issues severely limit the current use of VR applications.

© IFIP International Federation for Information Processing 2018
Published by Springer International Publishing AG 2018. All Rights Reserved
L. M. Camarinha-Matos et al. (Eds.): DoCEIS 2018, IFIP AICT 521, pp. 18–27, 2018.
https://doi.org/10.1007/978-3-319-78574-5_2

In this work, the use of semantic techniques for modelling of possible user interactions is proposed. On the VR system part, there is an ontology describing available interaction channels and techniques in a particular VR system configuration. On the VR environment side, an ontology describing interaction capabilities in the virtual environment is provided. Based on these two ontologies, an automatic mapping can be performed, to match the available interaction techniques in the VR system with the interaction capabilities of the environment. Because, the system-side ontology includes a semantic taxonomy of typical uses for particular interaction techniques, intuitive use of interaction channels can be automatically proposed – for example, a "Cancel" action in a VR environment can be mapped to the "X" button on a game controller, left-hand waving in gesture-based system, or back-button of a head-mounted display (HMD).

In the presented implementation, an authoring tool based on the Unity 3D editor is used. The tool enables describing interactive objects in a virtual environment with semantic metadata. At the runtime, a user can the control the virtual environment by the use of a mobile application. Mobile devices are currently widely available and provide advanced user-interface features, including high-resolution touch screen displays and various types of built-in sensors, such as gyroscope and accelerometer. In an interaction context, interaction metadata are sent to the mobile application to automatically generate a contextual personalized user interface. The interface depends on the collection of objects the user can interact with in a particular context. User interaction data are then sent from the mobile application to the environment, enabling the user to conveniently use all interactive functions of the virtual environment. The system has been tested on a Powerwall VR system.

The remainder of this paper is structured as follows. Section 2 situates the work in the context of resilient systems. Section 3 provides an overview of the current state of the art in the domain of methods of implementing interaction in VR environments. Section 4 describes the proposed method of semantic modelling of interactions in VR. Section 5 presents a reference implementation of the proposed method. Finally, Sect. 6 concludes the paper and indicates the possible directions of future research.

2   Relationship to Resilient Systems

With the constant progress in technology and hardware performance, virtual reality becomes a part of our everyday life. Activities that have been undertaken in the real world so far are now often replaced by activities in a virtual world. VR techniques allow us to meet our needs – both the basic ones (such as shopping) and those of a higher order (such as contact with art). Moreover, there are no physical barriers in the virtual world – virtual reality offers the possibility of "travelling" to distant places in no time, no cost and without all the dangers associated with traveling and staying in crowded places.

One of the main barriers to using virtual reality is the need for specialized presentation and interaction devices. There is a great variety of VR devices currently available. The offered systems differ significantly in their characteristics and capabilities, making it difficult to build applications that could be widely used. Even if multiplatform development tools (such as game engines) are being used, the diversity of the available input and output channels and their characteristics undermine the cross-platform use of VR applications.

The approach proposed in this paper enables dynamic adaptation of VR applications to the specific characteristics of interaction channels available in a particular VR system, enabling an application to run in multiple different hardware setups. As a result, the proposed solution may contribute to easier deployment and consequently wider dissemination of VR applications.

3   State of the Art

In this section, the most common approaches to user navigation and interaction in VR environments are described. Beyond the use of standard user interfaces, methods based on the use of specialized interaction equipment, natural user interaction, contextual approaches, and dedicated remote interfaces are presented.

3.1   Generic Input Device Approach

Interaction and navigation in VR environments can be implemented with the use of standard input devices (i.e., mouse and keyboard). Indirect mapping of 2D mouse interaction into 3D space can be implemented with the use of mouse or keyboard buttons [1]. However, the use of a mouse and a keyboard is problematic in immersive environments, such as HMD and CAVE. Moreover, although a keyboard enables relatively rich interaction, it is limited to two states only (button pressed or released), which may not be sufficient in modern virtual environments. In addition, these devices are generic, and therefore mapping of interface actions (mouse moves or pressing buttons) is often not intuitive and has to be memorized by users. Moreover, these devices do not provide any meaningful form of reverse communication, which prevents from implementing hints or facilitators for users. This makes the whole interaction complicated, inconvenient and, as a result, ineffective. An advantage of this approach is that users do not need additional devices. Generic input device approach is not recently a popular subject of research, however, methods of adapting this approach to touch-screens have been elaborated [2].

3.2   Specialized Input Device Approach

Specialized input device approach focuses on the use of specialized equipment – gaming input devices, such as joysticks and pads, or dedicated VR devices, such as haptic arms and flysticks – to navigate and interact in virtual environments. A significant advantage of this approach is higher user comfort and good control and accuracy in properly designed and configured environments [3]. A general disadvantage is the natural limitation of the number of available buttons and other interaction elements, which – in addition – do not allow creating user-friendly interfaces in context-based applications.

Device-based approach is often the basis for further research associated with virtual reality [4, 5].

3.3 Natural Interaction Approach

Natural interaction is a widely used method of interaction with virtual environments. This approach is based on a projection of natural human behaviour within virtual reality; it includes techniques such as motion capture (using marker tracking [6] or marker-less tracking, e.g., with Kinect sensor system [7]), gesture recognition [8], eye tracking [9], and verbal/vocal input [10]. All these techniques focus on providing an intuitive natural interface, which is user-friendly even for non-experienced users. Main problems encountered when using natural interaction are lack of precision and users' fatigue. This approach also requires specific sensor equipment, which registers and analyses users' behaviour.

3.4 Context-Based Approach

Context-based approach is an interaction technique invented long time ago [11]. This approach is popular in computer games, in particular simulations (e.g., "The Sims" and "SimCity" series) and adventure games. In practice, after initialization of interaction with a specific game object, the current context (e.g., time, position, current object state) is analysed and proper user interface is dynamically generated and presented within the virtual scene. In modern virtual environments, standard context-based approach is often used [12]. However, this approach is uncomfortable due to difficulty in navigation and the mismatch between classical UI elements (buttons, menus, charts) and the 3D virtual environment.

3.5 Dedicated Control Interface Approach

Dedicated control interface approach uses separate devices with their own CPUs to navigate and interact within virtual environments. Due to necessity of data exchange between the control device (the client) and the environment (the server), the communication aspect appears. Two most popular technologies used for such communication are Bluetooth [13] and WiFi [14]. This approach is used not only for VR interaction, but also for controlling vehicles (e.g., drones) and robots [14].

In the dedicated control interface approach, a predefined user interface is implemented on a control device. A user can manipulate the interface, which sends events (like pushing buttons) to a server, where appropriate actions are performed. Currently, application of this approach does not longer require implementing a dedicated client application due to the availability of generic software packages (e.g., PC Remote application by Monect [15]) that allow to configure interfaces depending on the user's requirements or to choose one of predefined interfaces (e.g., a TV remote control).

3.6 Discussion of Existing Approaches

Five common approaches to navigation and interaction within virtual environments have been described above. Mixing different techniques is also a commonly used solution. For example, specialized device approach and natural interaction approach can be mixed. In this case, comfort and functionality of a specialized device is combined with intuitiveness of natural interaction [16]. However, this type of combination requires even more devices, which makes it highly specific, expensive, and difficult to implement in practice.

There are also specific techniques that cannot be simply assigned to any of the above categories, e.g., usage of specific props for object manipulation within a CAVE-type system [17], which however are out of the scope of this paper.

4 Semantic Modelling of User Interactions

Currently, interaction channels are strongly related to the specific VR system configuration. Applications are usually created for a specific setup, which has a predefined set of user interface channels. This reduces the applicability of VR technology and makes users dependent on hardware.

However, the use of semantic techniques allows building resilient applications that can be dynamically adapted to the available input and output devices. This requires the creation of several ontologies. The first one describes VR systems and their configuration. This ontology classifies the interface elements, taking into account the senses associated with them, such as physical touch, sight, hearing, smell and balance, and particular actions. Next, the interaction channels available in the current system configuration are recognized. For example, if a user is in possession of an HMD interface, all data input techniques for the application will be recognized, such as reading the position of the gyroscope and the user's physical contact with the button located on the device.

The second ontology describes interactions foreseen in a VR environment. It includes all kinds of activities related to user's navigation and interaction in the environment. The basis of this ontology is a general model that takes into account navigation in many degrees of freedom and any possible interaction with virtual objects. Subsequently, the ontology is constrained by application-specific constraints. They may concern both the reduced number of degrees of freedom (e.g., if the application does not allow a user to "fly"), as well as limited interaction with objects (e.g., when the user cannot influence the arrangement of objects on stage or change their size). Both these ontologies are independent of each other (Fig. 1).

Additional ontologies and taxonomies can be used to represent typical uses of interaction elements (e.g., escape or return button to represent cancelling actions) and specific user preferences (e.g., a user prefers to use an interaction device instead of using body movements). Taking this information into consideration enables creation of more user-friendly personalized interfaces.

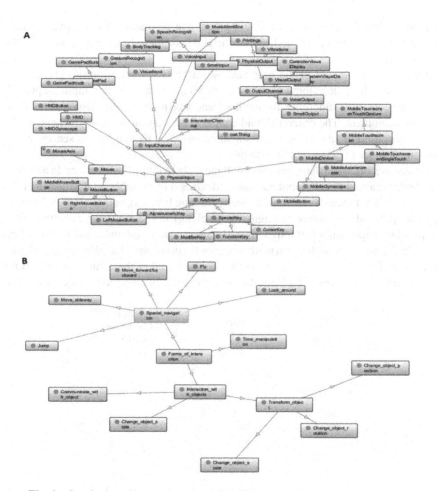

Fig. 1. Ontologies of interaction channels – VR system (A), VR environment (B)

Based on the available ontologies, an appropriate mapping of actions in the VR environment to the available interaction channels of a particular VR system can be generated. The mapping is performed in such a way that at least one interaction option is assigned for each possible action in the VR environment. If the number of interaction channels exceeds the number of activities, it is possible to leave unused channels of interaction. These channels can also be combined for making application interface more intuitive. For example, moving just the knob on a gamepad can be used for the movement of a virtual character, while moving this knob in combination with holding down the action button can be used for looking around the scene.

The semantic description of possible actions within a VR environment enables the mapping algorithm to assign intuitive interaction methods to actions. It enables faster adaptation of non-expert users to a VR environment thanks to their experience with similar applications and it allows smooth application transfer from one VR system to another. Intuitive use of interaction channels can be automatically proposed – for example, a "Jump" action in a VR environment can be mapped to the spacebar on a keyboard, a sudden lifting of the head on the HMD or literally a jump in the case of a tracking system.

An important element of the proposed approach is the selection of classification criteria for ontology describing capabilities of VR systems. As new input/output VR devices are constantly emerging, the method cannot be based on the classification of the available equipment. The criteria should enable classification of each new device. The solution proposed in this article is classification based on natural human senses, which limits the number of classification groups.

VR devices often use many senses simultaneously to interact with a user. In addition, most devices can act as both input and output devices. For example, a large part of HMD devices offer input interaction through touch (pressing buttons), balance (using the built-in gyroscope), and sometimes also sight (eye tracking) and hearing (voice control). They also provide output stimuli through sight (image), hearing (sound), and touch (vibrations). The proposed classification method forces the division of functions of each device into a limited number of categories. Such a solution makes these functions repetitive between different devices, which allows the interaction channels to replace each other.

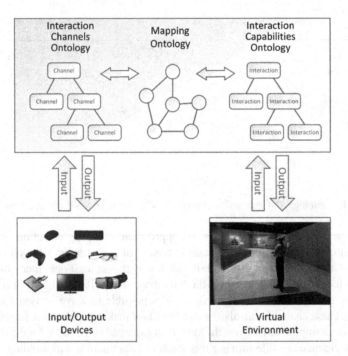

Fig. 2. System architecture

Figure 2 depicts the overall architecture of a system implementing the presented method. The architecture consists of three main elements. The first element is a group of interaction devices. Since many devices act as both input and output devices at the same time (e.g., with a mobile device it is possible to enter data into the system as well as display the application content) all these devices are treated as one element of the architecture. The second element is a VR environment represented by any of the specific applications. These applications contain sets of foreseen interactions and – depending on their nature – enable a user to interact with the environment in a variety of ways. The third architectural element is the VR system. With the use of a mapping ontology the system assigns available interaction channels to the interaction capabilities of the VR environment. After completing the mapping process, a user is able to interact with the application in the real time with the use of the available equipment.

5   Implementation

A prototype of the system implementing the proposed method has been developed using the Unity 3D engine. An interface that allows describing objects in virtual environments using semantic metadata directly in the Unity editor is presented in Fig. 3 (top-right). The described objects are presented in a virtual scene (Fig. 3, bottom). A user can

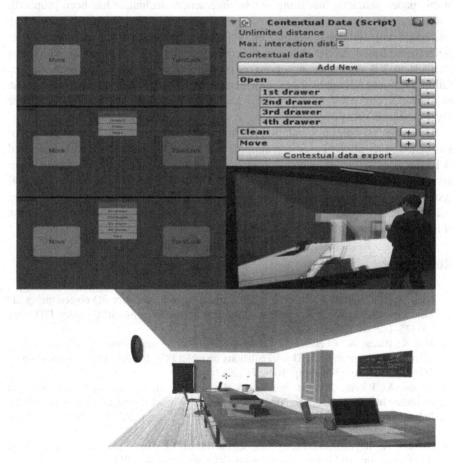

Fig. 3. Prototype implementation

navigate through the scene and initiate interaction with objects by directing towards them. When interaction with a specific object becomes possible, semantic metadata related to this object are sent to the mobile device. Since in the current version of the prototype the only input device is a mobile device, all interaction options associated with the currently active object are presented on the screen (Fig. 3, top-left).

Selecting one of the interaction options triggers appropriate action on the server side. In this way, a two-dimensional user interface is extracted from the three-dimensional presentation environment. WiFi connection has been used for communication between devices due to its speed, safety and sufficient range [14]. The popular JSON format has been used for structuring exchanged messages. All of this has been tested on a Powerwall setup with a 3.5-m screen and 4 projectors providing stereoscopic projection with 4 K resolution at 120 Hz (Fig. 3, middle-right).

6   Conclusion

In this paper, semantic modelling of user interactions technique has been proposed. Ontologies related to the VR system configuration and the VR environment are used in connection with additional mapping ontologies describing possible matching and user preferences. Based on the semantic descriptions, automatic mapping of interaction channels to VR environment actions is possible. The presented approach enables building resilient VR applications, in which the interface of a virtual environment can be adapted to any VR system. This may lead to simplification and acceleration of the development of VR applications.

This paper provides a foundation for future work on creating a general model for a semantic description of user interaction in VR. The prototype application is limited to interactions with one device – a smartphone or a tablet. Extending it to other devices requires the use of an extensive and well-structured ontology on the VR system side. In addition, not only the interaction channels, but also the presentation channels should be also taken into consideration in the general semantic description of user interactions in VR.

References

1. Nielson, G.M., Olsen Jr., D.R.: Direct manipulation techniques for 3D objects using 2D locator devices. In: Proceedings of the 1986 Workshop on Interactive 3D Graphics, I3D 1986, pp. 175–182. ACM, New York (1987)
2. Wu, S., Ricca, A., Chellali, A., Otmane, S.: Classic3D and single3D: two uni-manual techniques for constrained 3D manipulations on tablet PCs. In: 2017 IEEE Symposium on 3D User Interfaces (3DUI), pp. 168–171 (2017)
3. Kitson, A., Riecke, B.E., Hashemian, A.M., Neustaedter, C.: Navichair: evaluating an embodied interface using a pointing task to navigate virtual reality. In: Proceedings of the 3rd ACM Symposium on Spatial User Interaction, SUI 2015, pp. 123–126. ACM, New York (2015)
4. Alshaer, A., Regenbrecht, H., O'Hare, D.: Investigating visual dominance with a virtual driving task. In: 2015 IEEE Virtual Reality (VR), pp. 145–146 (2015)

5. Thomann, G., Nguyen, D.M.P., Tonetti, J.: Expert's evaluation of innovative surgical instrument and operative procedure using haptic interface in virtual reality. In: Matta, A., Li, J., Sahin, E., Lanzarone, E., Fowler, J. (eds.) Proceedings of the International Conference on Health Care Systems Engineering. Springer Proceedings in Mathematics & Statistics, vol. 61, pp. 163–173. Springer, Cham (2014). https://doi.org/10.1007/978-3-319-01848-5_13

6. Cortes, G., Marchand, E., Ardouinz, J., Lécuyer, A.: Increasing optical tracking workspace of VR applications using controlled cameras. In: 2017 IEEE Symposium on 3D User Interfaces (3DUI), pp. 22–25 (2017)

7. Roup, M., Bosch-Sijtsema, P., Johansson, M.: Interactive navigation interface for virtual reality using the human body. Comput. Environ. Urban Syst. **43**(Suppl. C), 42–50 (2014)

8. LaViola Jr., J.J.: Context aware 3D gesture recognition for games and virtual reality. In: ACM SIGGRAPH 2015 Courses, SIGGRAPH 2015, pp. 10:1–10:61. ACM, New York (2015)

9. Piumsomboon, T., Lee, G., Lindeman, R.W., Billinghurst, M.: Exploring natural eye-gaze-based interaction for immersive virtual reality. In: 2017 IEEE Symposium on 3D User Interfaces (3DUI), pp. 36–39 (2017)

10. Zielasko, D., Neha, N., Weyers, B., Kuhlen, T.W.: A reliable non-verbal vocal input metaphor for clicking. In: 2017 IEEE Symposium on 3D User Interfaces (3DUI), pp. 40–49 (2017)

11. Sobeski, D.A., Andrew, F.G., Smith, M.D.: Context-based dynamic user interface elements. US Patent 6,633,315 (2003)

12. Gebhardt, S., Pick, S., Voet, H., Utsch, J., al Khawli, T., Eppelt, U., Reinhard, R., Bscher, C., Hentschel, B., Kuhlen, T.W.: flapassist: How the integration of VR and visualization tool fosters the factory planning process. In: 2015 IEEE Virtual Reality (VR), pp. 181–182 (2015)

13. Steed, A., Julier, S.: Design and implementation of an immersive virtual reality system based on a smartphone platform. In: 2013 IEEE Symposium on 3D User Interfaces (3DUI), pp. 43–46 (2013)

14. Lutovac Banduka, M.: Remote monitoring and control of industrial robot based on android device and wi-fi communication. Automatika – J. Control Meas. Electron. Comput. Commun. **56**(3) (2015)

15. Monect: Monect PC remote. https://www.monect.com/

16. Young, T.S., Teather, R.J., MacKenzie, I.S.: An arm-mounted inertial controller for 6DOF input: design and evaluation. In: 2017 IEEE Symposium on 3D User Interfaces (3DUI), pp. 26–35 (2017)

17. Zielinski, D.J., Nankivil, D., Kopper, R.: Specimen box: a tangible interaction technique for world-fixed virtual reality displays. In: 2017 IEEE Symposium on 3D User Interfaces (3DUI), pp. 50–58 (2017)

Towards Collaborative Virtual Power Plants

Kankam O. Adu-Kankam[1,2(✉)] ⓘ and Luis M. Camarinha-Matos[1]

[1] Faculty of Sciences and Technology, UNINOVA - CTS,
Nova University of Lisbon, Campus de Caparica,
2829-516 Monte Caparica, Portugal
kankamadu@gmail.com, cam@uninova.pt
[2] School of Engineering, University of Energy and Natural Resources (UENR),
P. O. Box 214, Sunyani, Ghana

Abstract. To promote flexible integration of distributed energy resources into the smart grid, the notion of Virtual Power Plants (VPPs) was proposed. VPPs are formed by the integration of heterogeneous systems, organizations and entities which collaborate to ensure optimal generation, distribution, storage, and sale of energy in the energy market. The collaborative nature of VPPs gives the semblance of collaborative business ecosystem, constituted of a mix of highly interdependent relationship among stakeholders. The systematic literature review methodology is used to summarize research evidence of emerging convergence between the Collaborative Networks (CN) and VPP domains. It is observed that, various strategic and dynamic collaborative alliances are formed within a VPP which are similar to various CN organizational forms like: Virtual Breeding Environments (VBE), grasping opportunity driven-networks etc. CN principles like: virtual organization creation, operation and dissolution, negotiation, broker services, etc., are also found. Emerging collaborative forms like hybrid collaborations between known traditional CN forms were also visible.

Keywords: Collaborative Networks · Virtual Power Plants
Distributed energy resources · Energy market · Smart grid

1 Introduction

The VPP concept is rapidly transforming the way we think, design and plan the development of future energy grids. This is because VPPs have the capability of increasing the chances of integrating renewable energy sources into the conventional power grid. The advantage of this development is the enhancement of sustainable energy generation and its subsequent decline in the use of fossil-based energy sources. It also enhances decentralization of the energy grid from a single supply system to a more liberal and diverse supply, based on multiple sources [1]. The VPP concept will also enable small-scale energy producers to participate in the electricity market and will eventually help to overcome the stochastic nature of distributed energy resources (DERs), resulting in a more stable power grid [2].

© IFIP International Federation for Information Processing 2018
Published by Springer International Publishing AG 2018. All Rights Reserved
L. M. Camarinha-Matos et al. (Eds.): DoCEIS 2018, IFIP AICT 521, pp. 28–39, 2018.
https://doi.org/10.1007/978-3-319-78574-5_3

The VPP concept is supported by a merge of different technological and managerial concepts, principles and ideas from diverse domains and fields of study. One of such fields is the domain of Collaborative Networks (CN) which embodies knowledge about collaboration amongst heterogenous organizations, systems, and entities which are autonomous in nature and are geographically dispersed. These CNs usually collaborate to achieve common and compatible goals [3].

The objective of this work is to perform a panoramic review of the area of VPPs to primarily identify common grounds where both domains converge. This is necessary because VPPs are constituted of integration of heterogenous systems, organizations, and entities which collaborate to ensure optimal generation, distribution, storage, and sale of renewable energy in the energy market. The main contribution of this paper is the extraction of the underlying collaborative mechanisms, organizational forms, motivation for collaborations, key collaboration agents/players as well as related technologies within the VPP area, using foreknowledge in CNs.

2 Relationship to Resilient Systems

VPPs are anticipated to help in overcoming the stochastic nature of DERs which will result in a more stable and smart power grid. However, with the current level of system integration which includes: high levels of artificial intelligence, smart cyber components, cloud computing, IoT, etc., incorporated in the power grid, coupled with frequent disruptive events around the globe which include: natural disasters, globalization, climate change, economic crisis, demographic shifts, fast technological evolution, terrorism and cyber-attacks, the power grid is becoming more and more susceptible to many forms of attacks which will eventually endanger the sustainability of the grid.

The collaborative and decentralized nature of VPPs however presents a good opportunity to incorporate resilience into the power grid. For instance in [4] a collaborative observation network consisting of multiple DERs within the power system which monitor the behaviors of all its neighbors, and collectively decide to isolate DERs suspected to be under attack is proposed. Again, Egbue et al. [5], concluded in a survey work that micro-grids, which also function collaboratively like VPPs have high potential to increase the reliability and resilience of the smart grid during a blackout or cyber-attack. This is because the decentralized nature of micro-grids/VPPs can provide direct cyber-security benefits if structured properly.

In relation to resilient systems, or towards a resilient power grid, the VPPs concept is hereby perceived as essential components of the grid which cannot simply be overlooked.

3 Survey Approach

To establish a credible correlation of the application of CN principles in the domain of VPPs, the systematic literature review (SLR) method was used. The motivation for this approach is that it provides a balanced and objective summary of research evidence, by evaluating and interpreting available research work, relevant to a particular research question, topic area or phenomenon of interest [6]. SLR is evidence based and has been used extensively by many researchers in other domains.

Systematic Mapping (SM) [7] is a variant of SLR which provides a well-defined structure for any area of research, by categorizing articles in that domain in a way that gives a visual summary or pictorial map of the area. According to [8] the main goals of SM are to provide an overview of a research area and also identify the quantity, type of research and results available within it. A secondary goal can be to identify the forums in which research in the area has been published.

Research Questions. The concept of VPP is found to be relatively new, therefore publications in this area are highly dispersed in terms content and organization. To establish a good synergy and a better synthesis of the area, five guiding research questions are developed to help define the scope of the survey. The research questions are as follow:

> *Research question 1:* What are the key drivers or motivation for collaboration in the domain of VPPs? [Seeking to identify motivation for collaborations]
> *Research question 2:* Which collaborative organizations are emerging in the domain of VPPs? [Seeking to identify organizational forms]
> *Research question 3:* Which collaborative principles are being applied in the emerging collaboration forms? [Seeking to identify CN principles]
> *Research question 4:* Which technological elements support collaboration in VPPs? [Seeking to identify collaborative technological]
> *Research question 5:* Who are the key players, agents or systems that participate in these collaborations? [Seeking to identity key players in the collaborations]

4   Focus Areas

4.1   VPP Aggregation

VPP aggregation is the process of collecting and merging capacities of diverse dispatchable and non-dispatchable DERs, energy storage systems, which may include electric vehicles, controllable loads and demand response programs, etc., to create a composite VPP, which capacity, characteristics and functions are equivalent to a physical power plant. The aggregation method is expected to ultimately impact on the performance and operation of the VPP, hence various approaches proposed by different researchers. Table 1 below summarizes the findings under VPP aggregation with emphasis on motivation for collaboration, collaborative principles that were observed, collaborative forms that were seen, and key players involved in the collaboration.

Table 1. Summary of VPP aggregation

Motivation for collaboration	Collaborative principles	Collaborative forms	Key players/systems
1. Exchange of energy services and information [9] Partner search and selection [9] 2. Facilitate energy trade among DER clusters [10, 11] 3. Shared values such as sustainability, common social cohesion, common geographical location or energy sharing [12] 4. Aggregate energy produced by multiple DER supplier agents and make them available to consumers [13] 5. Build proposals for market bids using meter and sensor information as well as forecast of expected load, production, and flexibility from participating DERs [14] 6. Optimization of profit for Wind Power Producers (WPPs) [15]	1. VO creation processes [9–11, 14] VO operation and dissolution [9–11, 14] VBE broker services [9–12, 14] Partner search and selection process [9–11, 14] VBE administrator services [9–11, 14] Negotiations [9–11, 14] VO alignment with opportunity, operation, and competence [9–11, 14] 2. VPP planer/Market broker [13] 3. Contracts formations [15] Consortium formation [15]	*Traditional CN forms* 1. VBE [9–11] 2. Grasping opportunity driven network [9–11, 14] *Emerging CN forms* 1. Hybrid between VBE and Goal Oriented Networks [9–11, 14]	1. Resource provider [9] DER aggregator [9] DER owners [9] 2. Market operator [10, 11] VDCAs [10, 11] 3. Consumers [12] Small-scale producers [12] 4. DER supplier agents and consumers [13] 5. Aggregator solution [14] Prosumer systems [14] Smart meters [14] Sensor networks [14] Forecasting systems [14] 6. Wind power producers [15]

4.2 VPP Architecture and Infrastructure

VPP architecture and infrastructure covers articles that make contributions to VPP architecture and associated infrastructural support. General software and hardware architecture of power systems and related ICT infrastructure are also considered. In Table 2, a summary of VPP architecture and infrastructure in the context of collaborative technology, collaborative infrastructure, motivation for collaborations and key collaborative players are also analyzed and presented.

Table 2. Summary of VPP architecture and infrastructure

Motivation for collaboration	Collaborative technology	Collaborative architecture	Key players/systems
1. Management of power system through hierarchical architecture that enables integration of high number of DERs [16, 17] 2. To allow direct mapping and implementations in various programming languages 3. Enable effective inter device communication and collaboration	1. System of system technology 2. Multi-agent system technology 3. Artificial neural network algorithms 4. The unified modelling language 5. Communication architecture, technologies and standards such as: IEEE 802.11ac, Long Term Evolution (LTE), IEC 61850, IEC 62746, IEC 61968, IEC 62325	1. Hierarchical, Modular and Scalable architecture [11, 16, 17, 19] 2. Smart-grid Architecture Model framework [20]	1. VPPs at local, regional and district levels [16, 17] 2. Producer agents, consumer agents, and flexible consumer agents [21]

(continued)

Table 2. (*continued*)

Motivation for collaboration	Collaborative technology	Collaborative architecture	Key players/systems
4. VPP control and management [18]	standards, TCP-IP based communication protocols, Modbus TCP over Ethernet, Modbus RTU over RS-485	3. Intelligent multi-agent system architecture [21, 22] 4. Hybrid and modular architecture [18] 5. Regionalized multi-agent self-organizing, hierarchical architecture [23]	

4.3   VPP Management

VPP management is the process of organizing and coordinating all resources and activities within the VPP to optimize generation, transmission, and distribution of energy. VPP management results in cost and loss minimization and ultimately maximizes profit. In this work, the authors considered publications that made contributions to various aspects of VPP management. In this focus area, VPP managerial techniques as well as collaborative principles were analyzed. Additionally, motivation for collaborations and key collaborative players/systems were also reviewed. Table 3 present the findings under this focus area.

Table 3. Summary of VPP management

Motivation for collaboration	VPP management technique	Collaborative principles	Key players/systems
1. Resource sharing [24] 2. Address supply and demand mismatches in the grid [25] 3. Minimization of electricity bill for micro-VPP participant [26]	1. Contractual agreement for resource sharing [24] 2. Smart Energy Aggregation Network (SEAN) [25] 3. A business model in the form of services [26] 4. Formation of dynamic groups to handle optimal dispatch of local resources [18] 5. Adoptation of social concept of trust to measure the quality of predictions in the market [23] 6. Sustainable energy micro-system inside a residential building [27] 7. Management through dynamic price mechanism [28] 8. A multi-agent system approach to energy resource management [22]	1. Principle of resource sharing as used in collaborative virtual laboratory is visible, Partner search and selection processes, Internal consortium formation, call for tenders [23, 24] 2. Principle of trust [23] 3. VBE formation [23, 28] 4. VO administrator and broker, VO creation and broker services, VO planner, VO manager and VO coordinator is also visible [22]	1. Network operator, DER owners, Controllable loads, storage systems [25] 2. Residential community of multiple apartments [26, 27]

4.4   VPP Market

The core objective of VPPs is to enable DERs participation in the energy market. The energy market is a trading system that enables purchases, through bids for buying or selling, or through offers and short-term trades, generally in the form of financial or obligation swaps. Bids and offers use supply and demand principles to set the prices. In this work, the authors considered publications that covered tariffs, remunerations and negotiations within the VPP market. The focus here was to identify collaborative technology and collaborative principles within the VPP market.

Table 4. Summary of VPP market

Motivation for collaboration	Collaborating technology	Collaborative principles	Key players/systems
1. Introduction of new market players to modify current energy ecosystem [2] 2. Simulation of a multi-level negotiation mechanism for VPP in the energy market [29] 3. Autonomous VPPs that can decide whether to aggregate into a VPP or negotiate energy prices alone and outside the VPP [30] 4. VPPs to offer optimum remuneration packages to both customers and producers in the cluster [31] 5. clustering approach for a fairer tariff group organization that considers the resource type and the importance of each participating resource in each specific scenario [32] 6. Strategic bidding for VPPs [33] 7. Remuneration and tariffs [34] 8. Intelligent VPP remuneration [35]	1. Electronic notary [36] 2. Coalition formation theory [29] 3. Agent based-modelling and simulation [29, 30, 36] 4. Dynamic strategies [2] 5. Game theory for scenario analysis 6. Resilient systems, [30] 7. Cluster formation and dynamic system [2, 31, 32]	1. VBEs and dynamic VO creation using dynamic aggregations processes [2] 2. Application of e-Notary concepts [36] VO administrator services [36] 3. Grasping-opportunity driven networks [36]	1. VIMSIN Prosumers (individual prosumers) [2] VIMSIN Micro-Grid Aggregators [2] Virtual Micro-grid [2] Telecom Provider [2] 2. Market operator agent, System operator agent, Market facilitator agent, Buyer agents, Seller agents, VPP agents, VPP facilitator agents [36] 3. Clusters of producers and consumers [31] 4. Distributed generation units, DR programs, and suppliers [32]

4.5   VPP Security

VPP networks are supported extensively by ICT infrastructure which are deployed to enable wide area monitoring, protection, and control of the grid. With this kind of integration, the traditional power system is gradually evolving into a cyber-physical entity that is constituted of distributed smart devices which will eventually subject the power grid to cyber related attacks. VPP security therefore a very critical components future grid. Under this section (Table 5), the authors considered collaborative technologies, motivation for collaboration and key collaborative players under VPP security.

Table 5. Summary of VPP security

Motivation for collaborations	Collaborative technology	Key players/systems
1. Multi-layered security approach that can repel attacks and also help better contain cyber intrusions [37] 2. Prompt detection of cyber related attacks [38] 3. To enhance cross-layered grid security against pervasive and persistent attacks [38] 4. Simulate a security system that supports scalability and also insure system security in the distribution network [39] 5. Deploy a security system to detect intruders who may impair proper operation of the grid [40] 6. Simulate an attack-resilient cooperative control strategy [4] 7. Proposes a holistic attack resilient framework to protect the integrated DER and the critical power grid infrastructure from malicious cyber-attacks [41] 8. Assist in circumnavigation of software defined network (SDN) substations that may come under attacks [42] 9. To incorporate the concept of "mutual suspicion" which enables peers to protect themselves and their neighbours in the network [5]	1. Integration of security in cyber physical systems [5, 37, 38, 40, 41] 2. Multi-agent approach to system security 3. Hierarchical systems [39] 4. Software defined network technology [42] 5. Machine leaning techniques for deploying system security [5, 37, 38, 40, 41] 6. Artificial intelligence for system monitoring and intrusion detection [5, 37, 38, 40, 41] 7. System of system technology for integrating security at various levels of smart-grid architecture [37–39]	1. Cyber-security layer Behaviour estimation layer, and a physical security layer [38] 2. Control Centre Cloud, Primary Substation Cloud, Secondary Substation Backbone Cloud, Secondary Substation Cloud [39] 3. Low Level Intruder Detection System (IDS), Medium Level IDS and High Level IDS [40] 4. Cyber-threat modelling framework, resilience analysis framework, attack prevention and detection framework, and collaborative response framework at the cyber, physical, and utility layers of the power system [41] 5. SDN substation, SDN gateway switch, global SDN controller [42]

4.6 VPP Policy and Roadmaps

Under this section, the authors considered short-term and long-term policy documents that support plans intended to guide the development of technologies that will enhance communication and collaborative technologies in the smart grid (SG).

Table 6. Summary of VPP policy and roadmaps

Motivation for collaboration	Collaborative technology/roadmaps	Key players/systems
1. Develop a set of baseline requirements for information security within the SG [43] 2. Proposed the need to agree on a minimum level of security requirements for all components and systems that interconnect to support SG communications [43] 3. Proposition of a sub-roadmap for DR programme using the city of Oregon and the Pacific Northwest as case study [44] 4. Introduction of disclosure policy about cyber vulnerabilities [45] 5. Adaptation of utilities towards various enterprise ICT-architectures to ensure systems reliability and optimize new business capabilities [46] 6. Incorporate policy actions to promote utility access to spectrum by enabling utilities to share public safety spectrum and also share federal spectrum [47] 7. Implementation a green-growth plan as a national policy task [47]	1. Technological roadmaps for DR programs [44] 2. Roadmap to address various cyber security and associated vulnerabilities concerns [45] 3. Roadmaps for the advancement of ICT architectures and infrastructure to enhance smart grid enterprises and different business models [46] 4. Wireless communication technologies through spectrum allocations and sharing [47] 5. Smart grid prototyping [47] 6. Fully functional smart grid society [47]	1. The energy supply value chain which comprises: generation, transmission, distribution and load [43] 2. Various smart appliances running demand response program in the smart grid [44] 3. All entities connected to the grid [45] 4. Utility companies, Federal agencies [47] 5. Smart grid society in Korea [47]

5 Conclusions

The following general remarks can be made based on the outcome of the survey

(a) Various CN organizational forms, principles, concepts and technology have significant level of penetration within the VPP concept.

(b) VPPs are found to form various strategic and dynamic collaborative alliances which are similar to various CN organizational forms.

(c) Prospects for new and hybrid collaborative forms and mechanisms are very high within VPP energy ecosystem.

(d) The two communities were found to use different languages or terminologies although referring to similar concepts. For instance, in the VPP domain, the process of accumulating DERs is called "aggregation". However, a similar process in the domain of CNs is referred to as a "VO creation". Another process in CNs called "partner search and selection process" occurs in the VPP domain, however, this process is named differently and adopts different functional approach. Some examples include the Common Active Registry [9] and Virtual DER Clustering Aggregator [10, 11]. This suggests the need for the two communities to engage in further interactions to develop inter-disciplinary knowledge-base.

(e) The discipline of CN constitutes a matured, well defined and clearly structured body of knowledge in various aspects of collaborations across diverse disciplines. By adopting CN body of knowledge within the VPP collaborative environment, the VPP concept and its associated technologies can greatly be enhanced. A merge between these two disciplines could forge a clear niche for a collaborative VPP ecosystem which is agile and highly resilient.

Acknowledgement. This work was funded in part by the Center of Technology and Systems of Uninova and the Portuguese FCT-PEST program UID/EEA/00066/2013.

References

1. Kramer, O., Satzger, B., Lässig, J.: Managing energy in a virtual power plant using learning classifier systems. In: Proceedings of the 2010 International Conference on Genetic and Evolutionary Methods, GEM, pp. 111–117 (2010)
2. Lyberopoulos, G., Theodoropoulou, E., Mesogiti, I., Makris, P., Varvarigos, E.: A highly-dynamic and distributed operational framework for smart energy networks. In: 2014 IEEE 19th International Workshop on Computer Aided Modeling and Design of Communication Links and Networks, CAMAD, pp. 120–124 (2014)
3. Camarinha-Matos, L.M., Afsarmanesh, H.: On reference models for collaborative networked organizations. Int. J. Prod. Res. **46**(9), 2453–2469 (2008)
4. Kitchenham, B.: Procedures for performing systematic reviews. TR/SE-0401, NICTA Technical Report 0400011T.1 (2004). http://www.ifs.tuwien.ac.at/~weippl/systemicReviews SoftwareEngineering.pdf. Accessed 10 Aug 2017
5. Petersen, K., Feldt, R., Mujtaba, S., Mattsson, M.: Systematic mapping studies in software engineering. In: 12th International Conference on Evaluation and Assessment in Software Engineering, EASE 2008, vol. 17, pp. 68–77 (2008)
6. Dethlefs, T., Preisler, T., Renz, W., Hamburg, H.A.W., Tor, B.: A DER registry system as an infrastructural component for future smart grid applications. In: Proceedings of International ETG Congress, Die Energiewende - Blueprints for the New Energy Age, pp. 93–99 (2015)

7. Botsis, V., Doulamis, N., Doulamis, A., Makris, P., Varvarigos, E.: Efficient clustering of DERs in a virtual association for profit optimization. In: Proceedings - 18th Euromicro Conference on Digital System Design, DSD, pp. 494–501 (2015)
8. Rinaldi, S., Pasetti, M., Ferrari, P., Massa, G., Della Giustina, D., Unareti, S.A.: Experimental characterization of communication infrastructure for virtual power plant monitoring. In: 2016 IEEE International Workshop on Applied Measurements for Power Systems (AMPS), pp. 1–6 (2016)
9. Huang, Y., Warnier, M., Brazier, F., Miorandi, D.: Social networking for smart grid users. A preliminary modeling and simulation study. In: IEEE 12th International Conference on Networking, Sensing and Control, pp. 438–443 (2015)
10. Biswas, S., Bagchi, D., Narahari, Y.: Mechanism design for sustainable virtual power plant formation. In: IEEE International Conference on Automation Science and Engineering, pp. 67–72 (2014)
11. Siebert, N., et al.: Reflexe: managing commercial and industrial flexibilities in a market environment. In: IEEE Grenoble Conference PowerTech, POWERTECH, pp. 1–6 (2013)
12. Baeyens, E., Bitar, E.Y., Khargonekar, P.P., Poolla, K.: Wind energy aggregation: a coalitional game approach. In: Proceedings of 50th IEEE Conference on Decision and Control and European Control Conference, pp. 3000–3007 (2011)
13. El Bakari, K., Kling, W.L.: Development and operation of virtual power plant system. In: 2011 2nd IEEE PES International Conference and Exhibition on Innovative Smart Grid Technologies (ISGT Europe), pp. 1–5 (2011)
14. Bakari, K.E., Kling, W.L.: Fitting distributed generation in future power markets through virtual power plants. In: 2012 9th International Conference on the European Energy Market, pp. 1–7 (2012)
15. Han, X., Bindner, H.W., Mehmedalic, J., Tackie, D.V.: Hybrid control scheme for distributed energy resource management in a market context. In: 2015 IEEE Power & Energy Society General Meeting, pp. 1–5 (2015)
16. Kamphuis, R., Wijbenga, J.P., Van Der Veen, J.S., Macdougall, P., Faeth, M.: DREAM: an ICT architecture framework for heterarchical coordination in power systems. In: 2015 IEEE Eindhoven PowerTech, POWERTECH, pp. 1–4 (2015)
17. Messinis, G., Dimeas, A., Hatziargyriou, N., Kokos, I., Lamprinos, I.: ICT tools for enabling smart grid players' flexibility through VPP and DR services. In: 2016 13th International Conference on the European Energy Market (EEM), pp. 1–5 (2016)
18. Hernandez, L., et al.: A multi-agent system architecture for smart grid management and forecasting of energy demand in virtual power plants. IEEE Commu. Mag. 51(1), 106–113 (2013)
19. Raju, L., Appaswamy, K., Vengatraman, J., Morais, A.A.: Advanced energy management in virtual power plant using multi agent system. In: 3rd International Conference on Electrical Energy Systems (ICEES), pp. 133–138 (2016)
20. Oliveira, P., Pinto, T., Morais, H.: MASGriP—a multi-agent smart grid simulation platform. In: Power and Energy Society General Meeting, pp. 1–8 (2012)
21. Vale, Z.A., Morais, H., Khodr, H.: Intelligent multi-player smart grid management considering distributed energy resources and demand response. In: 2010 IEEE Power and Energy Society General Meeting, pp. 1–7 (2010)
22. Zehir, M.A., Bagriyanik, M.: Smart energy aggregation network (SEAN): an advanced management system for using distributed energy resources in virtual power plant applications. In: 3rd International Istanbul Smart Grid Congress and Fair, ICSG 2015, pp. 1–4 (2015)

23. Fu, H., Wu, Z., Li, J., Zhang, X.: A configurable μVPP with managed energy services: a malmo western harbour case. IEEE Power Energy Technol. Syst. J. **3**(4), 166–178 (2016). https://doi.org/10.1109/JPETS.2016.2596779
24. Brenna, M., Falvo, M.C., Foiadelli, F., Martirano, L., Poli, D.: From virtual power plant (VPP) to sustainable energy microsystem (SEM): an opportunity for buildings energy management. In: 2015 IEEE Industry Applications Society Annual Meeting, vol. 6, pp. 1–8 (2015)
25. Dagdougui, H., Ouammi, A., Sacile, R.: Distributed optimal control of a network of virtual power plants with dynamic price mechanism. In: Proceedings of the 8th Annual IEEE International Systems Conference, SysCon, pp. 24–29 (2014)
26. Morais, H., Pinto, T., Vale, Z., Praça, I.: Multilevel negotiation in smart grids for VPP management of distributed resources. IEEE Intell. Syst. **27**(6), 8–16 (2012)
27. Capodieci, N., Cabri, G.: Managing deregulated energy markets: an adaptive and autonomous multi-agent system application. In: Proceedings of the 2013 IEEE International Conference on Systems, Man, and Cybernetics, SMC 2013, pp. 758–763 (2013)
28. Spínola, J., Faria, P., Vale, Z.: Remuneration of distributed generation and demand response resources considering scheduling and aggregation. In: IEEE Power and Energy Society General Meeting, pp. 1–5 (2015)
29. Faria, P., João, S., Vale, Z.: Aggregation and remuneration of electricity consumers and producers for the definition of demand-response programs. IEEE Trans. Ind. Inform. **12**(3), 952–961 (2016)
30. Rahimiyan, M., Baringo, L.M.: Strategic bidding for a virtual power plant in the day-ahead and real-time markets: a price-taker robust optimization approach. IEEE Trans. Power Syst. **31**(4), 2676–2687 (2016)
31. Ribeiro, C., Pinto, T., Vale, Z.: Remuneration and tariffs in the context of virtual power players. In: Proceedings of the 23rd International Workshop on Database and Expert Systems Applications, pp. 308–312 (2012)
32. Ribeiro, C., Pinto, T., Morais, H., Vale, Z., Santos, G.: Intelligent remuneration and tariffs for virtual power players. In: 2013 IEEE Grenoble Conference PowerTech Towards Carbon Free Society Through Smarter Grids, POWERTECH, pp. 308–312 (2013)
33. Santos, G., Pinto, T., Vale, Z., Morais, H., Praca, I.: Balancing market integration in MASCEM electricity market simulator. In: Power and Energy Society General Meeting, pp. 1–8 (2012)
34. Enose, N.: Implementing an integrated security management framework to ensure a secure smart grid. In: Proceedings of the 2014 International Conference on Advances in Computing, Communications and Informatics, ICACCI, pp. 778–784 (2014)
35. Farag, M.M., Azab, M., Mokhtar, B.: Cross-layer security framework for smart grid: physical security layer. In: IEEE PES Innovative Smart Grid Technologies, Europe, pp. 1–7 (2014)
36. Hittini, H., Abdrabou, A., Zhang, L.: SADSA: security aware distribution system architecture for smart grid applications. In: Proceedings of the 2016 12th International Conference on Innovations in Information Technology, IIT, pp. 1–6 (2016)
37. Sedjelmaci, H., Senouci, S.M.: Smart grid security: a new approach to detect intruders in a smart grid neighborhood area network. In: 2016 International Conference on Wireless Networks and Mobile Communications (WINCOM), pp. 6–11 (2016)
38. Liu, Y., Xin, H., Qu, Z., Gan, D.: An attack-resilient cooperative control strategy of multiple distributed generators in distribution networks. IEEE Trans. Smart Grid **7**(6), 2923–2932 (2016)
39. Qi, J., Hahn, A., Lu, X., Wang, J., Liu, C.: Cybersecurity for distributed energy resources and smart inverters. IET Cyber-Physical Syst. Theory Appl. **1**(1), 28–39 (2016)

40. Aydeger, A., Akkaya, K., Cintuglu, M.H., Uluagac, A.S., Mohammed, O.: Software defined networking for resilient communications in smart grid active distribution networks. In: 2016 IEEE International Conference on Communications (ICC), pp. 1–6 (2016)
41. Egbue, O., Naidu, D., Peterson, P.: The role of microgrids in enhancing macrogrid resilience. In: 2016 International Conference on Smart Grid and Clean Energy Technologies (ICSGCE), pp. 125–129 (2016)
42. Line, M.B., Tøndel, I.A., Jaatun, M.G.: Cyber security challenges in smart grids. In: 2011 2nd IEEE PES International Conference and Exhibition on Innovative Smart Grid Technologies (ISGT Europe), pp. 5–7 (2011)
43. Cowan, K.R., Daim, T.U.: Integrated technology roadmap development process: creating smart grid roadmaps to meet regional technology planning needs in oregon and the pacific northwest. In: Proceedings of PICMET 2012: Technology Management for Emerging Technologies, pp. 2871–2885 (2012)
44. Hahn, A., Govindarasu, M.: Cyber vulnerability disclosure policies for the smart grid. In: 2012 IEEE Power and Energy Society General Meeting, pp. 1–5 (2012)
45. Danekas, C.: Deriving business requirements from technology roadmaps to support ICT-architecture management. In: 2012 International Conference on Smart Grid Technology (SG-TEP), Economics and Policies, no. Section II, pp. 1–4 (2012)
46. Kilbourne, B., Bender, K.: Spectrum for smart grid: Policy recommendations enabling current and future applications. In: 2010 First IEEE International Conference on Smart Grid Communications, pp. 578–582 (2010)
47. Kim, J., Park, H.-I.: Policy directions for the smart grid in Korea. IEEE Power Energy Mag. 9(1), 40–49 (2011). https://doi.org/10.1109/MPE.2010.939166

Decision Support Systems

Selection of Normalization Technique for Weighted Average Multi-criteria Decision Making

Nazanin Vafaei[✉][iD], Rita A. Ribeiro[iD],
and Luis M. Camarinha-Matos[iD]

Computational Intelligence Group of CTS/UNINOVA,
Faculty of Sciences and Technology, Nova University of Lisbon,
2829-516 Caparica, Portugal
{Nazanin.vafaei, rar, cam}@uninova.pt

Abstract. One of the main challenges when evaluating resilience frameworks is the aggregation of different criteria. To calculate the final level of resilience we have to measure different criteria and then fuse the information to obtain a score. Normalization is a crucial step in any decision making process of evaluation of alternatives or frameworks. Normalization transforms heterogeneous criteria data (qualitative, quantitative, different units, etc.) into numerical and comparable data to enable aggregation (fusion) of criteria to determine the rating of decision alternatives. In this study, we evaluate the effects of different normalization techniques on the most well-known multi-criteria (MCDM) method, called Weighted Average (WA) or SAW (Simple Additive Weighting). A small case study for selecting resilience frameworks illustrates our assessment process for selecting the suitable normalization technique.

Keywords: Normalization · MCDM · SAW · WA · Decision making
Pearson correlation · Spearman correlation · Data fusion · Aggregation
Resilient systems

1 Introduction

Usually, in multi-criteria decision-making (MCDM) problems, criteria have different scales (e.g. velocity, fuel consumption, design, etc., in selecting a car problem). As such, we should use some pre-processing to obtain a common scale, which will enable aggregation of numerical and comparable criteria to obtain a final score for each alternative [1, 2]. Therefore, the first step for modeling and applying any MCDM method is to choose a suitable normalization technique for the problem at hand. If the chosen normalization technique is not suitable the best decision solution may be overlooked [3]. As Chatterjee and Chakraborty [3] say: "*In fact, while the normalization process scales the criteria values to be approximately of the same magnitude, different normalization techniques may yield different solutions and, therefore, may cause deviation from the originally recommended solutions*". These considerations are the motivation for this article.

© IFIP International Federation for Information Processing 2018
Published by Springer International Publishing AG 2018. All Rights Reserved
L. M. Camarinha-Matos et al. (Eds.): DoCEIS 2018, IFIP AICT 521, pp. 43–52, 2018.
https://doi.org/10.1007/978-3-319-78574-5_4

In this work, the main research question that we address is: *Which normalization technique is more suitable for usage with the SAW method in the resilience systems?*

To this aim, we discuss an on-going assessment approach [4–6] for recommending the best normalization technique for the most common MCDM method, denoted Weighted Average or SAW (Simple Additive Weighting) [7]. WA/SAW method is the average of criteria rates per each alternative, where criteria can have relative weights [7]. Since SAW is a more formal designation, henceforth we will use this terminology. Preliminary studies, using parts of the mentioned assessment approach, were already applied to TOPSIS method [4, 5] and AHP method [6]. In addition, we expect to contribute to new ways of measuring and classifying resilience frameworks, since MCDM methods – here the SAW one - can provide the ranking for resilient systems [8].

To demonstrate the usefulness of the assessment process being developed, we use a small case of ranking 7 resilience frameworks with 3 criteria.

2   Relationship to Resilient Systems

It is believed that resilience is a succeeding paradigm after classical risk management because their aim is to improve the risk reduction by building systems, which can get back to nominal situation after being impacted by failures.

One of the main challenges in evaluating resilience frameworks is the aggregation of different criteria. MCDM methods are interesting for this kind of problem because they enable to structure and aggregate all information measured by different criteria (or sensors). To aggregate/normalize the different criteria and determine the resilience of a system there are three high level factors to be considered [9] (Fig. 1):

- Cognitive Resilience represents the organizational perspective of a system [10].
- Behavioral Resilience describes organizational behaviors and enable learning and facing uncertain, non-predictive and destructive situations.
- Contextual Resilience is important for measuring the collective reaction of the system to chocks or disasters [9].

Resilience approaches are more than risk management because their objective is to prepare systems for unpredictable situations or disruptions and due to the nature of faults should also be part of any systematic plan for catastrophe management [9]. The three factors mentioned above – usually denoted criteria in the MCDM arena - include

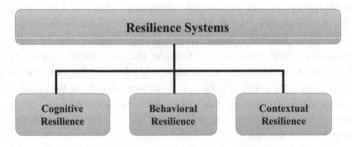

Fig. 1. Resilience system's factors [9]

heterogeneous values that need to be compared and fused to obtain a final score, therefore, normalization (same units) is crucial for enabling merging (fusing) data.

3   Normalization

There are several definitions for data normalization, depending on the study domain [11, 12]. Here we address normalization techniques for Multi-Criteria Decision Making (MCDM), specifically in the Simple Additive Weighting method (SAW). In general, normalization in MCDM is a transformation process to obtain numerical and comparable input data by using a common scale [4–6]. After collecting input data, we must do some pre-processing to ensure comparability of criteria, thus making it useful for decision modeling.

Several studies on the effects of normalization techniques on the ranking of alternatives in MCDM problems have shown that certain techniques are more suitable than others. For example, Chakraborty and Yeh [13] also analyzed four normalization techniques (vector, linear max-min, linear max, and linear sum) in the MCDM simple additive weight (SAW) method. They used a ranking consistency index (RCI) and calculated the average deviation for each normalization technique and concluded that the best normalization technique for SAW is the vector normalization. Here we

Table 1. Common normalization techniques.

Normalization technique	Condition of use	Formula
Linear: Max (N1) [16]	Benefit criteria	$n_{ij} = \frac{r_{ij}}{r_{max}}$
	Cost criteria	$n_{ij} = 1 - \frac{r_{ij}}{r_{max}}$
Linear: Max-Min (N2) [17]	Benefit criteria	$n_{ij} = \frac{r_{ij} - r_{min}}{r_{max} - r_{min}}$
	Cost criteria	$n_{ij} = \frac{r_{max} - r_{ij}}{r_{max} - r_{min}}$
Linear: sum (N3) [2]	Benefit criteria	$n_{ij} = \frac{r_{ij}}{\sum_{i=1}^{m} r_{ij}}$
	Cost criteria	$n_{ij} = \frac{1/r_{ij}}{\sum_{i=1}^{m} 1/r_{ij}}$
Vector normalization (N4) [2]	Benefit criteria	$n_{ij} = \frac{r_{ij}}{\sqrt{\sum_{i=1}^{m} r_{ij}^2}}$
	Cost criteria	$n_{ij} = 1 - \frac{r_{ij}}{\sqrt{\sum_{i=1}^{m} r_{ij}^2}}$
Logarithmic normalization (N5) [2]	Benefit criteria	$n_{ij} = \frac{\ln(r_{ij})}{\ln(\prod_{i=1}^{m} r_{ij})}$
	Cost criteria	$n_{ij} = \frac{1 - \frac{\ln(r_{ij})}{\ln(\prod_{i=1}^{m} r_{ij})}}{m - 1}$
Fuzzification (N6) [18] – trapezoidal function	Benefit & Cost criteria	Using membership functions. E.g. trapezoidal: $$f(x, a, b, c, d) = \begin{cases} 0 & x \leq a \\ \frac{x-a}{b-a} & a \leq x \leq b \\ \frac{d-x}{d-c} & c \leq x \leq d \\ 0 & d \leq x \end{cases}$$

advance this work by also applying Pearson and Spearman correlation coefficients to improve the on-going assessment process of recommending the most suitable normalization technique for MCDM methods.

We selected six (shown in Table 1) of the most promising normalization techniques [2, 4, 5, 14, 15] and analyzed their effect on the SAW method. In Table 1, each normalization method is divided in two formulas, one for benefit and another for cost criteria, to ensure that the final decision objective (rating of alternatives) is logically correct, i.e. for benefit criteria, high values correspond to high normalized values (maximization - benefit) and, for cost criteria, high values correspond to low normalized values (minimization - cost).

4 Assessment Approach for Normalization Techniques in MCDM Methods

There are already some interesting studies [2, 3, 13, 16, 19–21] to assess normalization techniques in MCDM problems. For example, Chakraborty and Yeh [13] used a ranking consistency index (RCI) and weight sensitivity as metrics in order to compare four normalization techniques in TOPSIS method. Also, Celen [16] used a consistency process for assessing banks performance in Turkey, which included using Pearson correlation, as a metric to assess normalization techniques. So far, most studies provide few metrics for solving specific problems and not a general framework to assess MCDM methods. Hence, the authors of this work are developing a general assessment approach to recommend the best normalization technique for well-known MCDM methods, and some preliminary results for the TOPSIS and AHP methods are already published [4–6]. At this stage, our proposed assessment process includes two steps:

A. Determining the Ranking Consistency Index (RCI) from [13].
B. Comparative study between ranking of alternatives using Pearson [16] and Spearman correlation [22] to determine the mean value.

Next, we apply the 2 steps to a small example of selecting resilience frameworks to describe the proposed process for choosing the best normalization technique.

5 Illustrative Example for Recommending Normalization Technique in SAW Method

In this section, we discuss the effect of normalization techniques using the SAW method on an illustrative example with 7 alternative resilience frameworks (RF1, RF2, …, RF7) and 3 criteria (C1 = Cognitive resilience, C2 = Contextual resilience, and C3 = Behavioral resilience). The alternatives and criteria are based on the characteristic of resilience systems, mentioned in Sect. 2 (see Fig. 1).

Table 2 shows the example's input data used for discussing the normalization techniques in the SAW method, where C1 and C2 are benefit criteria, i.e. the higher the raw values the better they should be on the normalization, and C3 is a cost criteria, where low normalized values are desirable.

Table 2. Input Decision matrix for resilience framework example.

	C1	C2	C3
RF1	171.3068	0.3176	3.9516
RF2	178.0288	0.3219	5.5274
RF3	179.3276	0.3263	5.5274
RF4	171.3068	0.3127	3.9516
RF5	179.3276	0.3171	5.5274
RF6	171.0295	0.3214	5.8126
RF7	162.0905	0.3079	10.6341

Following the SAW method procedure (See [7]), we first normalize the input data using the 6 different normalization techniques from Table 1 and then calculate their rating by summing the normalized criteria values per row. The normalized values and final ranking (ordering) of alternatives (resilience frameworks) using the six normalization techniques are depicted in Tables 3 and 4 respectively.

Table 3. Normalized values of alternatives with respect to the criteria using six normalization techniques for SAW method

	Max			Max-Min			Sum			Vector			Logarithmic			Fuzzificztion		
	C1	C2	C3	C1	C2	C3	C1	C2	C3	C1	C2	C3	C1	C2	C3	C1	C2	C3
RF1	0,96	0,97	0,63	0,53	0,53	1	0,14	0,14	0,19	0,37	0,38	0,76	0,14	0,14	0,15	0,83	0,44	1
RF2	0,99	0,99	0,48	0,92	0,76	0,76	0,15	0,14	0,14	0,39	0,38	0,66	0,14	0,14	0,14	0,96	0,44	1
RF3	1	1	0,48	1	1	0,76	0,15	0,15	0,14	0,39	0,39	0,66	0,14	0,14	0,14	0,99	0,45	1
RF4	0,96	0,96	0,63	0,53	0,26	1	0,14	0,14	0,19	0,37	0,37	0,76	0,14	0,14	0,15	0,83	0,43	1
RF5	1	0,97	0,48	1	0,5	0,76	0,15	0,14	0,14	0,39	0,38	0,66	0,14	0,14	0,14	0,99	0,43	1
RF6	0,95	0,98	0,45	0,52	0,73	0,72	0,14	0,14	0,13	0,37	0,38	0,65	0,14	0,14	0,14	0,82	0,44	1
RF7	0,9	0,94	0	0	0	0	0,13	0,14	0,07	0,35	0,37	0,35	0,14	0,15	0,13	0,64	0,42	0,96

Table 4. Ranking of alternatives using six normalization techniques for SAW method

	Max (N1)	Max-Min (N2)	Sum (N3)	Vector (N4)	Logarithmic (N5)	Fuzzification (N6)
RF1	1	4	1	1	2	5
RF2	4	2	4	4	4	3
RF3	3	1	3	3	5	1
RF4	2	6	2	2	1	6
RF5	5	3	5	5	3	2
RF6	6	5	6	6	6	4
RF7	7	7	7	7	7	7

Observing the rankings of the seven alternatives, it is impossible choose which normalization technique is more appropriate for using with the SAW method. Therefore, we applied steps A and B of the evaluation assessment we are developing, mentioned above in the Sect. 4.

For <u>step A</u>, we calculate the RCI (from [13]) with the number of similarity (e.g. T12 = Total number of times N1 and N2 produced the same ranking) or dissimilarity (e.g. TD123456 = Total number of times N1, N2, N3, N4, N5 and N6 produced different rankings) for the six normalization techniques (N1, N2, ..., N6). Further, we should define consistency weight (CW). Since we have 6 normalization techniques in CW is defined as:

1- If a normalization technique is consistent with all 5 techniques, then CW = 5/5 = 1.
2- If a normalization technique is consistent with 4 of other 5 techniques, then CW = 4/5.
3- If a normalization technique is consistent with 3 of other 5 techniques, then CW = 3/5.
4- If a normalization technique is consistent with 2 of other 5 techniques, then CW = 2/5.
5- If a normalization technique is consistent with 1 of other 5 techniques, then CW = 1/5.
6- If a normalization technique is not consistent with any other techniques, then CW = 0/5 = 0.

Now, illustrating the ranking consistency index (RCI) calculation for N1 [13]:

$$
\begin{aligned}
\text{RCI (N1)} = &[(\text{T123456} * (\text{CW}=1) + \text{T12345} * (\text{CW}=4/5) + \text{T13456} * (\text{CW}=4/5) \\
&+ \text{T12456} * (\text{CW}=4/5) + \text{T12346} * (\text{CW}=4/5) + \text{T12356} * (\text{CW}=4/5) \\
&+ \text{T1234} * (\text{CW}=3/5) + \text{T1235} * (\text{CW}=3/5) + \text{T1236} * (\text{CW}=3/5) \\
&+ \text{T1245} * (\text{CW}=3/5) + \text{T1246} * (\text{CW}=3/5) + \text{T1256} * (\text{CW}=3/5) \\
&+ \text{T1346} * (\text{CW}=3/5) + \text{T1356} * (\text{CW}=3/5) + \text{T1345} * (\text{CW}=3/5) \\
&+ \text{T1456} * (\text{CW}=3/5) + \text{T123} * (\text{CW}=2/5) + \text{T124} * (\text{CW}=2/5) \\
&+ \text{T125} * (\text{CW}=2/5) + \text{T126} * (\text{CW}=2/5) + \text{T134} * (\text{CW}=2/5) \\
&+ \text{T135} * (\text{CW}=2/5) + \text{T136} * (\text{CW}=2/5) + \text{T145} * (\text{CW}=2/5) \\
&+ \text{T146} * (\text{CW}=2/5) + \text{T156} * (\text{CW}=2/5) + \text{T12} * (\text{CW}=1/5) \\
&+ \text{T13} * (\text{CW}=1/5) + \text{T14} * (\text{CW}=1/5) + \text{T15} * (\text{CW}=1/5) \\
&+ \text{T16} * (\text{CW}=1/5) + \text{TD123456} * (\text{CW}=0))/\text{TS}] \\
= &[((1*1) + (1* 4/5) + (1* 4/5) + \ldots + (0))/1] = 24
\end{aligned}
$$

Where,

RCI (X) = Ranking consistency index for normalization procedure (X = N1, N2, ..., N6).
TS = Total number of times the simulation was run (in this study TS = 1).
TD123456 = Total number of times N1, N2, N3, N4, N5 and N6 produced different rankings.
T123456 = Total number of times N1, N2, N3, N4, N5 and N6 produced the same ranking.
T12345 = Total number of times N1, N2, N3, N4 and N5 produced the same ranking.

T1234 = Total number of times N1, N2, N3 and N4 produced the same ranking.
T123 = Total number of times N1, N2 and N3 produced the same ranking.
T12 = Total number of times N1 and N2 produced the same ranking.

The RCI for the other normalization techniques are also calculated and results are shown in Table 5.

Table 5. The result of RCI for SAW method

	RCI	Rank
Max (N1)	24	2
Max-Min (N2)	17,8	4
Sum (N3)	25,2	1
Vector (N4)	24	2
Logarithmic (N5)	21	3
Fuzzification (N6)	16,4	5

The results of RCI in Table 5 indicates that the best normalization technique is linear sum (N3). In second place, Vector (N4) and Max (N1) are recommended. Logarithmic (N5), Max-Min (N2) and Fuzzification (N6) do not seem appropriate normalization techniques for SAW method.

In Step B of evaluation assessment, we calculate Pearson and Spearman correlation and their mean ks values (average of correlation values for each normalization technique) for ranking the alternative (for calculation procedure please see [13, 16, 22]). Tables 6 and 7 show the results of Pearson and Spearman correlation.

Table 6. Pearson (P) correlation between rankings of alternatives and Mean ks values.

	Max	Max-Min	Sum	Vector	Logarithmic	Fuzzification	Mean k_s values	Rank
Max		0,321	1	1	0,821	0,107	0,650	1
Max-Min	0,321		0,321	0,321	0,107	0,929	0,400	3
Sum	1	0,321		1	0,821	0,107	0,650	1
Vector	1	0,321	1		0,821	0,107	0,650	1
Logarithmic	0,821	0,107	0,821	0,821		0	0,514	2
Fuzzification	0,107	0,929	0,107	0,107	0		0,250	4

The results of Spearman and Pearson correlation and their mean ks values (Tables 6 and 7) indicate that the best normalization techniques for SAW method is the Sum (N3) and vector (N4) because they have the same mean ks value. Max (N1) is the second most recommended technique because it is ranked in the first priority by Pearson correlation and second by Spearman correlation. Max-Min (N2), Logarithmic (N5) and Fuzzification (N6) are the less recommended techniques.

Table 7. Spearman (P) correlation between rankings of alternatives and Mean ks values.

	Max	Max-Min	Sum	Vector	Logarithmic	Fuzzification	Mean k_s values	Rank
Max		0,095	1	1	0,821	0,107	0,605	2
Max-Min	0,095		0,321	0,321	0,107	0,929	0,355	4
sum	1	0,321		1	0,821	0,107	0,650	1
Vector	1	0,321	1		0,821	0,107	0,650	1
Logarithmic	0,821	0,107	0,821	0,821		0	0,514	3
Fuzzification	0,107	0,929	0,107	0,107	0		0,250	5

Summing up, combining the result of RCI (Step A) and Correlation (Step B), we can see that both steps recommend Sum (N3) as the best normalization techniques for usage with SAW MCDM method and second best are Max (N1) and Vector (N4), thus we can conclude our proposed evaluation assessment process is on the right direction and this approach seems to be robust for recommending the best normalization techniques. We plan to add other measures in the near future, as explained in the conclusion.

6   Conclusions

This study's objective was to recommend which of six common normalization techniques is most suitable for SAW methods. In this paper we discussed the results obtained with an evaluation assessment process that is being developed, which now includes step A- calculating RCI and step B- calculating Pearson and Spearman correlation. To illustrate the approach an example of selecting resilience frameworks was used. With our results (see Tables 5, 6 and 7) the recommendation as best normalization techniques was: first Linear Sum (N3); second Vector (N4) and Max (N1). The other techniques are definitively not suitable for usage in the SAW method of MCDM.

As future work we plan to extend the assessment framework with other metrics and we also plan to perform simulations for testing and validation with specific case studies in resilient systems. With these additions we hope to generalize our preliminary conclusion about the most suitable normalization technique for well-known MCDM methods applied in resilient systems.

Acknowledgements. This work was partially funded by FCT Strategic Program UID/EEA/00066/203 of UNINOVA, CTS.

References

1. Triantaphyllou, E.: Multi-criteria decision making methods. In: Triantaphyllou, E. (ed.) Multi-criteria Decision Making Methods: A Comparative Study. Applied Optimization (APOP), vol. 44, pp. 5–21. Springer, Boston (2000). https://doi.org/10.1007/978-1-4757-3157-6_2
2. Jahan, A., Edwards, K.L.: A state-of-the-art survey on the influence of normalization techniques in ranking: improving the materials selection process in engineering design. Mater. Des. **65**, 335–342 (2015). https://doi.org/10.1016/j.matdes.2014.09.022
3. Chatterjee, P., Chakraborty, S.: Investigating the effect of normalization norms in flexible manufacturing sytem selection using multi-criteria decision-making methods. J. Eng. Sci. Technol. **7**(3), 141–150 (2014)
4. Vafaei, N., Ribeiro, R.A., Camarinha-Matos, L.M.: Importance of data normalization in decision making: case study with TOPSIS method. In: International Conference of Decision Support Systems Technology. Them: Big Data Analytic for Decision Making: An EWG-DSS Conference, [Abstract], Belgrade, Serbia (2015)
5. Vafaei, N., Ribeiro, R.A., Camarinha-Matos, L.M.: Data normalization techniques in decision making: case study with TOPSIS method. Int. J. Inf. Decis. Sci. **10**(N1) (2018, to appear)
6. Vafaei, N., Ribeiro, R.A., Camarinha-Matos, L.M.: Normalization techniques for multi-criteria decision making: analytical hierarchy process case study. In: Camarinha-Matos, L.M., Falcão, A.J., Vafaei, N., Najdi, S. (eds.) DoCEIS 2016. IAICT, vol. 470, pp. 261–269. Springer, Cham (2016). https://doi.org/10.1007/978-3-319-31165-4_26
7. Tzeng, G.-H., Huang, J.-J.: Multiple Attribute Desicion Making: Methods and Applications. Taylor & Francis Group, Boca Raton (2011)
8. Bhamra, R., Dani, S., Burnard, K.: Resilience: the concept, a literature review and future directions. Int. J. Prod. Res. **49**(18), 5375–5393 (2011). https://doi.org/10.1080/00207543.2011.563826
9. Jassbi, J., Camarinha-Matos, L.M., Barata, J.: A framework for evaluation of resilience of disaster rescue networks. In: Camarinha-Matos, L.M., Bénaben, F., Picard, W. (eds.) PRO-VE 2015. IAICT, vol. 463, pp. 146–158. Springer, Cham (2015). https://doi.org/10.1007/978-3-319-24141-8_13
10. Lengnick-Hall, C.A., Beck, T.E., Lengnick-Hall, M.L.: Developing a capacity for organizational resilience through strategic human resource management. Hum. Resourc. Manag. Rev. **21**(3), 243–255 (2011). https://doi.org/10.1016/j.hrmr.2010.07.001
11. Wiki3: Normalization. https://en.wikipedia.org/wiki/Normalization. Accessed 15 Oct 2015
12. Pavlicic, D.M.: Normalization affects the results of MADM methods. Yugosl. J. Oper. Res. **11**(2011), 251–265 (2011)
13. Chakraborty, S., Yeh, C.-H.: A simulation comparison of normalization procedures for TOPSIS. In: Computers and Industrial Engineering, pp. 1815–1820, IEEE, Troyes (2009). https://doi.org/10.1109/iccie.2009.5223811
14. Ross, T.: Fuzzy Logic With Engineering Applications, 2nd edn. Wiley, University of New Mexico, Chichester (2004)
15. Nayak, S.C., Misra, B.B., Behera, H.S.: Impact of data normalization on stock index forecasting. Int. J. Comput. Inf. Syst. Ind. Manag. Appl. **6**(2014), 257–269 (2014)
16. Celen, A.: Comparative analysis of normalization procedures in TOPSIS method: with an application to Turkish deposit banking market. INFORMATICA **25**(2), 185–208 (2014)
17. Patro, S.G.K., Sahu, K.K.: Normalization: a preprocessing stage (2015). http://arxiv.org/ftp/arxiv/papers/1503/1503.06462.pdf. Accessed 15 Aug 2015

18. Ribeiro, R.A.: Fuzzy multiple attribute decision making: a review and new preference elicitation techniques. Fuzzy Sets Syst. **78**(2), 155–181 (1996). https://doi.org/10.1016/0165-0114(95)00166-2

19. Chakraborty, S., Yeh, C.-H.: Rank similarity based MADM method selection. In: International Conference on Statistics in Science, Business and Engineering (ICSSBE 2012), Langkawi, Malaysia (2012)

20. Milani, A.S., Shanian, A., Madoliat, R., Nemes, J.A.: The effect of normalization norms in multiple attribute decision making models: a case study in gear material selection. Struct. Multidiscip. Optim. **29**(4), 312–318 (2004). https://doi.org/10.1007/s00158-004-0473-1

21. Chakraborty, S., Yeh, C.-H.: A simulation comparison of normalization procedures for TOPSIS. In: 2009 International Conference on Computers and Industrial Engineering, pp. 1815–1820, IEEE, Troyes (2009). https://doi.org/10.1109/iccie.2009.5223811

22. Wang, Y.-M., Luo, Y.: Integration of correlations with standard deviations for determining attribute weights in multiple attribute decision making. Math. Comput. Model. **51**, 1–12 (2010)

Residence Efficiency Based on Smart Energy Systems

André Monteiro[1], R. Pereira[1,2] (iD), and F. A. Barata[1(✉)] (iD)

[1] Instituto Superior de Engenharia de Lisboa, Instituto Politécnico de Lisboa,
R. Conselheiro Emídio Navarro 1, 1959-007 Lisbon, Portugal
a37768@alunos.isel.pt, {rpereira,fbarata}@deea.isel.ipl.pt
[2] LCEC, R. Conselheiro Emídio Navarro 1, 1959-007 Lisbon, Portugal

Abstract. The resources used to maintain all the activities in buildings are limited and, for this reason, it is critical the usage of energy efficiency measures. This paper is focused on a smart controller development able to automatically manage a building with the objective of improving its energy efficiency. The controller uses Artificial Neural Networks (ANN) as a forecast method and Fuzzy Logic Control (FLC) as a decision maker. The thermal house model is simulated using a MATLAB toolbox. One of the sub controllers is developed in this paper and is responsible for the HVAC control and optimization. The developed system uses renewable energy sources attached to an energy storage system to increase the overall energy efficiency. To demonstrate the effectiveness of the developed controller it is made an economic analysis of its influence in the build system.

Keywords: Energy efficiency · Artificial neural networks · Fuzzy logic
Smart control

1 Introduction

Data from 2015 shows that 39.8% of energy is consumed by buildings, where residences are responsible by 21.4% and commercial buildings by 18.4% [1]. This data shows that it is relevant to implement systems to reduce energy waste and, simultaneously should be able to provide the appropriate comfort to the habitants.

The intelligent system development that allows improving efficiency derived from technological evolution, resulting in numerous solutions that can be found in literature. In [2] several control intelligent solutions mainly based on predictive control, FLC and ANN are addressed. In [3] a solution for water heating and HVAC system based on ANN is developed as an energy efficiency improvement measure. This ANN is used for forecast and events evaluation. With the same purposed, in [4] a FLC based in MATLAB software is used to manage a building the thermal comfort. In [5] is presented a photovoltaic and energy storage system management resorting to FLC based strategies for commercial buildings. An ANN is developed in [6] for building consumption forecast and FLC is used to improve the system performance.

Considering this, and regarding that HVAC system, lighting and luminosity and standby consumed energy contributes for energy wasting, in this paper, an automatic

© IFIP International Federation for Information Processing 2018
Published by Springer International Publishing AG 2018. All Rights Reserved
L. M. Camarinha-Matos et al. (Eds.): DoCEIS 2018, IFIP AICT 521, pp. 53–61, 2018.
https://doi.org/10.1007/978-3-319-78574-5_5

load control, based in FLC are applied to individual control actions, leading to an improvement of global energy management.

This paper is organized as it follows. In Sect. 2 the relationship to technological innovation for resilient systems is addressed. The house characterization, the load diagram, the models for ANN and FLC for all developed FLC are described in Sect. 3. In Sect. 4 the results of the HVAC and storage FLC implementation are detailed and the overall results of the other sub controllers are summarized. In Sect. 5 there are discussed some of the most important results obtained by using this controller.

2   Relationship to Resilient Systems

Nowadays, the control application in large systems is moving toward a more decentralized approach. Applying this approach in a smart grid context, the interconnection of complex nature equipment brings up an opportunity and also a challenge to ensure more resilient system to threats.

In this paper an automatic load control, based in smart systems applied to individual control actions, leaded to an improvement of global energy management in the considered residence, showing to be a suitable solution supported by several FLC as described in Sect. 3.1. Thus, this model represents new algorithm that embraces all common electrical energy efficiency aspects that can be improved in a domestic residence such as, HVAC, light and luminosity, blinds, energy storage, standby energy and appliances management. The algorithm is used in a smart home context to limit and shift the daily peak demand (considering the peak and off-peak hours), and also to efficiently reduce the electricity consumption.

This approach resorts to the intercommunication between equipments, supported by the internet of things technology, allowing to accomplish more versatile and secure systems.

Due to buildings impact and function in everyone's life [7], they are key elements of cyberphysical systems. Additionally, buildings security and resilience aspects must be considered in energy management systems, mainly because mechanical and electrical systems are important components, in both local and global point of view. Locally, they allow guaranteeing people's comfort, and globally because buildings have impact in the power grid and in distributed energy systems [7].

Concerning to buildings resilience, there are several factors to be considered, such as: energy efficiency, construction aspects, ventilation and water management [8, 9]. However, energy resilience is pointed out as an element that allows improving buildings performance [10]. The use of renewable energy resources associated with energy storage systems, allow buildings to operate in island mode [7, 11], and also give them the possibility of being net-zero energy buildings [12].

Regarding energy management systems, intelligent control is a necessary component in modern buildings that contributes in a significant way for the buildings resilience achievement [7]. In this context, ANN, FLC and some evolutionary algorithms [13, 14] are some tools that support intelligent control actions, because they enable advanced control implementation and also the energy management system optimization, including energy storage system management [15]. These advanced control and optimization, not only contributes for buildings resilience, but also for the power grid resilience improvement [13, 14].

3 Domestic Smart Load Management

As referred, HVAC system, lighting and luminosity and standby consumed energy contributes for energy wasting, consequently it is necessary to characterize the system in order to identify which individual control actions should performed in order to contribute for a global energy management. Thus, the implemented smart load management strategy consists in using a global controller that can be adjusted to each controllable house devices.

3.1 Implemented System Description

The considered house has two bedrooms, one bathroom, one kitchen, one living room and one garage. The space has 100 m^2 and 2.5 m^2 height. Constructive aspects such as windows, walls, roof, among others were properly studied and the equivalent terms were set in the developed MATLAB/Simulink system. The developed thermal model of the house was based and adapted from [16, 17]. It's considered a charging point for an electric vehicle (EV), with 24 kWh battery capacity, is placed in the garage and the resultant charging load profile corresponds to 8 h during night, between 0.00 a.m and 8.00 a.m, with 3 kW.

In order to measure the house consumed energy it is necessary to consider an equipment scheduling and also to know equipment's load profile. Figure 1 shows a load profile example for a drying machine during a normal cycle.

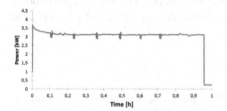

Fig. 1. Dryer consumption profile.   **Fig. 2.** Daily consumption diagram

To characterize the house global consumption, the load profile of other equipment was also considered, namely, washing machine, dishwasher, refrigerator, television, computer, router, lighting, HVAC equipment and electric vehicle charging. The sum of energy consumption of all equipment according to their usage schedule defines house's daily load diagram, as shown in Fig. 2. This diagram was built considering no energy efficiency measures.

From Fig. 2 analysis, can be stated that consumptions peaks require, in Portugal, a 10.35 kVA contracted power, which highly affects the final energy cost per kWh.

However, these energy peaks may be reduced if a better daily load distribution is considered. Hence, the implemented controllers aim also to reduce the average consumed energy and the contracted power needed to guarantee the service's continuity. Therefore, taking advantage of the characteristics for solving vague or uncertain problems, the fuzzy logic was chosen to develop the controllers. So, the smart controller is responsible for the equipment management and energy efficiency and consists in four sub controllers, namely, Light and Luminosity, Load, HVAC and Storage. The lighting and luminosity control associates the blinds opening or closing with lamps output power management. This means that the natural lighting usage is maximized and the electric energy used by artificial lighting minimized, always maintaining the overall illuminance as desired. Controller relates the time of the day with the luminosity inside the house to determine the lamps power output. The same FLC also uses the day time input and associates it with the blinds position to decide if these should open, close or maintain position.

The second sub controller is responsible for load management and verify if equipment is being used or just consuming energy passively. This verification requires the help of smart plugs that can measure the real time load consumed energy. From all studied load profiles analysis, 5 W maximum consumed standby energy is obtained. The controller was programmed to use the information given by the smart plug and after thirty seconds, and if the energy consumed is equal or lesser to 7 W the plug is turned off. Otherwise means the equipment is actively being used and the plug is left turned on. Therefore, the controller relates the inputs time and energy consumed, and decide to keep connected or disconnect the plug from the grid.

The storage sub controller is used in parallel with the implementation of photovoltaic panels. This FLC manage batteries using both energy from the power grid and from photovoltaic (PV) panels according to the time of the day and batteries charging charge. It was designed to maximize the renewable energy storage and if necessary fill the remaining storage space with grid's energy when it is cheaper. The implemented storage system is based on a Tesla Powerwall which allowed energy storage of 13.5 kW. The associated PV panels rated power is 2.7 kW.

3.2 House Control System

The HVAC FLC uses the number of people inside the house and indoor temperature as inputs and determines how much thermal energy is needed to heat or cool the building. The same analysis and methodology that is made for the HVAC controller was also used to build the other three controllers. The house thermal model, the ANN and FLC controllers' configuration, the appliances load profiles, the weather forecasts and the habitant schedule information are all integrated in the same model, as shown in Fig. 3.

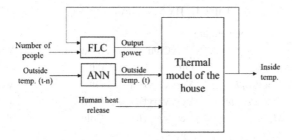

Fig. 3. HVAC control scheme

Regarding controller implementation, a feedforward backpropagation ANN, with 4-20-1 architecture, shown in Fig. 4, was used to forecast the outside temperature.

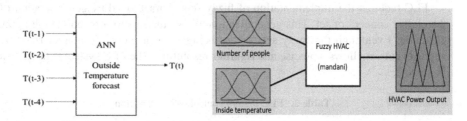

Fig. 4. ANN- Outside temperature forecast.   **Fig. 5.** FLC - HVAC control scheme.

Input data are given by a [182 × 4] matrix, where lines represent the considered days between October and March (2015/2016), and columns represents the four hours that most influence the fifth hour, obtained from a correlation analysis. This ANN has a performance given by a 0.6 mean square error and tests results shown a 2% deviation from the expected values.

In Fig. 5, information about the number of people inside house allows the system to pre-heat/cool the space and also enables knowing how much power the HVAC must provide to compensate peoples' heat release. The inside temperature is responsible by the HVAC control, it determines if the equipment is heating, cooling or if it can be turned off. The variable also defines the set-point and the grade of membership which indicates to the system how far is it from the desired set-point. Therefore, when the temperature is close to the set-point, the HVAC starts to generate less thermal power or, it can even turn off, reducing hereby the spent energy.

The HVAC power output defines quantitatively the needed power to reach the desirable inside temperature set-point in BTU. This variable limit is defined by the HVAC equipment and, in this particular case, 29300 BTU. This specification defines the universe of discourse of the variable that is also characterized by seven membership functions. After the variables fuzzification it is necessary to define a set of rules to control the FLC. Rules are presented in Table 1.

Table 1. HVAC controller rules.

HVAC power	Indoor temperature			
		VeryCold	Moderate	VeryHot
Number of people	*Empty*	Off	Off	Off
	FewPeople	VeryVeryHot	Off	VeryVeryCold
	SomePeople	VeryVeryHot	Cold	VeryVeryCold
	ManyPeople	VeryVeryHot	Cold	VeryVeryCold

All FLC's developed in this work uses the Mamdani inference algorithm, the centroid deffuzification method the maximum aggregation method, the minimum implication method, the "And" method is "min" and the "Or" method is "max". The Mamdani inference algorithm is known as min-max method because it uses minimum and maximum of area according to the fuzzy operator in use [18].

FLC requires the implementation of fuzzy sets characterized by generalization of the classic concept of set. This generalization allows the sets to incorporate the fuzzy character. Eventough the fuzzy sets theory being used in problems characterized by vague events, it allows a precise and rigorous resolution. The FLC system is described is Table 2.

Table 2. FLC system controller's description

Controller	Input/Output	Variable	Membership function	Type	Universe of discourse
Lighting	*In.*	Luminosity	5	Trap.	[0,1]
		Day Time	6	Trap.	[0, 23]
		Blind	3	Triang.	[0, 2.5]
	Out.	Ligthing	6	Triang.	[0, 1]
		Blind control	4	Triang.	[0, 3.5]
Load	*In.*	Equipment power	5	Trap.	[0,200]
		Timer	6	Trap	[0, 60]
	Out.	Plugger	2	Triang.	[−0.5, 1.5]
Storage	*In.*	Battery Status	2	Trap.	[0,1]
		Day Time	3	Trap.	[0, 23]
		PV Status	2	Triang	[0.5, 1.5]
	Out.	Charging	3	Triang.	[−0.5, 2.5]

To choose membership functions type, detailed in Table 2 and in HVAC controller, several empirical tests were performed resulting in the ones that led to better FLC results.

4   Results

The analysis of implemented intelligent controller impact in energy efficiency is addressed for all developed controllers and detailed for storage and HVAC system.

The result of storage controllers' implementation regarding the energy efficiency improvement can be observed in Figs. 6 and 7. From comparison between Fig. 2 and Fig. 6, can be observed an average power reduction, which resulted in a contracted power decreasing of 10.35 kVA to 6.9 kVA and consequently in a monthly energy bill reduction of 4.39€. The implement fuzzy controller allowed the storage system to support domestic consumptions between 6 p.m. and 0.00 a.m.

Fig. 6. House consumption profile.   **Fig. 7.** Global energy cost.

The result of HVAC controller implementation is shown in Fig. 8, where can be observed a 0,8€ daily energy cost, for the same operation period.

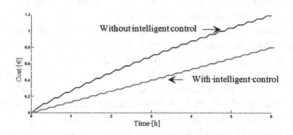

Fig. 8. HVAC with and without intelligent control

From comparison between the two lines shown in Fig. 8, the impact of intelligent control resulted in 0,4€ daily cost reduction. From these results is possible to empirically inferred that an annually energy cost reduction of 72€ can be obtained.

Comparing the consumptions without energy efficiency measures, each system controller have contributed for energy consumption reduction, namely 71% in the lightning and luminosity, 35% in the HVAC, 71% due to the PV panels and storage, and 93% in load passive energy consumption. The reduction in the contract power resulted in an annually energy bill reduction of approximately 53€.

5   Conclusions

The FLC and ANN implementation with the objective of consumed energy reducing is viable. The global developed controllers can be easily adjusted to fit users' needs and to overcome the restrictions provided by users. The variables and membership functions implemented allowed achieving good results. The implemented ANN is a reliable method for temperature forecasting and tests results show a 2% deviation from the expected values.

The storage system enables the energy consumption in off-peak periods and maximizes the photovoltaic energy usage, resulting in a less exigent load diagram. This not only represents lower energy peaks in the most expensive day periods, but also allows the consumer to contract less power to the energy distributor, which results in a lower energy bill.

This paper shows that, if consumer adopts energy efficiency measures, such as, replacing non efficient appliances by equivalent and more efficient ones, reducing standby consumption and improving HVAC control system, a significant energy cost reduction is achieved. Consequently, the global implementation of intelligent systems, namely FLC and ANN, allows contributing for a more resilient and efficient building.

It is intended to address a more detailed economic analysis of the implemented system in future work developments.

References

1. National Renewable Energy Laboratory (NREL). Renewable Energy Data Book (2015)
2. Alcal, R.: Tuning fuzzy logic controllers for energy efficiency consumption in buildings. In: EUSFLAT-ESTYLF Joint Conference, pp. 1–4 (1999)
3. Kolokotsa, D., Tsiavos, D., Stavrakakis, G.S., Kalaitzakis, K., Antonidakis, E.: Advanced fuzzy logic controllers design and evaluation for buildings occupants thermal-visual comfort and indoor air quality satisfaction. Energy Buildings 33(6), 531–543 (2001)
4. Ying-Guo, P., Hua-Guang, Z., Zeungnam, B.: A simple fuzzy adaptive control method and application in HVAC. In: Fuzzy Systems Proceedings, pp. 528–532 (1998)
5. Toshifumi, I., Masanori, K., Akira, T.: A hybrid energy storage with a SMES and secondary battery. IEEE Trans. Appl. Supercond. 15(2), 1915–1918 (2005)
6. Technische Universität Darmstadt. https://www.tracebase.org/
7. Manic, M., Wijayasekara, D., Amarasinghe, K., Rodriguez-Andina, J.J.: Building energy management systems: the age of intelligent and adaptive buildings. IEEE Ind. Electron. Mag. 10, 25–39 (2016)
8. Building resilience through sustainability. https://facilitydude.com/blog/buildingresilience-through-sustainability
9. 3 ways sustainability complements building resilience. http://www.facilitiesnet.com/emergencypreparedness/
10. Levite, B., Rakow, A.: Energy Resilient Buildings and Communities: A Practical Guide. Fairmont Press, Lilburn (2015)
11. Lamichhane, S., Nazaripouya, H., Mehraeen, S.: Micro grid stability improvements by employing storage. In: Proceedings of the IEEE Green Technologies Conference, pp. 250–258 (2013)
12. Kleissl, J., Agarwal, Y.: Cyber-physical energy systems: focus on smart buildings. In: IEEE Design Automation Conference, pp. 749–754 (2010)
13. Werbos, P.J.: Computational intelligence for the smart grid-history, challenges, and opportunities. IEEE Comput. Intell. Mag. 6(3), 14–21 (2011)
14. van Eck, N.J., Waltman, L., Berg, J., Kaymak, U.: Visualizing the computational intelligence field. IEEE Comput. Intell. Mag. 1(4), 6–10 (2006)

15. Kupzog, F., Sauter, T., Pollhammer, K.: IT enabled integration of renewables: A concept for the smart power grid. EURASIP J. Embed. Syst. **2011**(5), 1–8 (2011)
16. The MathWorks. https://www.mathworks.com/help/simulink/examples
17. Barata, F.A., Neves-Silva, R.: Distributed MPC for thermal comfort in buildings with dynamically coupled zones and limited energy resources. In: DoCEIS 2014, Technological Innovation for Collective Awareness Systems, IFIP Advances in Information and Communication Technology, vol. 423, pp. 305–312 (2014)
18. Quadrado, J.C.: Posicionadores eletromecânicos: Introdução à síntese de controladores inteligentes. Master Thesis (1994)

Next Day Load Forecast: A Case Study for the City of Lisbon

Svetlana Chemetova[1(✉)], Paulo Santos[1(✉)],
and Mário Ventim-Neves[2(✉)]

[1] Department of Electrical Engineering, EST Setúbal, Polytechnic Institute
of Setúbal, Rua Vale de Chaves Estefanilha, 2910-761 Setúbal, Portugal
{svetlana.chemetova,paulo.santos}@estsetubal.ips.pt
[2] Department of Electrical Engineering, Faculty of Sciences and Technology,
Universidade Nova de Lisboa, Quinta da Torre, 2829-516 Caparica, Portugal
ventim@uninova.pt

Abstract. Effective short-term load forecasting plays a crucial role in the operation of both traditional and deregulated power systems. Improving the accuracy of load forecasting can increase the appropriateness of planning and scheduling and reduce operational costs of power systems making them resemble resilient energy systems. In the present paper, we propose the regressive forecast model of the day ahead based on the artificial neural network. The electric load peaks were also calculated by the model. The data used were the time series of active power, recorded by EDP Distribution Telemetry System, collected in Lisbon. The results show that our approach provides a reliable model for forecast daily and hourly energy consumption, as well the load profile with accuracy.

Keywords: Load forecasting · Electric power system · Peak load
Neural networks · Load pattern

1 Introduction

Short-term load forecasting is an important basis for the reliable and economical operation of the electric power system. The accuracy of the forecast of power consumption directly affects the quality of dispatch control and the reliability of electricity supply [1]. With the development of driving assistance methodologies, according to the introduction of the Smart Grid concept, a short-term (24 h) load forecasting plays a fundamental role in the planning, management and control of electric power networks [2, 3]. The peak forecast also is very important since at peak times of consumption the price of energy in the electricity market is higher [4]. Thus, the creation of a suitable forecasting method that improves the accuracy of the electric load forecast is of a great practical importance.

The evolution of forecasting methodologies has made considerable progress. In the second half of the last century, methodologies were mainly based on regressive approximations. In the 1980s and early 1990s, the methodologies were related to the knowledge and techniques of artificial neural networks, hybrid systems and genetic algorithms. The limits and adaptability of each method depended on several factors,

© IFIP International Federation for Information Processing 2018
Published by Springer International Publishing AG 2018. All Rights Reserved
L. M. Camarinha-Matos et al. (Eds.): DoCEIS 2018, IFIP AICT 521, pp. 62–70, 2018.
https://doi.org/10.1007/978-3-319-78574-5_6

such as: the diversity and temporal depth of the information collected, the forecast horizon, the climate of the geographical region, and the degree of interdependence between consumption and meteorological variables [5].

Examples of statistical methods (traditional) include Multiple Linear Regression, Exponential Smoothing, Stochastic Time Series (Autoregressive Model, Autoregressive Moving-Average Model, Autoregressive Integrated Moving-Average Model) etc. Methods of Artificial Intelligence include Artificial Neural Networks (ANN), Fuzzy Logic, Genetic Algorithms, Knowledge-based Expert Systems etc. [6].

Statistical methods have the advantage of being well-developed and based on mathematical equations, although, they depend on a complex mathematical modelling. Those are hard to deal with rapid variations in factors such as climate, holidays, etc. The regressive model considers the load as a linear combination of functions, such as sinusoids, exponentials, etc. It consists in determining the curve that best fits the historical data of the load, using least squares method. Despite historical data incorporation, models based on time series can add other factors, such as ambient temperature, seasonality, and random effects.

Artificial intelligence-based methods allow a better handling with uncertainty, as well as non-linear functions. The prediction based on the ANN was widely accepted by the scientific and engineering spheres, becoming the most widespread technique for the load forecasting. There are several scientific publications that prove the quality and robustness of predictions based on neural networks [1, 2, 7–9].

This paper presents a methodology based on the ANN of daily load diagram and load peak forecast, using the time series of data of the EDP (Distribution System Operator in Portugal - DSO). To test and evaluate the load forecasting model, as the case study was selected the electric load demand of the capital of Portugal, Lisbon, in the years 2014–2015.

2   Relationship to Resilient Systems

The concept of resilience is a crucial element in shaping the strategies for establishing and developing the energy systems designed to withstand the widest range of external shocks. Resilience is defined as the ability of a system to absorb external shocks and restore itself while undergoing changes to maintain essentially the same function. Differentiating from the concept of sustainability, the emphasis of resilience is on how the system responds to non-linear dynamic perturbations.

Future resilient energy systems must provide, maintain sufficient and necessary customer services in the event of catastrophic events (for example, natural disasters, human error, and political instability) which could lead to the failure of the energy system [10].

From the definition presented, it is important to predict the load of the electrical system, the main function of the system is the power consumption of electric power. An electrical power system must be able to continuously meet the demand for electrical energy in an economical, reliable manner and in accordance with pre-established criteria for the risk of disruption. For these objectives to be achieved the operation of the system must be planned in advance. Better forecasting of the load diagram will allow planning of system reserves, lines and generation with a higher level of confidence to cope with any constraints and incidents that may occur.

Prediction methodology for the next day intends to increase the accuracy of the forecast, reducing its error, both in the temporal component of the diagram (load as a function of time) and in the peak (maximum value of the diagram). Thus, knowing better the expected load diagram, conditions of more effective response to the perturbations are created, increasing the resilience of the system.

The proposed model, aims to confirm the accuracy of the prediction in both aspects, enhancing the resilience of the operation of the electrical system.

3 Collection and Processing of Data

The proposed methodology is illustrated by a case study concerning the electric supply of the city of Lisbon (Fig. 1).

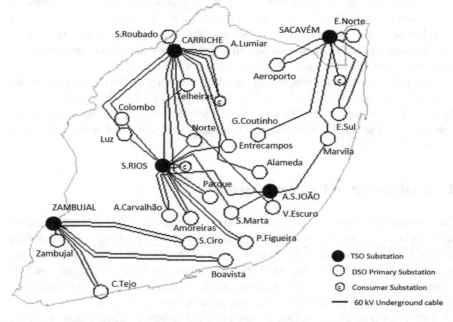

Fig. 1. Simplified scheme of the substations and high voltage lines of the electricity supply of the city of Lisbon.

There are five primary high voltage substations 220 kV/60 kV operated by REN (the Portuguese Transmission System Operator - TSO) that feed the electricity grid of the city of Lisbon: Sacavém, A.S. João, Carriche, Sete Rios, Zambujal [11]. The main sectors of energy consumption in Lisbon are domestic, services and commercial.

As an example of the time series of collected data, Fig. 2 shows the evolution of electric consumption in Lisbon in 2014.

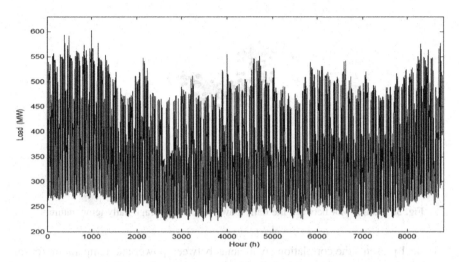

Fig. 2. Temporal series of load of Lisbon, 2014.

The source of the load data is the telemetry data acquisition system which records the power value which is equal to the average power of the instantaneous power during a time interval of 15 min. The data collection period was between 1 January 2014 and 31 December 2015.

After obtaining the load data it is important to evaluate the daily and weekly load profiles. This analysis helps to understand the behavior of demand for electricity over time, especially in the short-term horizon.

The data series were processed to discovering missing data, removing aberrant data or correcting outliers in system logs.

The discrepant data were replaced by the respective average values calculated considering the seasonality in the load curve in hourly basis, that is, daily, weekly and monthly.

The meteorological data were collected from the Portuguese Institute of the Sea and Atmosphere (IPMA) for the same period and for the same geographical area.

It is very important to choose the correct variables to compose the input vector in the forecast model. Load values (active power) are defined as endogenous variables and the meteorological factors (temperature, relative humidity, wind speed etc.) as exogenous variables.

After testing several interdependencies between the endogenous and exogenous variables, it was concluded that there is no strong correlation between them. The correlations obtained were weak. As an example, Fig. 3 shows the correlogram between daily average power and average daily temperature recorded in the city of Lisbon in 2014. There is an increase in power during cold (3–10 °C) and hot (25–30 °C) periods of the day.

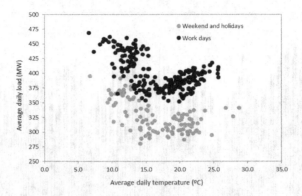

Fig. 3. Scatter plot between the daily power and the average daily temperature.

Table 1 presents the correlation coefficients between power and temperature for the same period.

Table 1. Correlation coefficients corresponding to the correlogram in Fig. 3.

All days of the year	Work days	Saturdays	Weekends and holidays
−0.31	−0.53	−0.61	−0.56

Studies of other Portuguese researchers have also shown that the meteorological factors (among which temperature is the most significant) did not affect the forecast of short-term load [8, 12]. Portugal is a country with a Mediterranean climate, where sudden temperature changes are rare. The forecast model can shape and adapt to small temperature changes in a short time. Thus, in the composition of the input vector, only the endogenous variables were considered, neglecting the exogenous variables.

4 Load Forecasting Methodology

In the proposed methodology, the ANN is used as a computational tool for forecasting.

By autocorrelation analysis of the Lisbon load data series of 2014, it was chosen to include in the input vector the homologous values of active power for the previous day of the forecast and the previous week, as well as the adjacent values: $P(t-23)$, $P(t-24)$, $P(t-25)$, $P(t-167)$, $P(t-168)$, $P(t-169)$, where t is the time of day of the forecast load. These adjacent values, taken together, indicate the trend of evolution of the electric load that is related to the derivative term concept [13].

To improve the performance of the forecast model, we added the values of load levels in the input vector of the model, as one more endogenous variable. The load levels are average values of load at intervals of several consecutive hours. In general, they serve to simplify the amount of information used in the analysis and procedures of phenomena that occur in the operation of electrical systems. The values of the levels can be identified as the average loads of Peak, Half-peak, and Off-peak. These names are the same as those used in the Portuguese tariff system [4], although the temporal intervals of the levels in our model are different, approaching the "natural" behavior of the load diagram. The time intervals in the Portuguese tariff system are determined considering the specific electrical characteristics of each region, in particular as regards to the evolution of its load diagram [14].

Three patterns of diagrams were established, one for weekdays, one for Saturdays and one for Sundays and holidays. The season (summer, winter) also influences electricity consumption. Thus, the average daily load diagrams were obtained for each season of the year and for each set of weekdays, Saturdays, Sundays and holidays. The duration of each level in each diagram is obtained by minimizing the sum of the differences between the average load diagram and the primary level diagram. As an example, Fig. 4 illustrates the profile of Lisbon city load levels, based on the calculation of the average winter power value on the working days of 2014. Similarly, the profiles of the summer load levels were constructed.

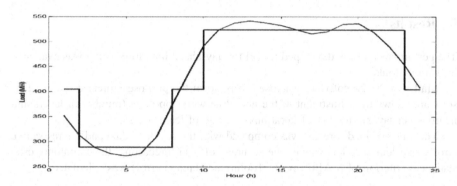

Fig. 4. Average daily load diagram with corresponding levels. Lisbon 2014, Winter, work days.

Figure 5 depicts the final input vector configuration. Schematically the inputs of the forecast model are represented on the left side, and outputs - on the right side. In this figure, the variables of historical load data are the power values used in the construction

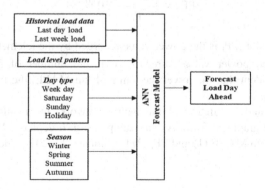

Fig. 5. Input vector configuration for day ahead load forecast.

of the input vector. Load level pattern variables correspond to the load profile level of a season of the year as well as to the type of day of the week.

The method based on the backpropagation algorithm, Levenberg-Marquardt, was used to train the three-layer feed-forward neural network. This algorithm is very efficient with fast and robust convergence. It is based, for the training acceleration, on the determination of the second-order derivatives of the quadratic error in relation to the weights, differing from the traditional backpropagation algorithm that considers the first-order derivatives. Thus, for a moderate amount of data it has an excellent performance with reduction of the number of iterations, but with a longer processing time.

The time power series were divided into three sets of data: training set (with 50% of data), validation set (with 25% of data) and test set (with 25% of data). The date for the year 2014 was used for the forecasting exercises, and simulation was performed with data from 2015. The training and validation set was used to adjust the parameters of the proposed model, while the test set is used to measure the performance of the model.

5   Results

The performance of the developed model for day-ahead load forecasting was evaluated by various trials.

MatLab R2012b software was used to perform the proposed forecast model. This software allows the adjustment of the neural network topology (number of layers and neurons per layer), number of iterations, the rate of learning, etc.

The value of load forecast was compared with the real load data and the respective error was calculated. To evaluate the accuracy of load forecasting, the commonly used statistical criterion of MAPE (mean absolute percentage error) was calculated:

$$\text{MAPE} = \sum_{t=1}^{n} \frac{|PE_t|}{n} \tag{1}$$

Where PE_t (percentage error) is defined as:

$$PE_t = \frac{(P_t - \hat{P}_t)}{\hat{P}_t} 100\% \tag{2}$$

In these expressions, P_t is the power value recorded by the telemetry system (real), $\hat{P}_t$ is the calculated power value (forecast), n is the number of registers (sample members). It is evident that the lower the value of the MAPE, the better the forecast performance [7, 8].

After training and simulation of the neural network, the results were obtained numerically and in graphical form. As an example, Fig. 6 depicts 24-hour-ahead load forecast together with MAPE (1) and PE_t (2) for January 22, 2015, and Fig. 7 - for July 2015.

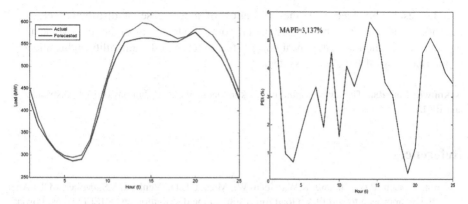

Fig. 6. Load diagram (real and forecast) and error calculated for January 22, 2015

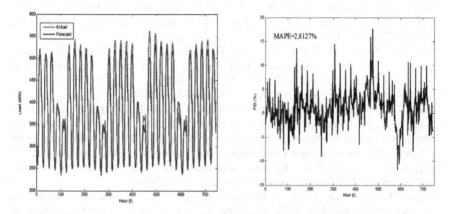

Fig. 7. Load diagram (real and forecast) and error calculated for July 2015

As MAPE, one of the most important statistical indicators of error, remains below 5%, and the results of the prediction model are satisfactory, the objective of the development of the accurate prediction model has been reached.

6 Conclusions

The present article describes a hybrid methodology for forecasting the load diagram for the next day, using the advantages of techniques based on ANN, combined with an innovative approach of composition of the input vector of these networks. The novelty introduced in the input vector that, in addition to active power variables, chosen according to the correlation techniques with meteorological and autocorrelation variables, includes the load levels.

The results obtained with the simulation of a real case of distribution electric network of the city of Lisbon, using the proposed model, confirm the improvement of the accuracy of the forecast, contributing to the increase in the reliability and resilience of the operation of the electric system.

Acknowledgments. The authors gratefully acknowledge the contributions of EDP Distribuição and IPMA.

References

1. Sun, X., Luh, P.B., Cheung, K.W., Guan, W., Michel, L.D., Venkata, S.S., Miller, M.T.: An efficient approach to short-term load forecasting at the distribution level. IEEE Trans. Power Syst. **31**, 2526–2537 (2016)
2. Rodrigues, F., Cardeira, C., Calado, J.M.F., Melício, R.: Family houses energy consumption forecast tools for smart grid management. In: Garrido, P., Soares, F., Moreira, A.P. (eds.) CONTROLO 2016. LNEE, vol. 402, pp. 691–699. Springer, Cham (2017). https://doi.org/10.1007/978-3-319-43671-5_58
3. Quilumba, F., Lee, W., Huang, H., Wang, D., Szabados, R.: Using smart meter data to improve the accuracy of intraday load forecasting considering customer behavior similarities. IEEE Trans. Smart Grid **6**(2), 911–918 (2015)
4. http://www.erse.pt/eng/Paginas/ERSE.aspx. Accessed Nov 2017
5. Gross, G., Galiana, F.: Short term load forecasting. Proc. IEEE **75**(12), 1558–1573 (1987)
6. Singh, A.K., Ibraheem, S.K., Muazzam, M., Chaturvedi, D.K.: An overview of electricity demand forecasting techniques. In: Proceedings of National Conference on Emerging Trends in Electrical, Instrumentation & Communication Engineering, vol. 3, no. 3, pp. 38–48 (2013)
7. Hippert, H., Pereira, C., Souza, R.: Neural networks for short-term load forecasting: a review and evaluation. IEEE Trans. Power Syst. **16**(1), 44–55 (2001)
8. Santos, P., Martins, A., Pires, A.: Designing the input vector to ANN-based models for short-term load forecast in electricity distribution systems. Electr. Power Energy Syst. **29**, 338–347 (2007)
9. Giacometto, F., Cárdenas, J., Kampouropoulos, K., Romeral, L.: Load forecasting in the user side using wavelet–ANFIS. In: IECON, 38th Annual Conference of the IEEE Industrial Electronics Society, pp. 1049–1054 (2012)
10. Kishita, Y., McLellan, B.C., Giurco, D., Aoki, K., Yoshizawa, G., Handoh, I.C.: Designing backcasting scenarios for resilient energy futures. Technol. Forecast. Soc. Chang. **124**, 114–125 (2017)
11. https://www.ren.pt/. Accessed Dec 2017
12. Fidalgo, J.N., Lopes, J.A.P.: Load forecasting performance enhancement when facing anomalous events. IEEE Trans. Power Syst. **20**, 408–415 (2005)
13. Lourenço, J.M., Santos, P.: Short term load forecasting using a Gaussian process model. Intell. Decision Technol. **6** (2012). ISSN 1872-4981/12
14. Apolinário, I., Felizardo, N., Leite Garcia, A., Oliveira, P., Trindade, A., Verdelho, P.: Determination of time-of-day schedules in the Portuguese electric sector. In: Power Engineering Society General Meeting. IEEE, 18–22 June 2006

Supervision Systems

Resilience Supported System for Innovative Water Monitoring Technology

Koorosh Aslansefat[(⊠)] ⓘ, Mohammad Hossein Ghodsirad ⓘ, José Barata ⓘ,
and Javad Jassbi ⓘ

Faculty of Science and Technology, Nova University of Lisbon,
UNINOVA - CTS, Campus de Caparica, 2829-516 Monte Caparica, Portugal
k.aslansefat@campus.fct.unl.pt

Abstract. The level of intelligence in monitoring & controlling systems are increasing dramatically. The critical issue for an autonomous resilient system is detecting the anomalous behavior through standard patterns to react properly and on time. In cyber-physical systems with the interaction of humans and machines, this will be more complicated. Deceptive alarm is a common dilemma in real systems which could reduce awareness and readiness and accordingly resilience of the system. In this paper, Markov modeling technique is used to predict human behaviors patterns to distinguish between human anomalous behavior and system failure. The data is from the real experience of implementing innovative monitoring system in a five-star hotel which was part of the project of gamification for changing guests' behavior. The idea was to develop Resilience Supported System to decrease the fault error and alarms and to increase the reliability and resilience of the system.

Keywords: Performance assessment · Markov modeling
Behavior-change systems · Resilience supported system · Deceptive alarm

1 Introduction

In commercial buildings, Building Automation Systems (BASs) are responsible for considerable savings in water and energy consumptions [1, 2], while the performance of the system is affected by consumption behavior of users.

An example can be made in the hospitality industry. In hotels, the execution of automatic energy management system is tied to heating, cooling, and light comfortability of guests including their consumption behavior. On the other hands, hotel managers are always concern about the experience of guests in rooms and never accept novel BAS solutions which threaten the comfortability of customers during their stay. So, more complexity will be added to BAS when it is affected by different elements at the same time.

The Optishower as a technology is an electricity, gas and water consumption monitoring system in hotels. The monthly, weekly, daily and real-time consumption graphs in the application show the consumption performance of the building and can be implemented in cost reduction strategies.

© IFIP International Federation for Information Processing 2018
Published by Springer International Publishing AG 2018. All Rights Reserved
L. M. Camarinha-Matos et al. (Eds.): DoCEIS 2018, IFIP AICT 521, pp. 73–80, 2018.
https://doi.org/10.1007/978-3-319-78574-5_7

Optishower applies both hardware and software technologies in a smart integrated platform. The ultra-low power consuming data transfer technology makes it possible to take advantage of installing smart Internet of Things (IoT) sensors measuring water and electricity consumption without any destructive intervention in water and electricity infrastructure of buildings. All measured consumption data is sent to a data transmission gateway to be delivered to the data analysis and decision-making platform in the cloud. Finally, this data will be shown to related technicians, managers, and individuals to consider them via interactive and smart online platform.

This service aims to increase the profit margin of the hotel by reducing water and electricity consumption. It works by creating awareness about environmental impacts of consuming water and electricity and motivating hotel guests to use water and energy in a wiser and eco-friendly manner.

Long-term influence prediction of educational interventions on adolescents' development based on the evolutionary causal matrices (ECM) and the Markov Chain has been developed through MATLAB by [3]. They created a computational model predicting longitudinal influences of different types of stories of moral exemplars on adolescents' voluntary service participation and verified the algorithm through surveyed data. The adaptive and automated decision-engine for improving the inherent resilience of autonomous systems has been presented in [4]. Reference [5], proposed new concepts on grid operation considering unexpected extreme disturbances and energy resources with leveraging distribution. In [6], low-cost detection methods have been identified, and novel guidelines for recovery and diagnosis has been provided with focusing on hard faults. Exploring the literature review it shows that there is no previous research work to employ Markov model in case of probabilistic BSC modeling. In this paper the main objective is to use Markov model to solve this challenge in the resilience support topic.

The organization of this paper is as follows. In Sect. 2, the contribution of this study in resilience systems will be presented. Section 3 introduces the innovative technology for water metering. Section 4 proposes the probabilistic solution based on Markov chain for Behavior Change Systems (BCSs). Hypothetical and experimental results will be addressed in Sect. 5. The use of the proposed solution for resilience improvement of a system will discuss in Sect. 6. Finally, the paper terminates with conclusions and future works.

2   Relationship to Resilient Systems

Resilience is widely recognized as a new paradigm and system designers, inspired by nature, are trying to improve the performance of the human-made systems to provide the capability of dealing with any disorder. The resilient system could be seen as the next generation of the robust or agile system. The resilient system works based on simple fact that it could recognize the attack or abnormal behavior and try to recover using its capability. This will help to reduce the level of the vulnerability of the system and to make sure that they could work effectively in an uncertain environment. Although the main challenge in resilient systems is how to recover and to react according to the input of the system but fault detection and recognizing unacceptable behavior is the trigger of the process which is a common challenge in all types of self-organizing system. In Cyber-physical systems, in which the output is the combination of both machine and human behavior, it would be difficult to distinguish the abnormal behavior by its cause. The challenge of the autonomous and intelligent system is how to detect a deviation and avoid fault error or alarm. In case of frequent deceptive alarms, the sensitivity of the operators will be decreased, and this will reduce the readiness and awareness of the system operators and increase the risk. Resilience support system, which is introduced in this paper, is a kind of decision support system based on Markov modeling technique to separate human behavior of system failure from machine failure. In this work, we use the experience and the database from the implementation of innovative technology which was selected by a five-star hotel as a solution for monitoring water and energy. Figure 1 illustrates innovative water consumption measurement called Optishower. The main mission was to use gamification and by producing feedback for hotel guests, help to change the behavior which means here decrease the amount of water and energy used by them.

Fig. 1. Optishower system demonstration

The sensors which are monitoring each room of the hotel are sending the data online, and separately. Aay strange behavior could be because of one of these reasons:

(a) Problem in water system such as water leakage
(b) Problem in monitoring system such as calibration of the sensors
(c) Strange human behavior

The first two problems could be categorized as a system failure which needs immediate recovery while the third one is just a result of unexpected guest behavior. Output analysis of the data will just show that something is wrong, but it was important to distinguish between human behavior and system failure. Sensors represent aggregation of both human and machine behavior, and it is important to understand that in case of alarm, what is the probability of machine failure. The challenge of using this new technology is to distinguish between the anomalous behavior of the system and humans. The importance of the proposed resilient support system increases simultaneously with the increase in the number of sensors/rooms, and it is very helpful when we are talking about thousands of rooms/sensors.

3   Probabilistic Modeling

This section studies the probabilistic modeling and calculation of water consumption behavior in case of applying the "Optishower" BCS on five-star hotel guests.

Consider the random discrete signal $x(t)$ as the water consumption measurement with sampling time h, and it is associated threshold x_{tp}. For the primary classification mechanism, a person will be considered as a non-eco-friendly guest if $x(t)$ exceeds x_{tp}. Otherwise, he/she will be regarded as an eco-friendly guest.

After categorizing eco-friendly and non-eco-friendly consumptions data, the probability density function of them can be obtained as illustrated in Fig. 2.

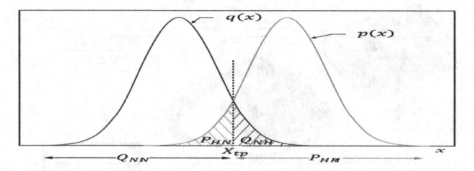

Fig. 2. Both pdfs of eco-friendly and non-eco-friendly water consumption

Having separated pdfs of eco-friendly and non-eco-friendly parts of the signal related to water consumption, the probability of becoming a non-eco-friendly for an eco-friendly type person can be calculated as follows:

$$P\_EF\_to\_NEF = \int_{x_{tp}}^{+\infty} q(x)dx \tag{1}$$

where q(x) is the probability density function of eco-friendly part of water consumption signal x(t) and *xtp* is a simple threshold that classifies the eco-friendly and non-eco-friendly consumptions. Similarly, the probability of becoming a EF for an NEF type person can be computed with the following expression.

$$P\_NEF\_to\_EF = \int_{-\infty}^{x_{tp}} q(x)dx \tag{2}$$

where p(x) is the probability density function of non-eco-friendly part of water consumption signal x(t).

Consider a stochastic process X that takes on a set of M which is finite and countable. The set of M has two elements engaged (E) and not engaged (NE). Having applied gamification or awareness strategy, the hotel guest will have eco-friendly consumption in case of being engaged and non-eco-friendly consumption in case of not being engaged with the strategy. Given an observed sequence of engagement states, the transition frequency $F_{N,NE}$ in the sequence can be found. Each element of one-step transition frequency matrix can be obtained by counting the number of changes from state E to NE in one step. The constructed frequency transition matrix for the sequence of the hotel guest engagement is as follows:

$$F = \begin{bmatrix} F_{N,N} & F_{N,NE} \\ F_{NE,N} & F_{NE,NE} \end{bmatrix} \tag{3}$$

Using (4), the probability of each transition in equivalent Markov model can be estimated [7]. In this section, the two-states Markov model is considered (m = 2).

$$P_{N,NE} = \begin{cases} \dfrac{F_{N,NE}}{\sum\limits_{N=1}^{m} F_{N,NE}} & \text{if } \sum\limits_{N=1}^{m} F_{N,NE} > 0 \\ 0 & \text{if } \sum\limits_{N=1}^{m} F_{N,NE} = 0 \end{cases} \tag{4}$$

Having estimated probabilities of Markov model transitions, the transition matrix can be achieved as (5).

$$P = \begin{bmatrix} P_{N,N} & P_{N,NE} \\ P_{NE,N} & P_{NE,NE} \end{bmatrix} \tag{5}$$

In order to calculate the transient probability from the Markov model, Eq. (6) can be recursively solved if $P(0)$ is known.

$$P(n\Delta t) = P^n . P(0) \tag{6}$$

If the Markov model satisfies the limiting probabilities in (7), then it will be irreducible ergodic Markov chain. Being independent of the initial state, from (8) the steady-state probabilities can be achieved.

$$\pi_{NE} = \lim_{l \to \infty} P^{(l)}_{N,NE} > 0 \tag{7}$$

$$\Pi = \Pi P, \sum_{NE} \pi_{NE} = 1 \tag{8}$$

where $\Pi = \begin{bmatrix} \pi_1, \pi_2, \dots \end{bmatrix}$.

4 Real Case Study

In this section, a real case study in a five-star hotel we are considered. To protect the data privacy of the hotel the name and the city will not be mentioned. In this case study, five different rooms are considered, and data of electricity and water consumption are stored in a cloud-based database. Figure 3-a shows the typical three-state Markov model of behavior change system in water consumption. This model has three states of behavior; (a) Low consumption, (b) Moderate consumption and (c) High consumption.

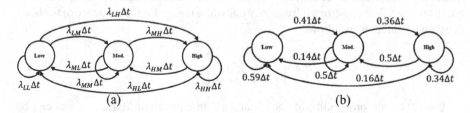

Fig. 3. Three-states Markov model of BCS in water consumption; (a) Typical, (b) Real model.

Based on recorded data and provided theory in the previous section, the three-state Markov model of BCS in cold water consumption is illustrated in Fig. 3-b. The model provided from 35 guests in five different rooms, and the average duration of using a room in the hotel was two days. Having modeled BCS in cold water consumption through Markov model as Fig. 3-b, the Fig. 4 can be provided. In this figure, the green curve shows the probability of "Low" state in the cold-water Markov model and the pink and red curves are showing the probability of states "Mod." Moreover, "High" respectively. As can be seen in this figure, from the beginning the probability of low consumption of cold water after applying the BCS will increase continuously, and the probability of high consumption of cold water will decrease continuously. The probability of state "Mod." Increases in two first days and after that will decreases. All probabilities will be steady-state after ten days.

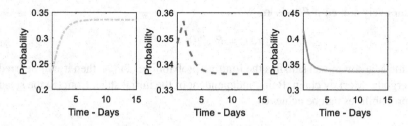

Fig. 4. Prediction based on Markov chain illustrated in Fig. 3-b (Color figure online)

5   Resilience-Supported System

In this section, the question of how the Markov model can be used as the resilience-supported system will be explained.

Figure 5 illustrates the block diagram of the proposed resilience supported system based on Markov model. As it mentioned in the previous section, through the Markov modeling, the behavior-change of a hotel guest in water consumption can be estimated. In the proposed system, real-time monitoring of a hotel guest water consumption is available. Any change in water consumption of the hotel guest can be detected by the "behavior change detector" block. It also can categorize the range of water consumption of a guest into three levels; (a) Low consumption, (b) Moderate consumption and (c) High consumption. In parallel, the time-variate Markov-based estimation of behavior-change in water consumption is available. Having counting clock, real-time state of consumption and estimated state of consumption, the final block can be used to formulate those inputs and generate a wise decision.

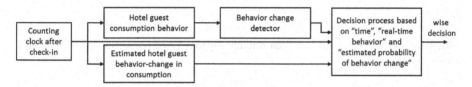

Fig. 5. Markov-based resilience supported system

For example, if the consumption is increasing in the second day of the stay of a person and it is not what we are expecting due to the promotion and gamification process, the first assumptions could be unexpected human behavior, water system failure or monitoring system failure. To understand better the system and the probability of the machine failure (case 2 & 3), the result should be compared with "estimated hotel guest behavior change in consumption" to see if the pattern is following the estimation or not. This is the key factor to recognize the type of failure.

If our estimation which is based on the model shows that the probability of unexpected human behavior is high, it means that the probability of machine failure is low and vice versa as they act unlike each other. This is the first step for any resilient system as the detection is a key factor for any recovery reaction. In case of having thousands of rooms, we have first to check the data comparing with our estimation, the probability of unexpected behavior coming from Markov model, and in case of lack of evidence to show that it could be the result of human behavior, we should immediately check if the monitoring system has a problem or it could be leakage problem in pipes.

6   Conclusion

In this paper, supported resilience system was developed based on Markov model to reduce the problem of Deceptive Alarm in human-made systems while the interaction of machine and human could mislead monitoring system operators. The basic assumption in this work is the fact that unexpected behaviors could be a result of human behavior so before reaction, we should make sure about the probability of the failure of the machine. In innovative IoT based system, this will help not to lose the concentration due to the frequency of alarms which is a common dilemma in monitoring systems while we are talking about thousands of sensors in different geographical hotels. The first step in all resilience system is to detect the disorder to react, and Resilience Supported System helps to increase the efficiency of the resilient process by detecting Deceptive alarms. The system was developed for an Innovative Water Monitoring Technology, and proposed model was tested in a real 5-star hotel to evaluate its applicability.

This work is ongoing research and the next step is to use "Fuzzy Inference System" to aggregate different criteria including the result from Markov model to determine the level of the risk of machine failure vs human unexpected behavior. This will help the operators to act according to the result of the intelligent support system using the analysis of Markov model. Also considering large amount of data (big data) from different rooms and over time, effective variables could be recognized to increase the accuracy of the system.

Acknowledgement. This work has been partially supported by "Temptation Keeper" through Optishower project.

References

1. Doukas, H., Patlitzianas, K.D., Iatropoulos, K., Psarras, J.: Intelligent building energy management system using rule sets. Build. Environ. **42**(10), 3562–3569 (2007)
2. Agarwal, Y., Balaji, B., Gupta, R., Lyles, J., Wei, M., Weng, T.: Occupancy-driven energy management for smart building automation. In: Proceedings of the 2nd ACM Workshop on Embedded Sensing Systems for Energy-Efficiency in Building, Zurich, Switzerland (2010)
3. Han, H., Lee, K., Soylu, F.: Predicting long-term outcomes of educational interventions using the evolutionary causal matrices and Markov chain based on educational neuroscience. Trends Neurosci. Educ. **5**(4), 157–165 (2016)
4. Marshall, C., Roberts, B., Grenn, M.: Intelligent control & supervision for autonomous system resilience in uncertain worlds. In: 3rd International Conference on Control, Automation and Robotics (ICCAR), Nagoya, Japan (2017)
5. Arghandeh, R., von Meier, A., Mehrmanesh, L., Mili, L.: On the definition of cyber-physical resilience in power systems. Renew. Sustain. Energy Rev. **58**(1), 1060–1069 (2016)
6. Li, M.L., Ramachandran, P., Sahoo, S.K., Adve, S.V., Adve, V.S., Zhou, Y.: Understanding the propagation of hard errors to software and implications for resilient system design. ACM SIGARCH Comput. Architect. News **36**(1), 265–276 (2008)
7. Ching, W.K., Fung, E.S., Ng, M.K.: A multivariate Markov chain model for categorical data sequences and its applications in demand predictions. IMA J. Manage. Math. **13**(3), 187–199 (2002)

Modelling Cyber Physical Social Systems Using Dynamic Time Petri Nets

Shabnam Pasandideh[1,2](✉) , Luis Gomes[1,2] , and Pedro Maló[1,2]

[1] Faculty of Science and Technology, NOVA University of Lisbon,
Campus da Caparica, 2829-516 Monte Caparica, Portugal
{lugo,pmm}@fct.unl.pt
[2] Centre of Technology and Systems - CTS, UNINOVA,
Campus da Caparica, 2829-516 Monte Caparica, Portugal
shabnam.pasandide@uninova.pt

Abstract. Performance evaluation models and techniques have been studied broadly in many fields. With the blooming of Cyber-Physical Systems (CPSs), Internet of Things (IoT) and other concepts in distributed systems, usage of a reliable and simple modelling strategy is becoming more important. This paper presents a Petri nets based strategy supporting behavioural modelling as well as performance analysis of Cyber-Physical-Social Systems (CPSSs) covering uncertainty situations when the social factor is also playing an effective role in the performance of these systems. The integration and interaction of system components including computation, physics and social factors as a challenging part of these systems are considered. Petri nets models, augmented with dynamic time dependencies associated with transitions, are applied in a case study, and validated as a promising tool for modelling and analysis of these kind of systems.

Keywords: Performance evaluation · Time Petri nets
Cyber Physical Social Systems

1 Introduction

In recent years the integration of computational or cybernetic systems with physical systems leaded to the concept of Cyber-Physical Systems (CPSs). CPSs are heterogeneous entities that cross the cyber and physical worlds, hardware and software, sensors and actuators, etc. [1]. As such systems are deployed mostly for dynamically-changing objectives, they must be highly adaptive and flexible to react within non-deterministic and changing environments with acceptable performance. Ultimately, the human perspectives as an entire and essential part of that should be studied to delicately counterbalance the design of such systems in accordance with human interactions. For this purpose, the concept of Cyber-Physical Social Systems (CPSSs) has recently evolved [2]. Smart cities, smart factories, smart healthcare, and public services can be named as some prevailing applications of CPSSs. The intrinsic complexity of CPSSs, as well as interconnection and interactions among components of the system, make their modelling

© IFIP International Federation for Information Processing 2018
Published by Springer International Publishing AG 2018. All Rights Reserved
L. M. Camarinha-Matos et al. (Eds.): DoCEIS 2018, IFIP AICT 521, pp. 81–89, 2018.
https://doi.org/10.1007/978-3-319-78574-5_8

more complicated. The characteristics of these kind of systems include concurrency, synchronisation, asynchronous, distributed, real-time, discrete and continuous features, additionally to the randomness commonly associated with human interactions. Regarding smart cities many correlated subjects should be considered and analysed according with the high performance of the systems, such as Intelligent traffic management, and Intelligent transportation systems, to mention a few. The aim of this paper is to introduce a proper formalisation to model CPSSs in uncertainly environment, using an intelligent traffic control system as validation example. This paper steps towards the design of a novel and adequate formalisation to model this kind of CPSSs.

Among modelling formalisms adequate to be used for specification, analysis and implementation of CPSSs, Petri nets (PN) [4] are well-suited to deal with the challenges of CPSSs, supporting a model-based development strategy, including component design, orchestration of components, as well as component and overall performance evaluation. They can also accommodate characterisation of stochastic environments [5] as PNs behavioural modelling is amenable to support modelling of both deterministic and non-deterministic characteristic of systems, which is an advantage for modelling essential features of CPSSs comparing with other formalisms (such as Markov chains).

In this paper, non-autonomous Petri nets modelling is used for the specification, analysis and implementation of CPSSs, where the PN model characteristics were augmented with the capability of dynamic time associated with the evolution of the model, as proposed in Sect. 5. Validation of the proposals is performed using an application example coming from an intersection traffic lights control system, where both cars' as well as pedestrians' arrival rate are considered to constraint the behaviour of the system, impacting on its performance.

2   Relationship to Resilient Systems

The notion of resilience is commonly associated with "the strength of a system to resist a significant disruption within satisfactory degradation parameters and to recover with a proper time and reasonable cost and risk" [6]. In this sense, resilience of a system, as one major performance evaluation criteria, is getting paramount importance in areas such as Internet of Things, Cyber Physical systems, Industry 4.0, as well as whenever misoperation or misconduct of users can affect overall performance, as in Cyber Physical Social Systems.

Evaluating resiliency in CPSSs needs to consider stochastic and non-deterministic environment, as well as automation and human decisions effects.

The use of modelling formalisms, such as Petri nets supporting different phases of the development, namely specification, and implementation, allowing a-priori verification of properties and anticipating the impact of some failures or misconduct, contributes to the improvement of systems' resiliency. In this extent, improving confidence in functional correctness in real-time operations, survivability during attacks, fault tolerance and robustness is supported by the adoption of a model-based development attitude, as the one proposed in this paper.

3   Summarised Overview of Related Literature

3.1   On Cyber-Physical Social Systems

The notion of CPS has a variety of definitions, but as a common one, a CPS is "an integration of computation, communication with monitoring or/and control of entities in the physical world" [7]. Information is a main part of the concept used for connecting computation, communication and control. Information can be generated from physical components, including sensors and actuators, or multi sources, such as societies, human operators, and embedded computers, as well as from networks monitoring and controlling the physical processes, usually with feedback loops where physical processes affect computations and vice versa, as shown in Fig. 1 [8].

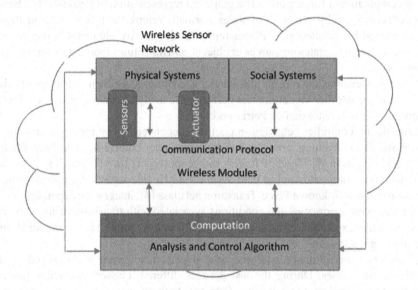

Fig. 1. CPSS overview

Many CPS operate in concert with human operators, and the human aspect of the design of such systems must be carefully considered. The challenge for modelling them is about heterogeneity, concurrency, and sensitivity of time. As CPSs have a discrete and dynamic nature, it is of interest to model them using qualitative and quantitative models. One popular formalisation for modelling CPS is via Petri nets.

Recently, definition of cyber-physical-social systems (CPSS) has been proposed, which integrates components/dependencies originated from social constraints into CPSs. In this new paradigm human has a centric role to design systems. Actually, there is lack of semantic design methodology in many existing systems.

3.2 Petri Nets

Petri nets (PN) are a graphical and mathematical modelling formalism, which are proper for analysis of behavioural properties and performance evaluation of discrete event systems [4].

Petri nets were initially introduced by Carl Adam Petri in the 1960s and have been receiving attention from different fields and areas of application. Petri nets are adequate to model core concepts of concurrent systems, namely concurrency itself, sequences, synchronisation, conflicts and resource sharing, benefiting from a rigorous execution semantics. Petri nets have a simple graphical representation, complemented by mathematical representation, as well as a text representation using the Petri Net Markup Language (PNML) defined in the standard ISO/IEC 15909 Part 2, allowing their usage within computational frameworks. The graphical representation of Petri nets is a bipartite graph composed by two types of nodes, normally referred as places and transitions, interconnected by directed arcs. Places represent the passive elements in the system, including conditions, states, resources or objects, and transitions model the dynamic part of the system.

Several extensions have been introduced, accommodating their usage for specific areas, namely controller modelling and performance analysis, among others, leading to the proposal of non-autonomous Petri nets classes.

Considering controller behaviour modelling, dependencies on the input and output signals and events coming from the environment under control need to be explicitly accommodated, as in [9]. The IOPT nets class proposed in [10, 11], which will be used as the underlying non-autonomous Petri net class used in this paper, is a non-autonomous extension of the well-known Place-Transition net class [4], integrating dependencies on signals and events, priorities and conditions associated with transition firing, allowing a deterministic execution, considering a cycle accurate single server maximal step execution semantics.

Considering performance analysis, time dependencies need to be added to the modelling capabilities. During the last decades different classes and definitions are introduced to expand application of the Petri nets to these areas. Time Petri nets (TPN) and Stochastic PNs (SPN) classes are extensions of PNs associating a random firing time to each transition, as in [12]. Remarkably, Generalized Stochastic Petri nets (GSPN) [13] were proposed considering two types of transitions: timed transitions (to accommodate the modelling and execution of time consuming activities) and immediate transitions (to describe transition firing not considering time dependencies), which are mostly used in performance evaluation where immediate (null delay) transitions are freely mixed with timed transitions [14]. Of particular relevance for this paper, Dynamic Time Petri net (DTPN) class, proposed in [12], on top of the static time interval associated with the firing of transitions, consider that the timing constraints are updated with a dynamic time interval mechanism.

4   Introducing a Traffic Light Control System

Traffic management, as a common and daily problem, took the attention for being studied in this work. In this paper, an intersection traffic light is modelled as a case study of a CPSS and will be used to illustrate the effectiveness of the presented proposals. Traffic light system is composed by one physical controller component, which is in communication with computation sections. In addition, social aspects are effective parameters affecting the performance of the system. Regarding modelling of the intelligent traffic light control system, several variables can be considered to constraint the time management of the system. Table 1 lists the more relevant ones.

Table 1. Traffic light control variables

No.	Variables	Description
1	Number of lights	Traffic; pedestrian
2	Number of vehicles on the road	
3	Tuning movements	Straight, right, left
4	Time and Days	Rush hours; morning; lunch time; mid-day; night; mid-night
		Weekdays, weekends and holidays
5	Determined time for Red light	Max time for stopping vehicles
6	Number of pedestrians	
7	Vehicles type	Cars; bus; trucks; tractor
8	Average age of pedestrians	Children, adults, elderlies

The objective of an intelligent traffic system is to improve performance of controlling the volume of traffic, reducing the queues and waiting times both for vehicles and pedestrians. Therefore, it is important to design a model which yield safe and efficient operation for the prevailing conditions.

Most of the previous studies found in the literature consider several operation modes (night, day, rush hours, etc.), each of them having pre-defined sequences and temporal behaviour (fixed or almost-fixed time periods). In this paper it is intended to allow that the time periods can be dynamically adjusted considering specific arrival profiles of vehicles and pedestrians.

For that end, the Petri nets class to be used will be augmented with a new attribute accommodating support for dynamically adjusted time delays, which will allow to take into consideration the variables presented in Table 1.

5   Extending IOPT Nets with Dynamic Time Delays

As referred in Sect. 3.2, the IOPT nets class will be used as the underlying non-autonomous Petri net class for this paper; additional information about IOPT nets can be obtained from [10, 11]. Formal definition is not presented here due to space limitations.

The extension proposed is based on the addition of a new characteristic, named as *TimeDelay*, which can be associated to each transition of the model having the same operational semantics as in time Petri nets [12] (which means that the firing of the transition will be delayed by a specific time interval after being enabled and ready to fire). In addition, this new characteristic is obtained as a result of the computation of an expression involving the values of dedicated input signals (this means that the value of *TimeDelay* is a non-autonomous characteristic, in opposition to other proposals in the literature where the time is fixed or computed based on the Petri net model characteristics). Formal definition is not presented here due to space limitations.

6   Modelling a Traffic Light Control System

The first step to model a traffic light control system is realising the transportation system and associated processes. The traffic light which is considered has five signal lights shown in Fig. 2(a) including Red light (R), Amber (A), left turn arrow on green (GL), straight arrow on green (GS), right turn arrow on green (GR), as in the example used in [15].

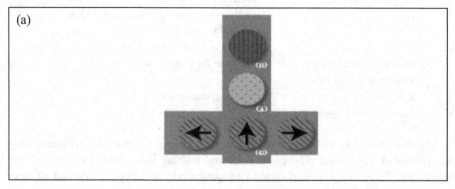

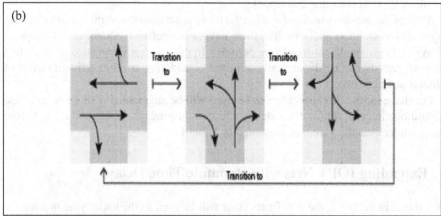

Fig. 2. (a) Traffic Light, (b) Three phase strategy for vehicle transitions (Color figure online)

Four traffic lights are placed at intersection entrances including north, south, east, and west. Several sequences for light activation are possible to be adopted, depending on specific characteristics of the intersection. For the application example, we will adopt a 3-phase strategy as shown in Fig. 2(b), and also referred in [15] (among other possible configurations).

Overall, the system can be divided into three components: one responsible for the vehicles' traffic lights activation (mainly actuators), other responsible for the pedestrians' lights activation (mainly actuators), and a third one responsible for the acquisition of information from the floor regarding vehicles' and pedestrians' arrivals (receiving information from a sensor network) and producing recommendations for the dynamic time delays to be used by the other two components. While the third component can be seen as the intelligent part of the system, which will take into consideration the social constraints, the other two can be seen as configurable controllers, which can be modelled using models expressed through IOPT nets extended with dynamic time delays.

The three components are modelled using Petri nets, as previously described. Due to space limitation, only the model for the component associated with the vehicles' traffic lights activation is presented in Fig. 3. A simple modelling strategy was adopted for Fig. 3, where a state machine representation with a Petri net notation is used, and where the input signals DTx (with x ranging from 1 to 9) are associated with the *Time-Delay* attribute associated with each transition. A different modelling attitude taking advantage of Petri nets modelling capabilities (namely concurrency and synchronisation

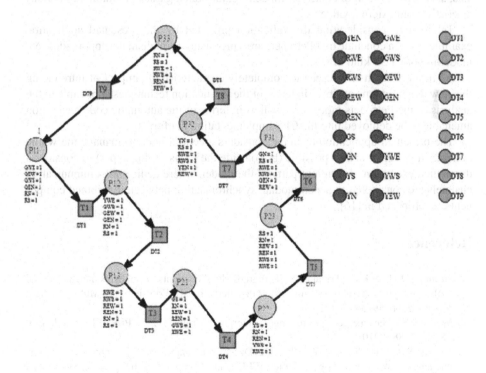

Fig. 3. Petri net model for intersection vehicle traffic lights control.

capabilities) could also be used, but for a simple intersection as the one under observation it is not worth to use.

To briefly explain the model of Fig. 3, we consider the 3-phase strategy as illustrated in Fig. 2b). Each of the phases referred in Fig. 2b) is associated with three places plus three-time transitions (with dynamic time attribute) in Fig. 3. There are nine places (P11, P12, P13, P21, P22, P23, P31, P32, P33), which represent nine states of the vehicles movement (active output lights are referred close to each place), and nine-time transitions (T1, …, T9), which defines firing sequences considering time delays (DT1, …, DT9) determined after computation occurred at the other component. As the computation associated with these time delays depends on different parameters according with Table 1, the behaviour of the system is dynamically tuned to optimise the performance of the system in terms of the waiting time for vehicles and pedestrians.

7   Conclusions and Future Work

In this work, the control part of a CPSS was modelled by Petri nets, where the list of attributes that can be associated with transitions firing was augmented with a dynamic time attribute (extending Time Petri nets common execution semantics). This attribute is the result of a calculation involving specific input signals. In this sense, this proposal put together and reuse time dependencies, normally considered for performance analysis, also in the context of controller modelling, introducing this non-autonomous characteristic as an external value.

Taking into consideration the validation provided by the presented application example, we can conclude that Petri nets are a promising formalism to support adequate modelling of CPSS kind of systems.

As future work it is foreseen to completely characterise the impact of introducing this new dynamic time attribute in terms of the verification techniques applicable to the analysis of the Petri nets models, as well as in terms of the automatic code generation amenable to be deployed into the CPS implementation platforms.

The presented application example considers a central node to manage the whole system. However, it is also possible (but outside the focus of this paper) to consider a distributed execution of different parts of the model, where dedicated communication channels are considered to accommodate synchronisation between distributed components, as proposed in [16].

References

1. Broman, D., Lee, E.A., Tripakis, S., Torngren, M.: Viewpoints, formalisms, languages, and tools for cyber-physical systems. In: 6th International Workshop on Multi-Paradigm Modeling, pp. 49–54 (2012)
2. Wang, F.Y.: The emergence of intelligent enterprises: from CPS to CPSS. IEEE Intell. Syst. 25(4), 85–88 (2010)
3. Seshia, S.A., Hu, S., Li, W., Zhu, Q.: Design automation of cyber-physical systems: challenges, advances, and opportunities. IEEE Trans. Comput. Des. Integr. Circ. Syst. 36(9), 1421–1434 (2017)

4. Girault, C., Valk, R.: Petri Nets for Systems Engineering - A Guide to Modeling, Verification, and Applications. Springer, Berlin (2003). https://doi.org/10.1007/978-3-662-05324-9
5. Mitchell, R., Chen, I.-R.: Effect of intrusion detection and response on reliability of cyber physical systems. IEEE Trans. Reliab. **62**(1), 199–210 (2013)
6. Denker, G., Dutt, N., Mehrotra, S., Stehr, M.-O., Talcott, C., Venkatasubramanian, N.: Resilient dependable cyber-physical systems: a middleware perspective. J. Internet Serv. Appl. **3**(1), 41–49 (2012)
7. Zeng, J., Yang, L.T., Lin, M., Ning, H., Ma, J.: A survey: cyber-physical-social systems and their system-level design methodology. Futur. Gener. Comput. Syst., 17 August 2016. In press
8. Zanni, A.: Cyber-physical systems and smart cities - Learn how smart devices, sensors, and actuators are advancing Internet of Things implementations, pp. 1–8 (2015). https://www.ibm.com/developerworks/library/ba-cyber-physical-systems-and-smart-cities-iot/. Accessed
9. Hanisch, H.-M., Lüder, A.: A signal extension for petri nets and its use in controller design. Fundamenta Informaticae **41**(4), 415–431 (2000)
10. Gomes, L., Barros, J.-P., Costa, A., Nunes, R.: The input-output place-transition petri net class and associated tools. In: INDIN 2007 - 5th IEEE International Conference on Industrial Informatics, Vienna, Austria, 23–26 July 2007
11. Gomes, L., Moutinho, F., Pereira, F., Ribeiro, J., Costa, A., Barros, J.-P.: Extending input-output place-transition petri nets for distributed controller systems development. In: ICMC 2014 - International Conference on Mechatronics and Control, Jinzhou, China, pp. 1099–1104 (2014)
12. Zilio, S.D., Fronc, L., Berthomieu, B., Vernadat, F.: Time petri nets with dynamic firing dates: semantics and applications. In: 12th International Conference, FORMATS, Florence, Italy, pp. 85–99 (2014)
13. Ajmone Marsan, M., Bobbio, A., Donatelli, S.: Petri nets in performance analysis: An introduction. In: Reisig, W., Rozenberg, G. (eds.) ACPN 1996. LNCS, vol. 1491, pp. 211–256. Springer, Heidelberg (1998). https://doi.org/10.1007/3-540-65306-6_17
14. Maione, G., Mangini, A.M., Ottomanelli, M.: A generalized stochastic petri net approach for modeling activities of human operators in intermodal container terminals. IEEE Trans. Autom. Sci. Eng. **13**(4), 1504–1516 (2016)
15. Huang, Y.-S., Chung, T.-H.: Modeling and analysis of urban traffic lights control systems using timed CP-nets. J. Inf. Sci. Eng. **24**(3), 875–890 (2008)
16. Moutinho, F., Gomes, L.: Asynchronous-channels within Petri net based GALS distributed embedded systems modeling. IEEE Trans. Industr. Inf. **10**(4), 2024–2033 (2014)

Supervisory Control System Associated with the Development of Device Thrombosis in VAD

José R. Sousa Sobrinho[1,2](✉) ⓘ, Edinei Legaspe[1], Evandro Drigo[2,3], Jônatas C. Dias[1], Jeferson C. Dias[1], Marcelo Barboza[1,2], Paulo E. Miyagi[1], Jun Okamoto Jr.[1], Fabrício Junqueira[1], Eduardo Bock[2], and Diolino J. Santos Filho[1]

[1] Escola Politécnica da Universidade de São Paulo, São Paulo, Brazil
ricardo1csousa@usp.br
[2] Instituto Federal de Educação, Ciência e Tecnologia de São Paulo, São Paulo, Brazil
[3] Instituto Dante Pazzanese de Cardiologia de São Paulo, São Paulo, Brazil

Abstract. Patients with advanced heart failure can use Ventricular Assist Device (VAD) to improve the quality of life and the rate of survival. However, the device thrombosis is responsible for serious complications and cause hospital return of patients. These complications may be result in surgical operation to change device and anticoagulant drugs are an alternative, but only inhibits thrombus formation and isn't a permanent solution. The levels of lactate dehydrogenase (LDH) should be monitored because interaction is intensive. Even if clinical features related to these complications are identified, patients can die when the diagnosis is delayed. The use of vibration analysis technique associated with clinical indicators has showed advanced results. This work proposes a supervisory control system to improve the resilience of the VAD through the development of a monitoring process to control the evolution of device and cannulas thrombosis. The objective is obtaining information in real time based on the system model and interactions of the specialist physician to promote control actions based on the clinical diagnosis.

Keywords: System model · Device thrombi · Control and supervision
Vibration analysis · Ventricular Assist Device

1 Introduction

Thrombogenesis is the process of natural blood clotting. The device thrombosis is responsible for serious complications and cause hospital return of patient to exchange device. Thrombus formation is an inherent failure associated with the use of Ventricular Assist Devices (VAD) that is potentiated by the presence of biomaterials, constructive forms of the blood pump or adverse factors that are investigated [1].

The technological advancement is responsible for replacement rudimentary devices. Therefore the using new devices there are a reduction in mortality rates and morbidity. However in Fig. 1 are showed examples of device thrombosis in the rotor of a commercial VAD, HeartMate II (Thoratec Corporation, Pleasanton, CA) [2].

© IFIP International Federation for Information Processing 2018
Published by Springer International Publishing AG 2018. All Rights Reserved
L. M. Camarinha-Matos et al. (Eds.): DoCEIS 2018, IFIP AICT 521, pp. 90–97, 2018.
https://doi.org/10.1007/978-3-319-78574-5_9

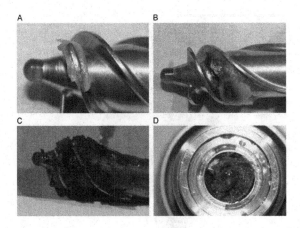

Fig. 1. Stages of devices thrombi on the HeartMate II. In (A) only fibrin, (B) fibrin and blood clot, (C) and (D) blood clot [2].

Blood clotting can be inhibit using anticoagulant drugs, but this therapeutic option isn't sufficient for permanent solution. The interaction between the device and blood is intense and the blood coagulation can be raising a risk factor for death to the patient. In periodic consultations for medical treatment the levels of Lactate Dehydrogenase (LDH) can be used to precede the presence of thrombus [3, 4].

The stagnant blood flow is may be favor the occurrence of device thrombosis. An alternative to solve this problem is increase the rotation of the blood pump. However, high rotations can result in ventricular suction events. The use of echocardiography techniques supports the observation of these complications and more: cannula position, dimension and ventricular volume, opening of the valves, blood flow and other indicators very important for clinical diagnosis [5, 6].

Another indicator of suspected device thrombosis is monitoring devices used in VAD as alarms indicating abnormal blood flow or high-power consumption. Between October 2009 and July 2015, a total of 524 patients suffered abnormal events classified as high power and low flow alarms. The use echocardiogram images combined with vibration analysis of the blood pump raises the possibilities for assertive diagnosis. The Fig. 2 show critical regions of thrombus formation: (1) pump inlet (pre-pump thrombosis); (2) rotor blood pump (intra-pump thrombosis); and (3) outlet cannula (post-pump thrombosis). In Fig. 2 is showed an illustration of a heart supported by a HVAD® System (HEART WARE®, Framingham, MA, USA) [7].

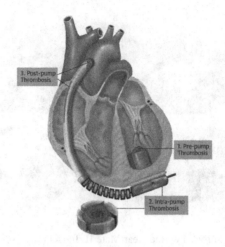

Fig. 2. Thrombus adherence in different levels of the HVAD® system [7].

1.1 State of the Art

The fault detection methodology that is applied in Dynamic Preventive Maintenance Scheduling of industrial critical equipments. The continuous flow VAD are a rotary machine and its dynamic characteristics facilitate the use of this methodology [9]. This method of vibration analysis is consolidated in the industrial environment for diagnose failures in rotating equipment. This methodology may be also used as an alternative for identifying defects or anomalies in VAD during operation [8].

Thrombus adherence in the blood pump rotor causes imbalance and critical stages that may change the rotor dynamics which may result in excessive stress on the bearings or the rotary pump actuator. This is a powerful feature that increases the risk of rotational element breakage and the thrombus may detach from the pump and cause great damage to the patient's health and increases the patient's risk of death [9].

A recent study showed promissory results with an industrial vibration analyzer to recognizing different patterns of pump behavior related to thrombus simulation. However the use of industrial sensors makes structural integration with the blood pump impossible based on biocompatibility aspects [10].

Micro-Electro-Mechanical System (MEMS) technology is a consistent approach to integrate devices into VADs. These applications are presents to monitoring bearings, gear box or dynamics equipments embedded for measurement of rotating elements in real time. Another alternative of measurement is observed in the study of an acoustic transducer MEMS of great relevance and may be outside the patient body [11–13].

2 Relation with Resilient Systems

The administration of anticoagulation drugs to inhibit blood clotting isn't a permanent solution because cannot avoid device thrombosis. The literature review showed knowledge of complications about thrombus formation to using with Ventricular Assist Devices (VADs). These problems reach many frontiers and involve two major areas of knowledge: engineering and medicine.

Thrombus formation is an inherent failure when used VAD the diagnosis and treatment represent a high complexity for clinical team. Advanced stages may require a surgical operation to exchange device or the patient can death without treatment for recover your health. Vibration analysis is an alternative for identification of thrombi, but the literature review isn't found applications of measurement systems to monitoring for thrombus formation severity or diagnosis of different stages of blood clotting.

The Standards IEC 61508 [14] and IEC 61511 [15] which address process safety in critical systems, state that one of the issues that needs to be considered is that any fabricated device is subject to failure. No matter how great the technological development, there will always be an inherent failure context that can endanger the life of the patient. For this reason, surveys are being conducted to improve the reliability of the device. One of the lines of action is to design a safety oriented control system that acts to keep the device functioning in case of failure, causing regeneration or degeneration of the device to a safe state [16].

Another important approach of action is associated with solutions developing for monitoring the evolution of failures while the patient is using device. Any late action for identification thrombus formation can inevitably induce death patient. Therefore it is important to consider the concept of resilience so that VAD and supervisory control system must be adapt to the changes in blood flow and is able to sustain a safe behavior for influence disorders [17].

3   Research Contribution and Innovation

The model of supervisory control system was idealized for uses a continuous flow blood pump. It will be considered that the Thrombogenesis is an inherent failure in the use of VAD and difficulty of identifying the region of adherence the thrombus. The objective is obtaining information in real time based on the system model and interactions of the specialist physician to promote control actions based on the clinical diagnosis. The specific objectives are:

A. The system should be efficient to monitor the evolution of thrombus formation;
B. Allow Remote device monitoring by Specialist Physician;
C. It has Resilience behavior;
D. Implement control actions based on clinical knowledge.

Therefore an initial proposal of the modular distributed architecture is presented in Fig. 3. With the evolution of the research our future objective is propose complementary characteristics so that concepts related to Health 4.0 are implemented in the supervisory control system architecture [18].

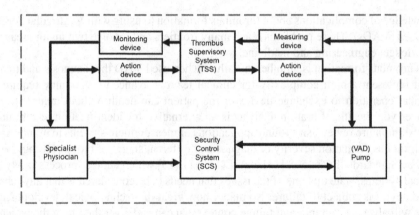

Fig. 3. Control and supervision system architecture.

4   Discussion of Results

The preliminary results showed the advances obtained during the thrombus detection tests. The thrombi were simulated at three critical points: (A) rotor base, (B) rotor flap and (C) rotor spiral.

Different speeds were used 1200 rpm, 1500 rpm and 1800. Normal spectrum was compared with signs vibration disturbed by the presence of thrombus. The yellow arrows represent multiples harmonics of fundamental frequency and red arrows indicate the peaks in frequency that represent anomaly harmonics frequencies.

The Fig. 4 show an experimental simulation realized with a blood pump prototype. A Micro-ElectroMechanical System (MEMS) accelerometer was inserted for vibration measurement of in closed hydraulic circuit to only behavior simulation of continuous flow blood pump.

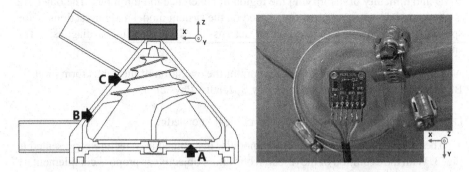

Fig. 4. The left image of the implantable centrifugal blood pump and experimental simulation on the right.

These results are based on experimental simulation data and Fast Fourier Transform (FFT) of vibration the centrifugal blood pump. The harmonics peaks showed amplitude severity stand out yellow or red arrows. More severity the red arrows but this study no realizes a qualitative analysis of harmonic.

The Fig. 5 compare measurements between excessive vibration signals by rotor base thrombus and normal behavior of the pump rotor. Figure 6 indicates the occurrence of rotor flap thrombus and Fig. 7 indicates rotor spiral thrombus.

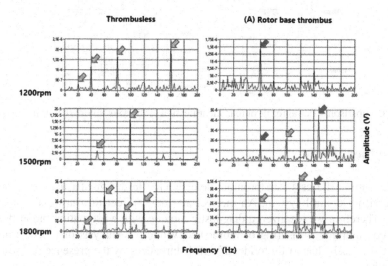

Fig. 5. Comparison spectral for thrombus identification at the rotor base. (Color figure online)

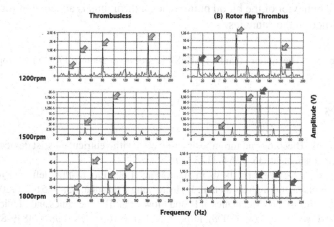

Fig. 6. Comparison spectral for thrombus identification at the rotor flap. (Color figure online)

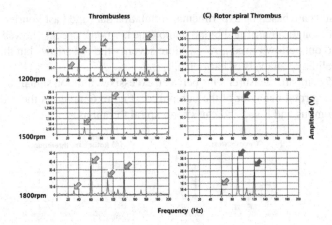

Fig. 7. Comparison spectral for thrombus identification at the rotor spiral. (Color figure online)

5 Further Work

We will continue the tests with thrombus detection for the design of a diagnostic module of supervisory control system. An indirect and other direct measuring are inputs of system. Therefore we idealized a diagnostic module with a estimation blood flow by power consumption measuring of motor and Vibration Analysis of a accelerometer Integrated for self-diagnostic to identify of disturbances by the presence or formation of thrombus.

Theses Some tests should be performed using computational simulation to map the dynamic fluid behavior of the blood pump when a thrombus is attached of the device or causing obstruction the cannulas.

Acknowledgment. The authors thanks FAPESP (Process no 2006/58773-1) for supporting this research.

References

1. Shah, P., et al.: Bleeding and thrombosis associated with ventricular assist device therapy. J. Heart Lung Transplant. **36**, 1164–1173 (2017)
2. Uriel, N., et al.: Device thrombosis in HeartMate II continuous-flow left ventricular assist devices: a multifactorial phenomenon. J. Heart Lung Transplat. **33**(1), 51–59 (2014)
3. Thiriet, M.: Biology and Mechanics of Blood Flows. Part II: Mechanics and Medical Aspects, 1st edn. Springer, New York (2008). https://doi.org/10.1007/978-0-387-74849-8. 468 p
4. Souza, H.L., Elias, D.O.: Fundamentos da Circulação Extracorpórea, vol. 1, 2nd edn. Alfa Rio, Rio de Janeiro (2006)
5. Stainback, R.F., et al.: Echocardiography in the management of patients with left ventricular assist devices: recommendations from the American Society of Echocardiography. J. Am. Soc. Echocardiogr. **28**(8), 853–909 (2015)

6. Ucar, M., et al.: Early thrombus formation in patient with HeartWare left ventricular assist device presenting with acute heart failure. J. Saudi Heart Assoc. **28**(1), 49–51 (2016)

7. Scandroglio, A.M., et al.: Diagnosis and treatment algorithm for blood flow obstructions in patients with left ventricular assist device. J. Am. Coll. Cardiol. **67**(23), 2758–2768 (2016)

8. Patil, M.S., Mathew, J., Rajendrakumar, P.K.: Bearing signature analysis as a medium for fault detection: a review. J. Tribol. **130**(1), 014001-1–014001-7 (2008)

9. Maruthi, G.S., Hegde, V.: Application of MEMS accelerometer for detection and diagnosis of multiple faults in the roller element bearings of three phase induction motor. IEEE Sens. J. **16**(1), 145–152 (2016)

10. Kawahito, K.: Transformation of vibration signals in rotary blood pumps: the diagnostic potential of pump failure. J. Artif. Organs **16**(3), 393–396 (2013)

11. Holm-Hansen, B.T., Gao, R.X.: Vibration analysis of a sensor-integrated ball bearing. J. Vib. Acoust. **122**(4), 384–392 (2000)

12. Luczak, S., Oleksiuk, W., Bodnicki, M.: Sensing tilt with MEMS accelerometers. IEEE Sens. J. **6**(6), 1669–1675 (2006)

13. Liu, Y., et al.: Mechano-acoustic detection of dysfunction in continuous flow VADs utilizing stretchable electronic systems. J. Heart Lung Transplant. **36**(4), S13 (2017)

14. International Electrotechnical Commission: Functional Safety of Electrical/Electronic/ Programmable electronic Safety-related Systems (2010)

15. International Electrotechnical Commission: Functional Safety - Safety Instrumented Systems for the Process Industry (2003)

16. Cavalheiro, A.C.M.: Control system for diagnosis and treatment of failures on ventricular assist devices. Doctoral thesis. University of São Paulo (2013)

17. Lamnabhi-Lagarrigue, F., et al.: Systems & Control for the future of humanity, research agenda: current and future roles, impact and grand challenges. Annu. Rev. Control **43**, 1–64 (2017)

18. Thuemmler, C., Bai, C.: Health 4.0: How Virtualization and Big Data are Revolutionizing Healthcare. Springer, Cham (2017). https://doi.org/10.1007/978-3-319-47617-9

Energy Management

Wind-PV-Thermal Power Aggregator in Electricity Market

I. L. R. Gomes[1,2], R. Laia[2], H. M. I. Pousinho[2], R. Melicio[1,2(✉)],
and V. M. F. Mendes[3,4]

[1] IDMEC, Instituto Superior Técnico, Universidade de Lisboa, Lisbon, Portugal
ruimelicio@gmail.com
[2] Departamento de Física, Escola de Ciências e Tecnologia, ICT,
Universidade de Évora, Évora, Portugal
[3] CISE, Electromechatronic Systems Research Centre,
Universidade da Beira Interior, Covilhã, Portugal
[4] Instituto Superior of Engenharia de Lisboa, Lisbon, Portugal

Abstract. This paper addresses the aggregation of wind, photovoltaic and thermal units with the aim to improve bidding in an electricity market. Market prices, wind and photovoltaic powers are assumed as data given by a set of scenarios. Thermal unit modeling includes start-up costs, variables costs and bounds due to constraints of technical operation, such as: ramp up/down limits and minimum up/down time limits. The modeling is carried out in order to develop a mathematical programming problem based in a stochastic programming approach formulated as a mixed integer linear programming problem. A case study comparison between disaggregated and aggregated bids for the electricity market of the Iberian Peninsula is presented to reveal the advantage of the aggregation.

Keywords: Aggregator · Day-ahead market
Mixed integer linear programming · Stochastic programming
Wind-PV-thermal units · Variable renewables

1 Introduction

The usage of non-renewable energy sources has been assumed as responsible for global warming and climate change, hovering a view of a worldwide sustainable development where renewable energy sources are the way of the future to be favored in the electricity industry [1]. But the variability of the available input energy in wind or photovoltaic (PV) renewable power units are doomed to fluctuate hastily, implying for the exploitation of the electricity industry uncertainty in energy delivering. If the rated power of these units is small in comparison with the power of non-fluctuating units in a power system, the impact of the fluctuation can be neglected. Else, the impact is of paramount significance and actions are to be taken. So, with the increasing penetration of wind or PV renewable power units, menaces to system reliability and security are to be expected [2, 3]. Additional ancillary service, i.e., more schedule of power of reserve units ready to be called if needed, must be incurred to recover an acceptable level of

© IFIP International Federation for Information Processing 2018
Published by Springer International Publishing AG 2018. All Rights Reserved
L. M. Camarinha-Matos et al. (Eds.): DoCEIS 2018, IFIP AICT 521, pp. 101–110, 2018.
https://doi.org/10.1007/978-3-319-78574-5_10

operation reliability and minimize the associated impact charge [4]. As the electricity industry follows in the way of Internet of Things (IoT) and of Energy 4.0, aggregation platforms are essential for distributed energy resources, storage, and multiple costumer loads like microgrids [5]. An aggregation platform for wind, photovoltaic and thermal units can be viewed as tool for an aggregator to act as facilitator for offering energy in an electricity market [6]. Strategic aggregation of power units is on an increasing interest for researchers and on demand from the electricity industry. The electricity industry not only requires adequate management information systems to improve profit in electricity market environment, but also to be in an electricity market environment is required to have enough capacity and ability to satisfy assumed compromised of delivering energy in due time [7].

Aggregation is able to harvest better bidding, accommodating fluctuation of variable renewables, carrying favorable imbalance while increasing the reliability and improving market performance. A convenient management of uncertainties due to the power of wind and PV units may reduce the deviations of energy and reduce the level of power required by thermal units in a wind-PV-thermal commitment. This convenient management for the wind-PV-thermal commitment is the issue in research in this paper in order to develop a computer procedure. The contribution is concerned with the: definition and advantages of a Wind-PV-Thermal power aggregator and the modeling in what concerns the strategy of bidding in electricity market. Market prices, wind and photovoltaic powers are data for the proposed computer procedure given by a set of scenarios. The thermal units modeling has continuous and integer decision variables in order to include a modeling for the start-up costs, variables costs and bounds due to technical operation constraints, such as: ramp up/down limits and minimum up/down time limits. The modeling is carried out in order to develop a management aggregation procedure based on a stochastic programming approach formulated as a mixed integer linear mathematical programming problem. A case study comparison between disaggregated and aggregated biddings shows the benefit of the aggregation proposed in this paper, particularly, the improvement in the profit.

2 Relationship to Resilient Systems

The electricity market approaches tend to maximize the economic efficiency, but may lead to deterioration on the resilience of energy systems [8]. United Nations has highlighted the importance of considering resilience in the designing of systems and infrastructures with the goal of being able to tolerate external shocks that might occur in the future [9]. Resilience of system is a capacity of being able of absorbing shocks and reorganizing while undergoing change to retain essentially the same function. Distinguished from sustainability, resilience is related to how the system responds to disturbance or non-linear dynamics [10]. A resilient energy system has the capacity to quickly recover from external shocks and to provide alternative means of maintaining a satisfactory level of services to consumers in the event of external disturbances [11]. In [12] is discussed the importance of ensuring the resilience of energy supply because the infrastructures have susceptibility to large-scale failures caused by unforeseen scenarios. Namely, natural disaster and technical failures. The future smart grid ambient and cyber physical systems have a layered architecture of a cyber infrastructure accessing resilient power applications able to give security and reliability, having the ability to act in order to maintain and correct infrastructure components without affecting the service [13]. Aggregated platforms have to be embodied with tools to allow resilience and contribute to improve the economic and technical performance in decision-making. This paper is a research on one of these tools customized for the Wind-PV-Thermal power aggregator, adding economic and technical value to the platform, giving the ability to manage the set of resources in coordination, harmonizing different functionality to assemble condition to be in an electricity market, achieving effectiveness, efficiency and efficacy. Considering the scenario of increasing penetration of variable-renewables the electrical system of Iberian Peninsula have to lead with some problems of resilience, since the fluctuations of variable-renewables may not be accommodated. To lead with this issue higher levels of interconnection with another regions should be verified according to European Union reports.

3 State of the Art

The former unit commitment in a view of the regulated electricity market is set with a paradigm formally given by an optimization problem related to the minimization of the operating costs while meeting the load demand and subjected to other constraints. This type of unit commitment is known as cost-based unit commitment. The deregulated electricity market paradigm brought new opportunities for power producers in an environment where the optimization is related to the maximization of profit. This new paradigm is known as price-based unit commitment [14]. The unit commitment problem can be divided into two sub-problems: (i) the unit on/off scheduling sub-problem period; and (ii) the economic dispatch sub-problem, which determines the power output of units in each period of the time horizon [15]. A line of research on variable-renewables and thermal unit commitment is raising up with the increasing penetration of wind or PV renewable power units. Some research is concerned with time resolution, i.e., the impact of considering sub-hourly periods in the commitment [16, 17]. In order to cope the uncertainties of variable-renewables is proposed the use with storage technologies [18] or financial options [19]. Another relevant research is concerned with the application of stochastic optimization methods and is expected to achieve a significant application in what concerns optimum bidding strategy for aggregation [20, 21]. Several of popular algorithms using stochastic optimization in unit commitment problems are based on decomposition techniques.

4 Wind-PV-Integrated Unit Commitment Problem

The Wind-PV-Thermal power aggregator owns wind turbines, PV modules and thermal units. The first two types of units have unpredictable and stochastic nature and are categorized as variable-renewable sources of energy, imposing challenges to the unit commitment problem. These challenges are linked with the determination of the optimal day-ahead bid under wind and PV power uncertainties and the decision for the optimal generation of thermal units. This paper addresses a two-stage stochastic programming approach that takes into account the uncertainties related to the variable renewables and the market prices by scenarios. The scenarios can be expressed by means of a scenario tree, but in the scope of the paper the scenarios are considered as given data attained by the use of an adequate tool. The stochastic programming is reformulated in an admissible way as a mixed-integer linear programming (MILP) approach to benefit from the available and well proved practical commercial optimization solvers. In a two-stage stochastic optimization decision variables are categorized into first and second stage decisions: (a) The first stage decisions, which are known as *here and know decisions*, are made before the realization of uncertainties which in this paper are: wind power, PV power and market prices, including imbalance prices. The thermal unit commitment and the hourly bids are the first-stage decisions. The objective is the maximization of the profit of the production aggregator in the first stage; (b) The second stage decisions, which are known as *wait and see decisions* are made after the realization of the aforementioned uncertainty decisions. Second stage decisions are related to the economic dispatch, i.e., the power output of units in each period of the planning horizon. The objective of the optimization problem is to maximize the realized expected profit of the production aggregator in the second stage. Normally is assumed that the operating costs of variable renewables are negligible in comparison with the one of thermal units. Therefore, the total operating cost to be considered for the aggregation is the cost due to the operation of the thermal unit. The total operation cost OP_{sit} is stated as follows:

$$OP_{sit} = GC_{sit} + SUC_{sit} + SDC_t z_{it} \qquad \forall s, \quad \forall i, \quad \forall t \qquad (1)$$

In (1) the three terms are measures for the generation cost GC_{sit}, start-up cost SUC_{sit}, and shut-down cost SDC_{sit}. The generation cost GC_{sit} is stated as follows:

$$GC_{sit} = A_i u_{it} + d_{sit} \qquad \forall s, \quad \forall i, \quad \forall t \qquad (2)$$

In (2) the two terms are the fixed cost and the variable cost of unit i. The generation cost is normally approximated by a quadratic function, but this paper follows the model for an approximation as a piecewise linear function stated as follows:

$$d_{sit} = \sum_{l=1}^{L} F_i^l \delta_{sit}^l \qquad \forall s, \quad \forall i, \quad \forall t \qquad (3)$$

$$p_{sit} = p_i^{min} u_{it} + \sum_{l=1}^{L} \delta_{sit}^l \qquad \forall s, \quad \forall i, \quad \forall t \qquad (4)$$

$$(T_i^1 - p_i^{min}) t_{sit}^1 \le \delta_{sit}^1 \qquad \forall s, \quad \forall i, \quad \forall t \qquad (5)$$

$$\delta_{sit}^1 \le (T_i^1 - p_i^{min}) u_{it} \qquad \forall s, \quad \forall i, \quad \forall t \qquad (6)$$

$$(T_i^l - T_i^{l-1})t_{sit}^l \le \delta_{sit}^l \qquad \forall s, \quad \forall i, \quad \forall t, \quad \forall l = 2, \dots, L-1 \tag{7}$$

$$\delta_{sit}^l \le (T_i^l - T_i^{l-1})t_{sit}^{l-1} \qquad \forall s, \quad \forall i, \quad \forall t, \quad \forall l = 2, \dots, L-1 \tag{8}$$

$$0 \le \delta_{sit}^L \le (p_i^{max} - T_{sit}^{L-1})t_{sit}^{L-1} \qquad \forall s, \quad \forall i, \quad \forall t \tag{9}$$

In (3) the variable cost function is given by the sum of the product of the slope of each segment, F_i^l, by the segment power δ_{sit}^l. In (4) the power generation of the unit is given by the minimum power generation plus the sum of the segment powers associated with each segment. The binary variable u_{it} ensures that the power generation is equal to 0 if the unit is in the state offline. In (5)–(9) are defined the limits of power generation in each segment. The start-up cost SUC_{sit} are normally represented by means of an exponential function. This exponential function is approximated by a stair wise function with a convenient selection of the number of intervals and is stated as follows:

$$SUC_{sit} \ge K_i^\beta \left(u_{it} - \sum_{r=1}^{\beta} u_{it-r} \right) \qquad \forall s, \quad \forall i, \quad \forall t \tag{10}$$

In (10) the right term is the imputed cost for a unit to have a start-up, i.e., if the unit is in the state online at hour t, and in the previous β hour has been in the state offline, the expression in parentheses is equal to 1, a start-up happens and the respective cost is has to be considered. The constraints to limit the power of unit i are stated as follows:

$$p_i^{min} u_{it} \le p_{sit} \le p_{sit}^{max} \qquad \forall s, \quad \forall i, \quad \forall t \tag{11}$$

$$p_{sit}^{max} \le p_i^{max}(u_{it} - z_{it+1}) + SDz_{it+1} \qquad \forall s, \quad \forall i, \quad \forall t \tag{12}$$

$$p_{sit}^{max} \le p_{sit-1}^{max} + RUu_{it-1} + SUy_{it} \qquad \forall s, \quad \forall i, \quad \forall t \tag{13}$$

$$p_{sit-1} - p_{sit} \le RDu_{it} + SDz_{it} \qquad \forall s, \quad \forall i, \quad \forall t \tag{14}$$

In (11) for unit i at hour t the variable p_{sit} represents the energy bid limited by the maximum power p_{sit}^{max} of unit i at hour t. The maximum power p_{sit}^{max} value takes into consideration the actual capacity unit i at hour t, considering start-up, shut-down ramp rate limits and ramp-up limit. In (12)–(14) the relation between the start-up and shut-down variables of unit i is imposed, using binary variables. The modeling for the minimum up and down time constraints is based on constraints imposed on the binary variables. The minimum up time constraints are stated as follows:

$$\sum_{t=1}^{N_i} (1 - u_{it}) = 0, \qquad \forall i, \quad \forall t \tag{15}$$

$$\sum_{t=k}^{k+UT_i-1} u_{it} \geq UT_i y_{it}, \qquad \forall i, \quad \forall k = N_i+1 \ldots T - UT_i + 1 \tag{16}$$

$$\sum_{t=k}^{T} (u_{it} - z_{it}) \geq 0, \qquad \forall i, \quad \forall k = T - UT_i + 2 \ldots T \tag{17}$$

$$N_i = \min\{T, (UT_i - U_{i0})u_{i0}\}$$

In (15)–(17) once the unit is on remain on until reaching minimum up time. A unit start-up up is kept on by UT_i hours by (16). Similarly, the minimum down time constraints are stated as follows:

$$\sum_{t=1}^{J_i} u_{it} = 0, \qquad \forall i, \quad \forall t \tag{18}$$

$$\sum_{t=k}^{k+DT_i-1} (1 - u_{it}) \geq DT_i z_{it}, \qquad \forall i, \quad \forall k = J_i+1 \ldots T - DT_i + 1 \tag{19}$$

$$\sum_{t=k}^{T} (1 - u_{it} - z_{it}) \geq 0, \qquad \forall i, \quad \forall k = T - DT_i + 2 \ldots T \tag{20}$$

$$J_i = \min\{T, (DT_i - s_{si0})(1 - u_{i0})\}$$

In (18)–(20) the unit once off stays minimum down time ours off. The start-up and shutdown coupling constraints between the binary variables and the total power of the thermal units are respectively stated as follows:

$$y_{it} - z_{it} = u_{it} - u_{it-1}, \qquad \forall i, \quad \forall t \text{ and } y_{it} + z_{it} \leq 1, \quad \forall i, \quad \forall t \tag{21}$$

$$p_{st}^g = \sum_{i=1}^{I} p_{sit} \qquad \forall s, \quad \forall t \tag{22}$$

In (21) is imposed a restriction on start-up and shutdown variables so that a unit is not allowed to start up or shut down simultaneously.

Imbalance prices, i.e., up and down-regulation prices, are given in the Iberian balancing market through a double price procedure, computing how the prices for deviations are accommodated in a view of both the system imbalance and producer imbalance as shown in Table 1. In Table 1 λ_t^+ and λ_t^- are positive and negative imbalance prices, respectively, λ_t^D is the day-ahead market price, λ_t^{UP} and λ_t^{DN} are the up and the down-regulation prices, respectively. λ_t^+ and λ_t^- can be expressed by means of price ratios pr_t^+ and pr_t^-.

Table 1. Imbalance prices

		System imbalance	
		Negative	Positive
Power producer imbalance	Negative	$\lambda_t^- = \max(\lambda_t^D, \lambda_t^{UP})$	$\lambda_t^- = \lambda_t^D$
	Positive	$\lambda_t^+ = \lambda_t^D$	$\lambda_t^+ = \min(\lambda_t^D, \lambda_t^{DN})$

The main objective of the production aggregator wind-PV-integrated unit commitment is to maximize the total expected profit in the day-ahead market. The objective function is stated as follows:

$$\sum_{s=1}^{N_S} \sum_{t=1}^{N_T} \frac{1}{N_S} \left[(\lambda_{st}^D P_{st}^{Aggregated} + \lambda_{st}^D pr_{st}^+ d_{st}^+ - \lambda_{st}^D pr_{st}^- d_{st}^-) - \sum_{i=1}^{I} OP_{sit} \right] \quad \forall s, \ \forall t \quad (23)$$

Subject to:

$$0 \leq P_{st}^{Aggregated} \leq \sum_{i=1}^{I} p_{sit}^{max} + p^{Wmax} + p^{PVmax} \quad \forall s, \ \forall t \quad (24)$$

$$d_{st} = \left(p_{st}^g + p_{st}^W + p_{st}^{PV} - P_{st}^{Aggregated}\right) \quad \forall s, \ \forall t \quad (25)$$

$$d_{st} = d_{st}^+ - d_{st}^- \quad \forall s, \ \forall t \text{ and } 0 \leq d_{st}^+ \leq p_{st}^g + p_{st}^W + p_{st}^{PV} \quad \forall s, \ \forall t \quad (26)$$

In (23) the four terms of the objective function are the per scenario profit of the day-ahead bid, the outcome incurred by the respective imbalances, and total thermal operating cost, respectively. Offer curves are normally subjected to a non-decreasing dependence stated as follows:

$$(P_{st}^{Aggregated} - P_{s't}^{Aggregated})(\lambda_{st}^D - \lambda_{s't}^D) \geq 0 \quad \forall s, \ s', \ \forall t \quad (27)$$

In (27) as the price is incremented the power increment is imposed as non-negative.

5 Case Study

The case study is for a Wind-PV-Thermal power aggregator owning a wind system with a power rated of 100 MW, a PV system with 100 MW and 2 thermal units with 285 MW. The day-ahead market prices and the price ratios pr_t^+ e pr_t^- are the ones reported in [22]. The bidding is for a 24 h horizon on an hourly basis. The scenarios considered are 5 for wind power, 5 for PV power, 5 for day-ahead market prices and 5 for price ratios. The data for the thermal units used in this research work are reported in [21]. The day-ahead market price scenarios, the wind powers scenarios and PV power scenarios are shown in Fig. 1.

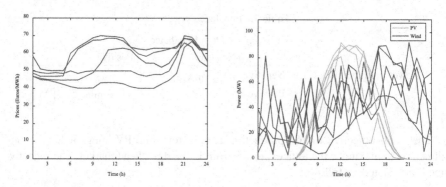

Fig. 1. Scenarios: Left, day-ahead market price; Right, PV (red lines), Wind (blue lines). (Color figure online)

The unit commitment of thermal units and the expected profit of disaggregated and aggregated coordination are shown respectively in Tables 2 and 3.

Table 2. Unit status

Hour	1	2	3	4	5	6	7	8	9	10	11	12
Unit 1 status	1	1	1	1	1	1	1	1	1	1	1	1
Unit 2 status	0	0	1	1	1	1	1	1	1	1	1	1
Hour	13	14	15	16	17	18	19	20	21	22	23	24
Unit 1 status	1	1	1	1	1	1	1	1	1	1	1	1
Unit 2 status	1	1	1	1	1	1	1	1	1	1	1	1

Table 3. Expected profit

Case	Expected profit [€]
Wind disaggregated	53 411.9
Photovoltaic disaggregated	33 515.0
Thermal disaggregated	130 217.2
Wind photovoltaic thermal - disaggregated	217 144.2
Wind photovoltaic thermal - aggregated	222 671.5

In Table 2 unit 1 is on online state (1) in all the 24 h, while unit 2 is offline (0) in hour 1 and 2, then has a start-up and stays online. Table 3 shows the expected profit for disaggregated and aggregated coordination. The aggregation allows obtain an increase in performance of about 2.5% in profit.

6   Conclusion

Aggregation platforms deliver an increase of resilience on the aggregated systems since better decisions are made in the bidding process while achieving an adequate technical performance. The wind and photovoltaic uncertainties and market uncertainties, including day-ahead market prices and imbalance prices are considered as input data given by scenarios for the proposed platform. A mixed-integer stochastic optimization approach is proposed to address the problem of aggregation in an optimum way. The result of the proposed approach is the optimum commitment of the units and bidding. The aggregation can improve the aggregator's profit in 2.5% as shown by the case study. The aggregation of wind systems, PV systems and thermal units using stochastic optimization allows to manage uncertainties in a more convenient way. Although this management is not able to ensure null imbalances, the management harmonizes the uncertainties to deliver imbalances in accordance with the aim of accessing optimum profit in the electricity market. This management with aggregation has proved to be better than the previous single stochastic optimization works. The consideration of energy storage devices in a future work can improve the reliability and the profit of the aggregated system proposed in this research work, in the way that energy can be stored in periods of low market prices and sold at periods of likely high prices. In addition stored energy can be used to compensate the fluctuations of the variable-renewables.

Acknowledgments. To thank the Millennium BCP Foundation for the financial support; and Foundation for Science and Technology-FCT project LAETA ref: UID/EMS/50022/2013.

References

1. Destek, M.A., Aslan, A.: Renewable and non-renewable energy consumption and economic growth in emerging economies: evidence from bootstrap panel causality. Renew. Energy **111**, 757–763 (2017)
2. González-Aparicio, I., Zucker, A.: Impact of wind power uncertainty forecasting on the market integration of wind energy in Spain. Appl. Energy **159**, 334–349 (2015)
3. Purvins, A., Zubaryeva, A., Llorente, M., Tzimas, E., Mercier, A.: Challenges and options for a large wind power uptake by European electricicty system. Appl. Energy **88**(5), 1461–1469 (2011)
4. Ummels, B.C., Gibescu, M., Pelggrum, E., Kling, W.L., Brand, A.J.: Impacts of wind power on thermal generation unit commitment and dispatch. IEEE Trans. Energy Convers. **22**(1), 44–51 (2007)
5. Asmus, P.: Microgrids, virtual power plants and our distributed energy future. Electr. J. **23**(10), 72–82 (2010)
6. Carreiro, A.M., Jorge, H.M., Antunes, C.H.: Energy management systems aggregators: a literature survey. Renew. Sustain. Energy Rev. **73**, 1160–1172 (2017)
7. Wang, Q., Zhang, C., Ding, Y., Xydis, G., Wang, J., Ostergaard, J.: Review of real-time electricity markets for integrating distributed energy resources and demand response. Appl. Energy **138**, 695–706 (2015)
8. Lietaer, B., Ulanowicz, R.E., Goemer, S.J., McLaren, N.: Is our monetray structure a systemic cause for a finacial instability? Evidence and remedies from nature. J. Futures Stud. **14**(3), 89–108 (2010)
9. United Nations: Sustainable development goals (2016). https://sustainabledevelopment.un. org/topics/sustainabledevelopmentgoals
10. Folke, C.: Resilience: the emergence of a perspective for social-ecological systems analyses. Glob. Environ. Change **16**(3), 253–267 (2006)

11. Kishita, Y., McLaren, B.C., Giurco, D., Aoki, K., Yoshizawa, G., Handoh, I.C.: Designing backcasting secnarios for resilient energy futures. Tech. Forecast. Soc. Change **124**, 114–125 (2017)
12. Lovins, A.B., Lovins, L.H.: Brittle Power: Energy Strategy for National Security. Brick House Publishing Co., Inc., Andover (1982)
13. Gomes, I.L.R., Pousinho, H.M.I., Melicio, R., Mendes, R.: Bidding and optimization strategies for wind-PV systems in electricity markets assisted by CPS. Energy Procedia **106**, 111–121 (2016)
14. Shahidehpour, M., Yamin, H., Li, Z.: Market overview in electric power systems. In: Shahidehpour, M., Yamin, H., Li, Z. (eds.) Market Operations in Electric Power Systems. Wiley, New York (2002)
15. Wood, A.J., Wollenberg, B.F., Sheblé, G.B.: Power Generation, Operation, and Control, 3rd edn. Wiley, New York (2013)
16. Deane, J.P., Drayton, G., Gallachóir, B.P.O.: The impact of sub–hourly modelling in power systems with significant levels of renewable generation. Appl. Energy **113**, 152–158 (2014)
17. Yang, Y., Wang, J., Guan, X., Zhai, Q.: Subhourly unit commitment with feasible energy delivery constraints. Appl. Energy **96**, 245–252 (2012)
18. Angarita, J.L., Usaola, J., Martinez-Crespo, J.: Combined hydro-wind generation bids in a pool-based electricity market. Electr. Power Syst. Res. **79**(7), 1038–1046 (2009)
19. Hedman, K.W., Sheble, G.B.: Comparing hedging methods for wind power: using pumped storage hydro units vs options purchasing. In: Proceedings of International Conference on Probabilistic Methods Applied to Power Systems–PMAPS 2006, Sweden, pp. 1–6 (2006)
20. Gomes, I.L.R., Pousinho, H.M.I., Melicio, R., Mendes, V.M.F.: Stochastic coordination of joint wind and photovoltaic systems with energy storage in day-ahead market. Energy **124**, 310–320 (2017)
21. Laia, R., Pousinho, H.M.I., Melicio, R., Mendes, V.M.F.: Bidding strategy of wind-thermal energy producers. Renew. Energy **99**, 673–681 (2016)
22. REE-Red Eléctrica de España (2018). http://www.esios.ree.es/web-publica/

Quantifying Potential Benefits from Flexible Household Storage Load

Beata Polgari[✉] 🔟, David Raisz, and Daniel Divenyi

Department of Electric Power Engineering, Budapest University of Technology and Economics,
18 Egry Jozsef Street, Budapest, Hungary
{polgari.beata,divenyi.daniel,raisz.david}@vet.bme.hu

Abstract. Household energy storage appliances offer a great potential for various direct load control (DLC) purposes. This paper investigates DLC schemes that could be realized by ripple control (RC) or long-wave radio control (LWRC). The increasing number of electric vehicles, heat-pumps or other household energy storage appliances are great candidates for DLC but in this paper only the currently available boilers have been considered. The authors prepared four case studies in order to quantify the potentials in the existing and future DLC infrastructure therefore analysed the distribution network loss reduction potential of DLC, the benefits of peak-shaving to reduce the procurement cost of the DSO, the attainable balancing energy (BE) reduction of the DSO and the possibility of offering ancillary services (AS) by directly controllable loads.

Keywords: Demand side management · Ripple control · Direct load control
Demand response · Household flexibility · Load management

1 Introduction

Demand Side Management (DSM), and especially Demand Response (DR) programs are gaining increasing interest in the power system control community [1]. RC and LWRC are worth to be considered for DLC as these are wide-spread technologies in Europe and Australia [2, 3]. In Hungary, Electric Storage Water Heaters (ESWHs) and partly storage space heaters are controlled currently with these technologies. The current control strategy was designed for peak shaving. This strategy can be called 'static', because there is no adjustment of the command sequence regarding the actual state of the system. However, there might be more profitable schemes to use the DLC infrastructure. Moreover, the LWRC infrastructure and some of the smart metering systems would be suitable for dynamic control methods with further investment while RC is only suitable for static applications.

Apart from helping to determine the future of the current control infrastructure that is approaching the end of its lifetime, the driving forces behind this investigation were to leverage the potentials in the smart metering infrastructure trying to solve balancing problems as well. This paper analyses four case studies to compare their potential benefits. However the RC and LWRC infrastructure are owned by the DSO currently, the

© IFIP International Federation for Information Processing 2018
Published by Springer International Publishing AG 2018. All Rights Reserved
L. M. Camarinha-Matos et al. (Eds.): DoCEIS 2018, IFIP AICT 521, pp. 111–119, 2018.
https://doi.org/10.1007/978-3-319-78574-5_11

benefits discussed below correspond to different market participants. Therefore the conclusion is only a techno-economic analysis disregarding market barriers.

The simulations are based on physical operation of ESWHs that calculate the power consumption in response to a control function. The control algorithm defines a control (setpoint) function which should be followed by the controlled loads. The models and the algorithm are already published in [4, 5]. The setpoint function is optimized to meet a certain goal defined by the case study and can be static or dynamic. In case of loss or procurement cost reduction the setpoint function is static because it can be defined day-ahead. Dynamic control options include BE reduction or offering system reserve as appliances are controlled according to the actual state of the system. Virtual energy buffer method is applied for dynamic setpoint function (P_{set}) determination: the setpoint function is determined so that the load consumes the same amount of energy daily as without control:

$$P_{set}(t) = P_{req}(t) + \frac{1}{T_{int}} \sum_{i=0}^{t} \left[P_{avg} - P_{set}(i) \right] \Delta t. \tag{1}$$

In Eq. (1), P_{req} corresponds to the required controlled load, P_{avg} is the constant setpoint value equal to the annual average of the DSO consumption, Δt is the time-step (5 min) and T_{int} is an integration time constant. Evaluating the simulations yields results that are suitable for a financial estimation of the achievable benefits.

2   Technological Innovation for Resilient Systems

RC and LWRC systems are robust and reliable. LWRC can switch hundreds of appliances in seconds while smart metering technologies can face difficulties to reach that many load at a time. These attributes can be useful even in case of emergency e.g. for load curtailment and storage control in extreme weather. An intelligent load management system is able to reduce the financial impact of an outage if fewer consumers fall out. Unexpected costs are avoidable with fast reaction to extreme prices.

Ripple or radio control technologies fit well to a smart grid concept applied for energy management and flexibility control. Their combination with smart meters can be useful to apply better tariff incentive, facilitate the settlement with the end-user and better estimate the balancing energy need. The proposed use cases lead to better capitalization of the digital infrastructure.

The proposed load management is flexible as control commands can be changed to suit possible new needs and the changing market environment. There is a great potential in the proposed use cases in many countries to enable residential flexibility because of the wide radio frequency coverage especially in Europe. The examined – mainly storage-type - loads are typically the biggest electric consumers in a household such as boilers, space heaters, the growing number of electric cars and batteries.

The proposed solutions contribute to the decrease of the gas-dependency as it promotes solutions for electric loads which have gas-fired alternatives (like space heating or water heating). Moreover, the resiliency of the system increases due to the non-gas-fired flexibility.

Resiliency aligns closely with sustainability: DSM adjusts the consumption closer to the intermittent renewable generation thus facilitating higher renewable share. In addition, the local intervention reduces the technical loss leading to less energy generation need either through local consumption of the local generation or by steadier load. The latter corresponds to the distribution loss reduction use case [6].

3 Use Cases

3.1 Flattening the Load Curve for Low Voltage Loss Reduction

Loss reduction problem is actually a load shifting issue since the loss is the smallest on a network if the energy is delivered by a constant current over time, for the reason that the loss is proportional to the square of the current. The load capacity factor is defined as follows showing the increase due to the time variant RMS value of the current:

$$LCF = \frac{W_{loss\,actual}}{W_{loss\,ideal}} = \frac{\int_0^T i^2(t)dt}{\frac{1}{T}\left[\int_0^T i(t)dt\right]^2} \tag{2}$$

Measurements of several distribution transformer loads have shown that in Hungarian LV networks $LCF \approx 1.1$, thus approximately 10% of the LV line loss is due to the time-variant consumption (base is the total fed-in energy into the LV network).

Two approaches were used to estimate the potential earning from a DLC with the purpose of loss reduction. In the first approach standard load profiles were used (available from DSO measurements at a statistically representative number of LV consumers). Residential load profiles without DLC were scaled up to the total number of households, then the upscaled DLC load profiles were added with ESWHs present in only 40% of the Hungarian households [7, 8]. Residential load profiles without DLC have a LCF of 1.082, the total load profile has 1.02, so this would suggest that the loss component that is due to the time-variant consumption can be reduced by a factor of almost four.

In the second approach a selected LV network model was used with measured load data and a valley-filling has been modelled. The results were approximately the same as above. It has to be noted, that a great part of this reduction is already achieved by the currently used DLC system, thus further optimization potential is low. Generally smaller loss reduction rates can be expected in the LV transformers and in the MV network than in the LV network, since greater load aggregation and thus smoother load curves can be observed on transformer and MV line level than on LV line level.

One has to note that in order to make full use of DLC for network loss reduction, a DLC management system has to be built that is network topology aware, i.e. theoretically the load curves and the DLC potential for each LV feeder have to be known, and the load has to be controlled separately for each feeder. If comparing these results to the results of the dynamic use-cases, there is no incentive to start such investments. Paper [9] comes to a similar conclusion.

The effect of the transformer peak load reduction has also been analysed and found to be 26...36%. Supposing that the actual peak load of a transformer without DLC is 0.7 p.u. (base: rated power of the transformer), and that the annual peak load increase is 1.5%. Furthermore, the peak load can be reduced by a constant 0.132 p.u. (40% of the peak load is ESWH load and 33% of this ESWH load can be eliminated). Thus the peak load exceeds 1.2 p.u. in the 38th year without DLC and in the 45th year with DLC. This means that the investment in a new transformer could be postponed by 7 years. However, this has no relevance in Hungary as most of the distribution transformers are by far not overloaded.

3.2 Reducing Procurement Costs

The energy procurement cost reduction of the DSO has been translated to minimizing the peak/base product demand ratio. Only the year-ahead futures trade has been considered in the simulations in a way that the average of the annual DSO load is procured as base and the rest is bought as peak product. First of all, the authors calculated the annual average cost of electric energy per kWh for a cost comparison, by determining the base and peak quantities from the total load curve of the DSO without load control. Then the average cost per unit energy in the following situations have been computed: ESWHs without control (simulated by constant ON command) plus the uncontrollable load, ESWHs controlled by the current command program plus the uncontrollable load, ESWHs controlled by an optimal peak-shaving algorithm plus the uncontrollable load and ESWHs controlled by a sub-optimal peak-shaving algorithm plus the uncontrollable load. The results are summarized in Table 1.

Table 1. Procurement cost reduction potentials.

	Average energy cost reduction [%]	ΔT_{avg} of ESWHs [°C]
Without control	Base	0
Current	–0.74	–2.30
Optimal	–2.61	–9.15
Sub-optimal	–1.46	–5.87

The results show that the average energy cost would be 0.74% higher without DLC than with the current command schedule. The optimal procurement cost reducing algorithm achieves an even better saving, though the average water temperature in the ESWHs would be decreased significantly (by 9.15 °C). Further experiments have shown that another set-point function can be created that yields less decrease of the comfort level (the average water temperature in the ESWHs [4]) while bringing an additional decrease of 0.72% in the procurement cost.

3.3 Balancing Energy Reduction

These simulations investigate how much the controlled ESWHs could help to reduce the BE need of the DSO: in case of consumption surplus (consumption larger than the schedule) the directly controllable loads have to be switched off, and vice-versa. It must be emphasized that the quantity of the BE has been minimized in the simulations instead of minimizing the cost of BE. These two approaches do not provide the same saving in cases when the BE price is different for surplus and deficit as it is usual. The balancing cost is determined posteriorly for every 15-min interval of a day by accumulating the deviation for that interval.

Simulations were carried out for one year. In the original dataset, the mean value of the BE over all 15-min intervals was 0.23 MWh with a standard deviation of 7.51 MWh. The simulation results are shown in Table 2, where deviations smaller than 2.5 MWh are considered to be zero.

Table 2. Distribution of BE sign changes - share from the total number of 15-min. cases.

| | | Resulting | | |
		Negative	Zero	Positive
Original	Negative	23.16%	9.97%	1.61%
	Zero	7.22%	18.75%	6.85%
	Positive	1.62%	9.82%	20.99%

From Table 2 it can be observed, that in approx. 10% of all intervals with originally negative BE and 10% of the originally positive intervals were changed to almost zero by DLC. The results of the balancing energy costs are summed up in Table 3.

Table 3. The balancing energy cost change by DLC.

| Original balancing energy need | Yearly balancing energy costs (M€) | | |
	Original cost	Simulated cost with optimal DLC	Cost decrease
Negative	–0.252	0.246	–0.497
Nearly zero	0.655	1.500	–0.8407
Positive	9.514	5.546	3.968
SUM	9.918	7.292	2.630

A large number of originally negative BE needs turned over to be positive due to the DLC, thus the BE cost turned from negative (income) to positive (expenditure). Similarly, in several cases when the BE was approx. 0, the DLC turned the BE to positive, thus the costs increased. These two impacts result in 1.34 M€ more expenditure, but the originally positive BE was decreased in several cases, resulting in 3.97 M€ cost decrease. The total balance is 2.63 M€ decrease of BE costs.

Using a dynamic type of DLC brings also some disadvantages. First, it could be more difficult for the balance group responsible party to provide a day-ahead schedule, since the controlled load is not known in advance. Secondly, the benefit of valley-filling is lost with an online modification of the switching schedule. Thirdly, the real-time balancing requires the online estimation of the value of BE or remote online meter reading at all consumers.

An analysis of DLC for balancing is presented in [10] by freezers and their application is found to be promising for balancing e.g. renewable generation.

3.4 Offering Ancillary Services

DR as control reserve for the TSO promise great business opportunity also in [11]. For residential DR, secondary reserve (SR or also called frequency restoration reserve - FRR) is adequate since time is needed for activation while the reserve cannot be guaranteed long enough to offer tertiary reserve. The control commands have to be modified in real-time here as well. The Code of Commerce (CC, [12]) can be hardly applied for DR. The CC distinguishes 4 qualifications. Their interpretation is as follows (Table 4).

Table 4. Reserve qualification.

Qualification	Description
Not available	In case of an upward request, the total DLC load is greater than $P_{average}$; In case of a downward request, the total DLC load is smaller than $P_{average}$
Available	If in case of an upward ramping request the controlled load decreases, or the power of the controllable loads increases after a downward ramping request
Follower	If DLC is available & its load follows the request with max. 5% error
Partial follower	If DLC is available & its load follows the request with greater than 5% error

Only followers are entitled for capacity price while partial followers are not. Yet both of them get energy price proportionally to the provided reserve energy. Constant availability has been assumed in the simulations, because the operational limits are hardly interpretable for a portfolio consisting of thousands of boilers and the ability of the system is highly dependent on the previous reserves (e.g. after a high power consumption period it is difficult to achieve further increase in the power consumption).

The SR demand in the simulations (35.75% up and 63,84% down) come from the netting of the secondary up and down need of the TSO during a whole year proportionated to the DSO [13]. Simulations were conducted with 30 MW, 60 MW and 90 MW offered capacity symmetrically. Table 5 contains the result of this case study.

Table 5. Annual rate of entitlements for energy and capacity price.

		±30 MW	±60 MW	±90 MW
Upward 35.75%	Entitled for capacity price	71.10%	69.48%	69.34%
	Entitled for energy price – follower	6.85%	5.23%	5.09%
	Entitled for energy price – partial follower	23.45%	22.77%	22.65%
Downward 63.84%	Entitled for capacity price	40.73%	39.12%	38.97%
	Entitled for energy price – follower	4.56%	2.96%	2.81%
	Entitled for energy price – partial follower	47.37%	43.07%	39.84%

The capacity and energy prices that were used in the simulations can be seen in Table 6 while the resulted capacity and energy fee incomes for a year are given in Table 7.

Table 6. The unit fee of the capacity and the energy fee based on the year-ahead procurement for the selected year in Hungary [13].

Secondary up		Secondary down	
Capacity fee (€/MW/h)	Energy fee (€/MWh)	Capacity fee (€/MW/h)	Energy fee (€/MWh)
41.2	187.8	39.8	2.52

Table 7. The total income of energy and capacity revenue according to the simulations.

		±30 MW	±60 MW	±90 MW
Up	Capacity fee [M€]	7.70	15.04	22.52
	Energy fee [M€]	8.47	11.47	12.29
	Sum [M€]	16.16	26.51	34.81
Down	Capacity fee [M€]	4.26	8.19	12.24
	Energy fee [M€]	−0.13	−0.19	−0.22
	Sum [M€]	4.14	8.00	12.02
	SUM [M€]	**20.30**	**34.51**	**46.83**

It should be treated with criticism that the *capacity fee* forms a significant part of the total income thanks to the model assumptions. Capacity fee was gained in most cases for the non-requested direction (which cannot be checked), not for a follower qualification. Furthermore, the upward reserve brings more income because the downward ramping demand (~64%) was twice as frequent as the upward (~35%) as in Table 5 and the availability fee was almost the same. The DR received capacity fee in only 71% of the year according to Table 3 because it was a partial follower in the rest of the time. It is questionable how high a capacity fee can be if the concept of availability is so uncertain. However, even considering the quarter of the above capacity, the saving is still 25…200% higher than by optimizing for procurement cost reduction.

The negative energy fee means that the DLC provider pays for the system operator. Although, in case of downward ramping in fact the power of the ESWHs is increased, costing more contrary to the conventional power plant ramping.

The idea of applying DLC for SR is quite unique. The authors in [14] report encouraging initial simulation results by a virtual power plant concept aggregating small controllable loads. Koch et al. also prove in [15] by an entirely different method that thermostat-controlled appliances are good candidates for following a dynamically variable control signal. Paper [16] concluded that 33.000 ESWHs (4.5 kW each) would allow 2 MW SR offer 24 h a day or 20.000 pieces between 6:00 and 24:00. Scaling up this result for our 430.000 ESWHs case (considering the controlled built-in power) would mean 10.8 MW nonstop reserve or 17.8 MW excluding the 0:00 to 6:00 interval. This is comparable to the power ranges that were simulated.

4   Conclusions

The authors found that there is a great potential in the DLC of ESWHs at present and in the future, too. However, an infrastructural development is necessary to realize most of the benefits. The highest income could be realized by offering AS, namely SR. For this purpose the older RC system should be replaced entirely by LWRC infrastructure. Secondary control must be realized in a closed-loop control system and it would be necessary to install an online metering system. Further significant potential lies in BE reduction. Smaller cost saving can be achieved by static load control programs used for peak-base ratio optimization to decrease energy purchase costs or distribution system losses. The approximate ratios of simulated financial potentials are as follows (Table 8):

Table 8. Comparison of the results – ratio of financial benefits for different DLC use cases, relative to the most promising one.

Secondary reserve	Minimizing BE	Decreasing year-ahead procurement cost	Reducing LV network loss
1	0,788	0,700	0,318

Further work could extend the analysis with other domestic controllable loads like EVs and heat-pumps covering also a supporting market environment.

Acknowledgement. Project no. FIEK_16-1-2016-0007 has been implemented with the support provided from the National Research, Development and Innovation Fund of Hungary, financed under the FIEK_16 funding scheme. The authors are indebted to Zsolt Bertalan and Péter Márk Sőrés for their valuable comments.

References

1. Palensky, P., Dietrich, D.: Demand side management: demand response, intelligent energy systems and smart loads. IEEE Trans. Ind. Inf. **7**(3), 381–388 (2011)
2. Dán, A., Divényi, D., Hartmann, B., et al.: Perspectives of demand-side management in a smart metered environment. In: International Conference on Renewable Energy and Power Quality (2011)
3. Polgári, B., Raisz, D., Hartmann, B.: Overview on demand side management. In: International Conference on Deregulated Electricity Market Issues in South-Eastern Europe Proceedings (2013)
4. Polgári, B., Divényi, D., Raisz, D.: Control response analysis of aggregated energy storage household appliances. In: IEEE Industry Applications Society Annual Meeting, pp. 1–9 (2016)
5. Raisz, D.: Network Loss Calculation and Demand Side Management: Optimization Problems for Cost Efficient Distribution System Operation, 128 p. Lambert Academic Publishing, Berlin (2012). ISBN: 978-3-8484-2036-0
6. EPRI: Electric Power System Resiliency, challenges and opportunities (2016)
7. Hungarian Central Statistical Office: Electricity consumption of households (2006)
8. Energy Center Not-for-profit Ltd.: Electricity consumption of households (2009)

9. Shaw, R., Attree, M., Jackson, T., Kay, M.: The value of reducing distribution losses by domestic load-shifting: a network perspective. Energy Policy **37**, 3159–3167 (2009)
10. Vande Meerssche, B.P., Van Ham, G., et al.: General and financial potential of demand side management. In: 9th International Conference on the European Energy Market (2012)
11. Talari, S., Shafie-khah, M., Hajibandeh, N., Catalão, J.P.S.: Assessment of ancillary service demand response and time of use in a market-based power system through a stochastic security constrained unit commitment. In: Camarinha-Matos, L.M., Parreira-Rocha, M., Ramezani, J. (eds.) DoCEIS 2017. IFIP AICT, vol. 499, pp. 233–241. Springer, Cham (2017). https://doi.org/10.1007/978-3-319-56077-9_22
12. Commercial Code of MAVIR (Hungarian TSO), March 2015
13. MAVIR, Secondary up and down reserve need data publication and the selected bidders for the secondary up and down direction, 30 March 2015
14. Ruthe, S., Rehtanz, C., Lehnhoff, S.: Towards frequency control with large scale virtual power plants. In: 3rd IEEE PES ISGT Europe (2012)
15. Koch, S., Mathieu, J.L.: Modelling and control of aggregated heterogeneous thermostatically controlled loads for ancillary services. In: Power Systems Computation Conference (2014)
16. Kondoh, J., Lu, N., Hammerstrom, D.J.: An evaluation of the water heater load potential for providing regulation service. IEEE Trans. Power Syst. **26**(3), 1309–1316 (2011)

Energy Efficiency in Buildings by Using Evolutionary Algorithms: An Approach to Provide Efficiency Choices to the Consumer, Considering the Rebound Effect

Ricardo Santos[1(✉)], J. C. O. Matias[2,3,4], and Antonio Abreu[5]

[1] University of Aveiro, Aveiro, Portugal
ricardosimoessantos84@ua.pt
[2] Department of Economics, Management, Industrial Engineering
and Tourism (DEGEIT), University of Aveiro, Aveiro, Portugal
jmatias@ua.pt
[3] C-MasT, University of Beira Interior, Covilhã, Portugal
[4] Govcopp, University of Aveiro, Aveiro, Portugal
[5] CTS Uninova, ISEL - Instituto Superior de Engenharia de Lisboa,
Instituto Politécnico de Lisboa, Lisbon, Portugal
ajfa@dem.isel.ipl.pt

Abstract. Energy efficiency can be achieved, by making optimal choices of household appliances, based on specific rules for consumption and use. However, it's not always possible to achieve good solutions, since in general, an efficient equipment, with an economic consumption savings during his life cycle, is usually an expensive one, with a high initial investment. Additionally, the interaction of these choices, associated with consumer behavior, could lead toward to efficient losses during the lifecycle of the equipment, and then to a situation of indirect rebound effect. In this work, it is presented an approach, applied to the residential buildings, by using evolutionary algorithms to support consumer decisions. The approach presented here, could promote energy efficiency by providing the consumer with several optimal and feasible solutions, and at the same time, with information about the impact of his choices made on future.

Keywords: Energy efficiency · Evolutionary algorithms
Life cycle cost analysis · Indirect rebound effect

1 Introduction

The reduction of energy consumption is a priority to sustainability [1], with buildings, accounting for about 30–45% of the energy consumed in most countries [2].

Regarding the residential sector, the share of electrical energy consumption, represents about 13,9% of the final energy consumed in the world in 2012 [3], while in Portugal, this value is about 18% [4] representing thus an important area and an opportunity to increase energy efficiency.

© IFIP International Federation for Information Processing 2018
Published by Springer International Publishing AG 2018. All Rights Reserved
L. M. Camarinha-Matos et al. (Eds.): DoCEIS 2018, IFIP AICT 521, pp. 120–129, 2018.
https://doi.org/10.1007/978-3-319-78574-5_12

Recently, there has been some energy efficiency improvements, regarding electrical household appliances, by establishing some measures (e.g. mandatory labelling) to reduce energy services cost and their equivalent CO_2 emissions [3, 5]. This allows to inform the consumer about relevant issues, regarding each appliance (e.g. energy consumption, capacity in litters (fridge), clothe capacity (washing machine)) by adjusting the use of each appliance to consumer needs [6].

However, and given the several options available on the market, it's difficult to analyze what's the best solution to adopt, regarding the issues referred before [7]. Additionally, and related with that choices, there may be a situation of a Rebound Effect, which one way to reduce it, could be by improving the consumer behavior [8, 9], through awareness for the efficiency that he could earn or lost in a long term.

Recently, several entities, including governments, associations and manufactures, have tried to promote energy efficiency in the household's sector [5, 10, 19]. However, these measures don't provide efficient electrical appliances, that seek to satisfy both consumer and environment needs, according to what was referred before.

Traditional optimization techniques, allows to tackle the problem, although, they usually begin with a single potential solution, which is iteratively manipulated, until finding a unique and final one [11], which is undesirable to consumer.

The evolutionary algorithms, like Genetic Algorithms (GAs), allows to obtain optimal, feasible and alternative solutions [12–14] to attend the consumer needs.

Several authors have presented methods, that try to exploit some of these needs by including some criteria (annual primary energy consumption, initial investment cost) [7], although without considering the appliances from the market and its features.

Additionally, they don't include the possibility of Indirect Rebound Effect, regarding the optimal choices, although the proof of its existence on literature [8].

The Indirect Rebound Effect can be estimated through the monetary savings from energy efficiency improvements and by assuming that this increase of disposable income is "re-spent" in other services that needs energy to be produced [15]. According to [9], rebound effect can be mitigated, through an improvement on consumer behavior.

The purpose of this work, is to present a methodology to support the consumer with efficient choices, considering the indirect rebound effect as well.

Research Question

1. How to support a consumer who wants to buy energy services in their household, through a set of diversified and optimal solutions that simultaneously:
   i. Promote economic welfare;
   ii. Promote social welfare;
   iii. Promote environment welfare;
   iv. Minimize the possible (negative) impacts of its choices on future.

Hypothesis

H1: Optimization of economic savings with investment and consumption;
H2: Optimization of thermal, light and noise comfort;
H3: Optimization of carbon footprint;
H4: Use of rebound effect to estimate energy efficiency loss.

2 Relationship to Resilient Systems

The integration of resilience into the systems, is relevant, due to the increased rate of disruptive events regarding energy area (e.g. climate change, economic crisis, technological evolution, etc.) challenging the way that systems are designed.

This occurs particularly with energy services, due to the technology innovation, economic crisis (consumer surplus), and climate change.

To be resilient to that changes, the system should include the capability to adapt and preserve at the same time its original functions.

This work, presents a methodology, to provide the agent decision (consumer) with a set of optimal solutions (energy services), adapted to the technology development and extended to other energy services, although preserving the same functions.

3 Research Method

3.1 Problem Dimensions and Case Study

The problem presented here, will consider a decision-agent (consumer), who wants to buy a set of electrical appliances (energy services), as described next.

The dimension of the problem (j) corresponds to the number of different types of appliances (option i) that can be purchased by the consumer.

The number of combinations depends on the number of available options regarding each dimension, thus corresponding to a high number of possible combinations for the user (about 22 million in this case) (Fig. 1).

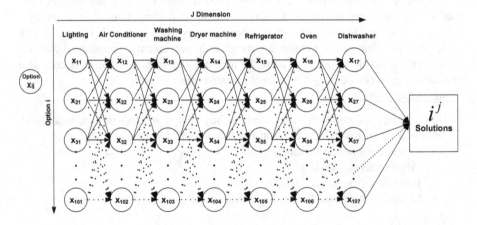

Fig. 1. Consumer's decisions space

This number can be reduced, if we consider that the consumer cannot make any choice x_{ij}, since he has a limited budget available, thus reducing the feasible solutions.

The available solutions were pre-selected, according to a set of criteria (Table 1), and based on number of occupants/consumers, to improve the use of the efficient appliances. In this work, it was considered four (e.g. family) occupants.

Table 1. Adopted criteria based on [17, 18]

Dimension	Criteria used	Quantity
Air conditioner	• Types of air conditioner considered: • Zone to be heated/cooled by the air conditioner • Minimum capacity required	• Wall (mono split) wall (multisplit) Portable • Living room • 9905,6 BTU
Washing machine	• Capacity, based on the number of household's occupants [18]	• 7 kg
Dishwasher	• Load capacity	• 12 cutlery
Oven	• Useful volume, available for cooking according to the nr. of occupants [18]	• 47 cm × 68 cm
Dryer machine	• Type of dryer machines • Load capacity from [18]	• By exhaust • 7 kg
Lighting	• Technology	• halogen CFL Fluorescent
Refrigerator	• Capacity of the refrigerators [18] • Type of refrigerator according to the number of occupants [18]	• 150 L • Refrigerator combined type

Additionally, the consumer wants to explore the consequences of his choices in future, regarding those made on present, and by predicting an eventual situation of indirect rebound effect, between light and air conditioner appliances.

This can be estimated by using scenario simulations during the air conditioner life cycle (previously selected), and by making changes around different light appliances (Table 2) according to a set of scenarios (Table 3).

Table 2. Set of considered scenarios

Year/Scenario	1..4	5..6	7..9	10
1	Opt. Std - CFL	Opt. C – Halogen	Opt. A – CFL	Opt. E – Halogen
2	Opt. Std - CFL	Option D – Fluorescent		
3	Opt. Std - CFL	Option F – LED		

Table 3. Set of considered options and heating loss calculations per lamp

Opt.	Technology	P [W]	Eff. Flux [lm/W]	q_{ele} [W]	Nr. lamps same E[lux]	q_{ele} [W] Equiv.	Air Cond. [BTU/h]
Std	CFL	11,0	60,0	32,18	5	160,92	10055,0
A	CFL	13,0	60,0	38,04	4	152,14	10025,0
B	Halogen	35,0	20,1	93,09	5	465,47	11252,0
C	Reg. Fluorescent	10,0	80,1	31,92	4	127,67	9905,6
D	Halogen	29,0	24,2	77,13	5	385,67	10953,0
E	Led	7,0	57,8	21,78	8	174,27	10116,0

3.2 Problem Formulation and Implementation by Using GAs

Based on the diagram of Fig. 1, the decision variables can be defined as:

$$x_{ij} : j(dimensions) = \{1..7\} \wedge i(options) = \{1..n_i\} \tag{1}$$

Where the objective is to maximize the savings for the consumer:

$$\max V_R(x_{ij}) : V_R \in \mathbb{R} \wedge x_{ij} \in \{0,1\} \tag{2}$$

According to [6], it was considered the following objective function (V_R):

$$V_R = \sum_{i=1}^{20} \left(\frac{P_{cons_{i,1}}(x_{i1})}{P_{inv_{i,1}}(x_{i1})} \right) \cdot \frac{I_{inv.ef_{ij}}(x_{i1})}{I_{total}} \cdot x_{i1} + \sum_{i=1}^{11} \left(\frac{P_{cons_{i,2}}(x_{i2})}{P_{inv_{i,2}}(x_{i2})} \right) \cdot \frac{I_{inv.ef_{ij}}(x_{i2})}{I_{total}} \cdot x_{i2}$$
$$+ \sum_{j=3}^{7} \sum_{i=1}^{10} \left(\frac{P_{cons_{i,j}}(x_{ij})}{P_{inv_{i,j}}(x_{ij})} \right) \cdot \frac{I_{inv.ef_{ij}}(x_{ij})}{I_{total}} \cdot x_{ij} \tag{3}$$

Where $P_{cons_{i,j}}(x_{ij})$ and $P_{inv_{i,j}}(x_{ij})$, are variable parameters associated with the objective function, regarding the correspondent savings obtained in consumption and investment respectively and for a given solution i, regarding to a dimension j.

These parameters were obtained by making previously a life cycle cost analysis (LCCA) regarding each selected appliance, and other assumptions made.

Since, only one option, can be selected from each dimension, the constraints are:

$$R_1(x_{i1}) : \sum_{i=1}^{20} x_{i1} = 1 \wedge x_{i1} \in \mathbb{N}, \forall i = \{1..20\} \wedge x_{i1} = \{0,1\} \tag{4}$$

$$R_2(x_{i2}) : \sum_{i=1}^{11} x_{i2} = 1 \wedge x_{i2} \in \mathbb{N}, \forall i = \{1..11\} \wedge x_{i2} = \{0,1\} \tag{5}$$

$$R_k(x_{ij}) : \sum_{i=1}^{10} x_{ij} = 1 \land k = \{3..7\} \land x_{ij} \in \mathbb{N}, \forall i = \{1..10\}, j = \{3..7\} \land x_{ij} = \{0,1\}$$

(6)

The constraint associated with the total investment made (budget) is:

$$R_8(x_{ij}) : I_{total}(x_{ij}) = \sum_{i=1}^{20} x_{i1} * I_{ef_{i1}} + \sum_{i=1}^{11} x_{i2} * I_{ef_{i2}} + \sum_{j=3}^{7} \sum_{i=1}^{10} x_{ij} * I_{ef_{ij}} \quad I_{total} \in \mathbb{R} \land x_{i1} = \{0,1\}$$

(7)

The GAs individual framework is presented as follows on Fig. 2:

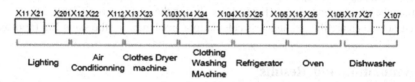

Fig. 2. GA's individual codification

The codification used, was binary, given the complexity with the use of real formulation to tackle this problem, mainly in terms of the additional number of constraints.

The initial population was 100 individuals, and the parents were selected by using Roulette method, with a single crossover point method and crossover rate of 0,45.

The selected mutation operator, was Normal Random with a mutation rate of 0,01.

3.3 Indirect Rebound Effect and Its Estimation

When the effective energy saving, is different from the one initially expected, it occurs an efficiency loss, and therefore a Rebound Effect [15]:

$$Rebound\,Effect(\%) = \frac{Expected\,Savings - Actual\,Savings}{Expected\,Savings} \times 100$$

(8)

One type of Rebound Effect is the Indirect Rebound Effect, which, and according to [9, 15], these phenomena results from the additional income, freed up by saving energy costs, which in this case it will be studied the impact from the consumer behavior (through is choices), by assuming that he chooses different light appliances, during the life cycle of the air conditioner, considered on this work (10 years).

In this work, the indirect rebound effect, was estimated, by using LCCA with focus on the interaction between Air Conditioner and Lighting, during the "usage" phase.

By considering the acquisition in the future of different options available on the market (Table 3), it was performed a set of possible scenarios (Table 2), predicting the impact of consumer actions in terms of investment and consumption savings.

On Table 2, it's presented the scenarios, considered for this work, by assuming an air conditioner life cycle of 10 years.

According to [16], the internal loads from light-emitting, can be calculated using the following expression:

$$q_{el.} = 3,41.W.F_{ul}.F_{sa} \tag{9}$$

Where q_{el}, is the correspondent heat gain [Btu/h], W the total light power [W], F_{ul}, the lighting use factor and Fsa, a lighting special allowance factor, obtained from [16].

The total light wattage values, were obtained for all options, by considering an illuminance maintenance level (Em) of 300 lx for the living room (Table 3).

These values are the heating loss calculations per lamp, for the same Em (approx.).

4 Simulation and Results

To perform the following simulations, it was adopted the following parameters:

- Average number of generations/run: 157
- Average CPU time: 12,3 s
- Average N° Runs/budget value: 12

For the model implementation, it was used the Risk Solver Platform in an *Excel VBA* spreadsheet for GAs implementation, and *GAMS* software for the *Simplex* implementation to assess the GAs quality solutions, by comparing those to *Simplex*.

4.1 GAs Behavior and a Solution Example

The following results presented on this section, regards the best solution found, concerning each value of the available budget, varying from 1800 up to 3000 euros.

The model was simulated for different budget scenarios, by using GAs and Simplex methods.

The results and behavior of GAs, are shown on Fig. 3(a) and (b) respectively, regarding the solutions with investment supported by the family for each scenario/budget considered, and the behavior of GAs through the evolution of its fitness value, and considering a budget constraint of 1900 euros.

The results presented (Fig. 3(a)), allows to show the good quality of solutions, obtained from GAs, compared to those from Simplex, and its correspondent behavior (Fig. 3(b)).

On Table 4, it is presented an example of a feasible solution obtained from GAs, as well as its savings in terms of CO_2 emissions, for a budget of 2600 Euros.

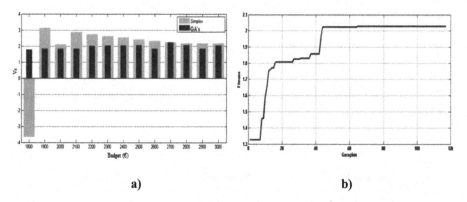

Fig. 3. (a) Final value of the investment for each value of available budget (b) Evolution of fitness, corresponding to the best individual of each generation, for a budget of 1900 euros

CO_2 savings were calculated by using a carbon footprint indicator (emission factor), obtained from [20].

Table 4. Example of an efficient solution obtained from this approach

Dimensions	Stand. Sol. Inv. (€)	Effic. Sol. Inv. (€)	Invest. Sav. (€)	Consum. Sav. (€)	CO_2 Sav. (kg)
Lighting	0,6	5,0	5,8	28,5	28,5
Air conditioner	368,0	299,0	69,0	1315,6	1315,6
Refrigerator	250,0	549,0	−287,0	2,7	8,5
Wash dishes	310,0	669,0	−320,0	6,2	6,2
Cloth wash machine	262,0	445,0	−195,0	94,8	94,8
Ovens	170,0	170,0	−429,0	−2,6	−2,6
Clothes dryer	349,0	385,3	−75,3	1,8	1,8
Total	1709,6	2522,3	−1231,5	1447,1	1452,9

According to Table 4, the consumer can save up to € 215.6 (€ 1,447.1 € − 1231.5), corresponding to a reduction of 1452,9 kg of CO_2 and during a period of 10 years.

4.2   Indirect Rebound Effect

The indirect rebound effect was estimated by using LCCA, by analyzing the interaction between air conditioner and lighting, during the usage phase. On the following figures (Figs. 4), are graphically presented the 3 scenarios, presented before.

Both indicators are proportional to the electric consumption, and the values are related to the overall electric energy consumption, accumulated with the years.

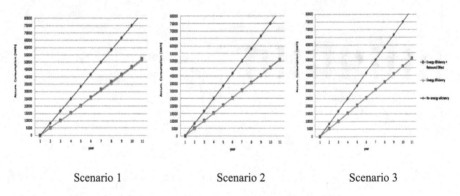

| Scenario 1 | Scenario 2 | Scenario 3 |

Fig. 4. Household electric appliances consumption

Through the obtained results, it is noted the relevance on the difference between the consumption patterns, mainly through observation of the numerical values, where according to the obtained solution (Option Std-CFL), the consumer would have better results for the next 10 years, if he chooses Scenario 2, where the rebound effect assumes negative values, suggesting that the energy efficiency could be improved with the new choice.

Through this method, the consumer (if unsatisfied), can choose the 2nd best Scenario, or even run again the model presented before to get a previous solution, where he can obtain new projections and therefore choose the best scenario.

5   Conclusions

In this work, it was presented an approach to provide efficiency choices to the consumer, regarding a set of household appliances needed to be acquired.

The quality of the obtained solutions, as well as GAs behavior on providing feasible solutions, was also considered, by comparing GAs results with those obtained from Simplex method.

Although Simplex has provided the best results for this problem, rather GAs, it is feasible to use Gas in this model, given the quality of the obtained solutions in comparison with those from Simplex.

GAs can also provide more feasible and different solutions instead of methods such Simplex, where only one solution can be achieved for the same budget scenario.

The estimation of Rebound Effect, by predicting the consumer behavior, through is choices on future, can be particularly interesting, to provide him with some information regarding the impacts of his choices in terms of consumption, allowing him to make the best options in the future.

Additionally, the methodology presented here, can be suitable to the consumer needs, by providing just the appliances to be acquired, eliminating or adding dimensions to the problem, and by using the same principles that were presented here.

Other dimensions could be explored on further work, such the electronic equipment's, or even a future application, applied into an industrial context, which could reduce the energy consumption allowing savings to the consumer, both in the consumption, as well as in the investment.

References

1. Matias, J., Devezas, T.: Socio-economic development and primary energy sources substitution towards decarbonization. Low Carbon Econ. **2**, 49–53 (2011)
2. IPCC: Climate Change 2014: Mitigation of Climate Change Summary for Policymakers and Technical Summary. Intergovernmental Panel on Climate Change (IPCC) (2015)
3. IEA: Energy Efficiency 2017 – Market Reports Series. OECD/IEA (2017)
4. Santos, C.: Reabilitação de edifícios para promoção do conforto e da eficiência energética. In: Net-Zero Energy Buildings, Conference (1st) LNEG, Lisbon (2012)
5. Krivošík, J., Attali, S.: Market surveillance of Energy Labelling and Eco Design product requirements. ADEME (2014)
6. Santos, R., Reis, F.: An approach to provide smart and energy efficient decisions using evolutionary algorithms. 4ª Escola Luso Brasileira de Computação Evolutiva (ELBCE), Departamento de Matemática, Universidade de Coimbra, Coimbra (2013)
7. Malatji, E.M., Zhang, J., Xia, X.: A multiple objective optimization model for building energy efficiency investment decision. Energy Build. **61**, 81–87 (2013)
8. IRGC: The rebound effect: implications of consumer behaviour for robust energy policies - a review of the literature on the rebound effect in energy efficiency and report from expert workshops, Lausanne (2013)
9. Ouyang, J., Long, E., Kazunori, H.: Rebound effect in Chinese household energy efficiency and solution for mitigating it. Energy Build. **35**, 5269–5276 (2010)
10. Fell, M.: Energy services: a conceptual review. Energy Res. Soc. Sci. **27**, 129–140 (2017)
11. Henggeler Antunes, C., João Alves, M., Clímaco, J.: Multiobjective Linear and Integer Programming. EATOR. Springer, Cham (2016). https://doi.org/10.1007/978-3-319-28746-1. ISBN 978-3-319-28744-7
12. Abreu, A., Calado, J., Pêgo, E.: Preventive maintenance planning using genetic algorithms Planeamento da manutenção preventiva usando algoritmos genéticos. In: International Conference on Engineering for Society, Covilhã (2015)
13. Goldberg, D.: Genetic Algorithms in Search Optimization and Machine Learning. Addison Wesley, Maryland (1989)
14. Randall, M., Rawlins, T., Lewis, A., Kipouros, T.: Performance comparison of evolutionary algorithms for airfoil design. Procedia Comput. Sci. **51**, 2267–2276 (2015)
15. González, J.: Methods to empirically estimate direct and indirect rebound effect of energy-saving technological changes in households. Ecol. Model. **223**, 32–40 (2011)
16. ASHRAE: ASHRAE Handbook: Fundamentals, American Society of Heating, Refrigeration and Air-Conditioning Engineers, Atlanta (1999)
17. ADENE: Manual da Etiqueta Energética, ADENE, Lisboa (2017). ISBN 978-972-8646-36-3
18. DGEG: Eficiência Energética em Edifícios – Programa E4, Direção Geral de Energia e Geologia, Lisboa (2002)
19. Wong, I.L., Krüger, E.: Comparing energy efficiency labelling systems in the EU and Brazil: implications, challenges, barriers and opportunities. Energy Policy **109**, 310–323 (2017)
20. IEA: CO_2 Emissions from Fuel Combustion 2017 - Highlights, OECD/IEA (2017)

Smart Grids

The Use of Smart Grids to Increase the Resilience of Brazilian Power Sector to Climate Change Effects

Débora de São José[1,2(✉)] and J. Nuno Fidalgo[2,3]

[1] Department of Industrial Engineering and Management,
University of Porto, Porto 4200-465, Portugal
up201106063@fe.up.pt
[2] Institute for Systems and Computer Engineering,
Technology and Science, Porto 4200-465, Portugal
[3] Department of Electrical and Computer Engineering,
University of Porto, Porto 4200-465, Portugal

Abstract. Climate change has been a much-commented subject in the last years. The energy sector is a major responsible for this event and one of the most affected by it. Increasing the participation of renewable is a way to mitigate these effects. However, a system with large share of renewables (like Brazil) is more vulnerable to climate phenomena. This article analyzes the implementation of smart grids as a strategy to mitigate and adapt the electricity sector to climate change. Different climate and energy sector scenarios were simulated using a bottom-up approach with an accounting model. The results show that smart grids can help save energy, increase network resilience to natural hazards and reduce operational, maintenance costs and investments in new utilities. It would also allow tariffs diminution because of generation and losses costs reductions.

Keywords: Resilience · Electricity · Smart grids · Climate change
Brazil · Adaptation · Mitigation

1 Introduction

The climate is always changing, as human activity has been intensifying this natural process by adding more greenhouse gas (GHG) into the atmosphere. The main concerns are related to the impact that this event can have on health, food production, well-being and energy generation.

The electricity sector is one of the main responsible for those changes and the renewable generation is seen as a tool to mitigate the associated effects. In fact, the energy usage often results in greenhouse gases emissions, affecting climate in such way that deteriorates the potential of renewables generation. As less renewables production means more emissions, this positive feedback loop should be broken.

© IFIP International Federation for Information Processing 2018
Published by Springer International Publishing AG 2018. All Rights Reserved
L. M. Camarinha-Matos et al. (Eds.): DoCEIS 2018, IFIP AICT 521, pp. 133–146, 2018.
https://doi.org/10.1007/978-3-319-78574-5_13

1.1 Climate Change

Many researches have been conducted in different countries, in order to assess the local/ regional impacts of global climate change in Brazil. Despite the diversity of these studies (see Sect. 1.3), all of them arrive to the same conclusion: Brazil is vulnerable to climate change because of its high share of renewable sources. Not only hydro production is expected to reduce but also the biofuels potential will be compromised.

To minimize these negative impacts, it is necessary to reduce the GHG emission and increase the system resilience through the implementation of adaptation strategies.

1.2 Electricity Sector in Brazil

Not many years had passed since the 2001 electric crisis in Brazil and the local specialists (GEE 2015; Losekann 2015a, b; Queiroz 2015) already talk about a new one. When analyzing the 2001 electric crisis, Lorenzo (2002) concluded that this event was not unexpected for those who have been following and analyzing the evolution of the sector, which has been always struggling to grow in a way to guaranty the security of supply and low tariffs.

The analysis of power sector history since 19th century shows that previous (and current) problems i were driven by poor administration decisions, lack of investment, depletion of water resources and difficulty in meeting the growing demand – common issues in a developing country (Centro da Memória da Eletricidade no Brasil 2015; Lorenzo 2002; Baer and McDonald 1998; Grupo de Economia da Energia 2015).

That shows the necessity of a broad planning strategy to better support the management of available and upcoming the resources and allow different solutions to attend the growing demand.

Table 1. Past studies in the area

Ref.	Region	Results
Perazzoli et al. (2013)	Brazil	Reduction in river flow of 39.2% and 41.2%, respectively
		Flood peaks could reach more extreme values in the future, especially in scenario A2
Krol et al. (2006)	Northeast	Water availability can be affected in a significant level impacting in long-term water policies
Carvalho et al. (2015)	Goiana and Itambé, Zona da Mata of Pernambuco	Zona da Mata region can be strongly affected by climate changes, reflecting on a future climate with increased temperature
		Reduced rainfall, which could reduce the potential sugarcane yield, and that could already be perceived in the near future

(continued)

Table 1. (*continued*)

Ref.	Region	Results
Pereira et al. (2013)	Brazil	Tendency for an 15% e 30% growth in wind power density in land for most of Northeast region
		Regional intensification of more than 100% mainly in the north sector of Northeast
		Decreasing trend particularly along the coast of Bahia state
		South region exhibited an average growth of 10%, peaking to more than 20% in some areas
Lucena et al. (2010)	Brazil	The wind power potential would not be jeopardized due to possible new climate conditions
Lucena et al. (2009)	Brazil	The lowest wind speeds planned for inside northeastern can reduce up to 60% in potential wind
Lucena et al. (2012)	Brazil	Reliability of hydroelectric generation in Brazil could be compromised
		Electrical transmission lines can be more vulnerable to potential high winds, especially in South
		In some places the average loss in electricity generation may exceed 80% (firm energy)

1.3 Vulnerability to Climate Change

Several studies report Brazil vulnerability to climate change. Some of these results can be seen in Table 1.

These researches pointed out that the country will suffer with the impacts of climate change, namely the availability of water will and the electricity generation in North and Northeast regions. Even with the projected increase in the potential for wind generation in coastal areas, the vulnerability is considerable, especially because the sector is not prepared to work with intermittent sources of energy.

This paper aims to present the potential of smart grids to increase the resilience of Brazilian power sector to climate change effects.

2 Climate Change Resilience

Resilience is defined by Nelson et al. (2007) as the "amount of change a system can undergo and still retain the same function and structure while maintaining options to develop".

The following mitigation and adaptations strategies are often found presented in the scientific literature:

- Increasing the participation of wind and solar. However, according to Ferraz (2015), the new renewable sources will not just add more energy into the grid; they cannot be treated as any other conventional source because of their flashing nature, and the system is not yet adapted to it;
- Increase the cooperation with neighbor countries for energy integration. There are already a few ones with Paraguay, Argentina, Uruguay and Venezuela, but must be extended and reinforced. Moreover, hydropower inventory studies are being performed in Guyana, whose potential is about 7.5 GW and it is possible "to negotiate a construction of one or two power plants in this country to import part of the electricity produced" (Soito and Freitas 2011);
- Use of smart grids. Some pilot projects were already developed in Brazil (see Sect. 2.1).

2.1 Smart Grids

According to International Energy Agency (2011) a smart grid is *"an electricity network that uses digital and other advanced technologies to monitor and manage the transport of electricity from all generation sources to meet the varying electricity demands of end-users. Smart grids coordinate the needs and capabilities of all generators, grid operators, end-users and electricity market stakeholders to operate all parts of the system as efficiently as possible, minimizing costs and environmental impacts while maximizing system reliability, resilience and stability."*

Studies of the U.S. Department of Energy's (*apud* Hamilton and Summy 2011) suggest that a full operational smart grid in the U.S. could lead to an 18% reduction in carbon dioxide emissions by 2030: "if the electrical grid were simply 5% more efficient, the efficiency gain could displace the equivalent of 42 coal-fired power plants, and that would equate to permanently eliminating the fuel and greenhouse gas emissions of 53 million cars" (Hamilton and Summy 2011).

Smart grids are also expected to reduce CO_2 emissions. In the Smart Grids Roadmap by IEA (International Energy Agency) (2011), the best scenario "estimates that smart grids offer the potential to achieve net annual emissions reductions of 0.7 Gt to 2.1 Gt of CO_2 by 2050" (IEA 2011, 29). According to this report, these reductions will occur through: feedback on energy usage; lower line losses; accelerated deployment of energy efficiency programs; continuous commissioning of service sector load; energy savings due to peak load management; smart grid support for the wider introduction of electric vehicles; deeper integration of intermittent renewable generation.

IEA (2010b, p. 154 apud Clastres 2011) found out that, compared to the baseline scenario in 2050, smart grids offer the potential to achieve savings of between 0.9 $GtCO_2$ and 2.2 $GtCO_2$ a year by direct reductions (from 0.2 to 0.85 $GtCO_2$ a year under the Blue Map scenario compared with the baseline scenario) and indirect cuts (from 0.65 to 1.31 $GtCO_2$ a year).

According to the IEA (2011) "peak demand will increase between 2010 and 2050 in all regions" and "smart grids deployment could reduce the projected peak demand growth by 13% to 24%". Moreover, according to Faruqui et al. (2007 *apud* Clastres 2011), automated management can reduce peak demand by 20 to 50%, and overall demand by 10 to 15%, but "the danger with automated management is that peak consumption may simply shift, reappearing when all the loads reconnect at the same time". However, this possibility is quite reduced under a smart grid environment.

Lastly, it is important to reduce peaks because electricity system infrastructure is designed to meet the highest level of demand. Therefore, during non-peak times the system is typically underutilized. Building the system to satisfy occasional peak demand requires investments in capacity that would not be needed if the demand curve were flatter (IEA 2011).

2.2 Smart Grids in Brazil

In Brazil, although the generation and transmission systems have a fairly advanced monitoring and controlling levels, the majority of the distribution system does not have monitoring, control or associated communications networks functions (Pelegrini and Vale 2014).

The main reasons why Brazil would implement smart grids is for reducing theft and fraud and increase the quality of the service. Some pilot projects are under development to study the available technologies.

In what concerns the cost-benefits analysis, Duarte et al. (2015) conclude that, for the three smart grids implementation scenarios – slow, moderate and fast – the minimum expected benefits exceed the anticipated costs.

3 Methodology

The goal of present research is to uncover the smart grids potential to mitigate the challenges caused by climatic change changes on the Brazilian energy sector.

Twelve scenarios were created to simulate the sector in 2030, resulting from the combination of three climate hypothesis and four power system states. The climate cases comprise a reference scenario and two scenarios with climate impacts based on IPCC SRES scenarios A2 and B2:

1. A2 scenario represents a very heterogeneous world, with self-reliance and preservation of local identities, fertility patterns across regions converge very slowly, continuously increasing global population, economic development is primarily regionally oriented and per capita economic growth and technological change are more fragmented and slower than in other storylines;
2. B2 scenario represents an emphasis is on local solutions to economic, social, and environmental sustainability, continuously increasing global population at a rate lower than A2, intermediate levels of economic development, less rapid and more diverse technological change than in other storylines, oriented toward environmental protection and social equity, focuses on local and regional levels.

The four power system cases involve one without smart grids, and three with different levels of smart grids technologies' penetration in the Brazilian grid, namely slow, moderate and fast, as in Duarte et al. (2015).

Table 2. Scenarios used in the case study

Economic, social and technological	Smart grids implementation	Year	Code
Current account		2015	CA
Without climate change	Without smart grids	2030	REF
	Slow	2030	SG_S
	Moderate	2030	SG_M
	Fast	2030	SG_F
A2 climate change scenario	Without smart grids	2030	REF_A2
	Slow	2030	A2_SG_S
	Moderate	2030	A2_SG_M
	Fast	2030	A2_SG_F
B2 climate change scenario	Without smart grids	2030	REF_B2
	Slow	2030	B2_SG_S
	Moderate	2030	B2_SG_M
	Fast	2030	B2_SG_F

The baseline scenario was developed assuming that current development trends continue, and no actions are undertaken to explicitly reduce the GHG emissions.

The Table 2 presents the combination of settings used to come up with each scenario, the years analyzed, and the code used for each of them.

The estimation of the installed capacity evolution took into consideration the projections presented by national energy plan 2030, the decennial energy expansion plan's projections and the real execution values, i.e., a comparison between what was planned and the observed capacity until 2015. The projections for the power sector in 2030 were based on a bottom-up approach, based on n energy accounting model, LEAP – Long-range Energy Alternatives Planning System (LEAP and Energy Community 2017), developed at the Stockholm Environment Institute. LEAP has become the *de facto* standard for countries undertaking integrated resource planning and for the assessment of greenhouse gas mitigation, especially in the developing world and "has been adopted by thousands of organizations in more than 190 countries worldwide: government agencies, academics, non-governmental organizations, consulting companies, and energy utilities. It has been used at many different scales ranging from cities and states to national, regional and global applications" (LEAP and Energy Community 2017).

The bottom-up approach was chosen as it helps understand the potential of technologies, in this case smart grids, to increase efficiency and reduce GHG emissions. This research followed the characteristic three-step structure presented by Sathaye and Ravindranath (1998):

1st.   Characterization of the options for mitigation;

2nd.   Development of a baseline scenario;

3rd.   Development and analysis of mitigation scenarios.

These authors also proposed the following relation to evaluate the factors contributing to CO_2 emissions, used the follow identity:

$$CO_2 \text{ emissions} = \text{population} * \text{GDP per capita} * \text{Energy Intensity} \\ * \text{Carbon Intensity.} \tag{1}$$

Note: GDP = Gross Domestic Product

4 Results and Discussion

The simulations predict a decrease in Hydro generation from the reference scenario to the A2 and B2 scenarios due to the reduction of capacity factor[1] and firm energy[2].

The demand can be separated into 3 groups (Fig. 1). The middle one is the Reference scenario. A2 group (lines in the top) show a much higher demand, as it anticipates a large increase of air conditioner use in residential and commercial sectors. B2 group (bottom lines) presents the smaller demand as it considers the use of more efficient electrical equipment even considering the increase in air conditioner use. According to these results, the transition to smart grids has a small impact on the total demand.

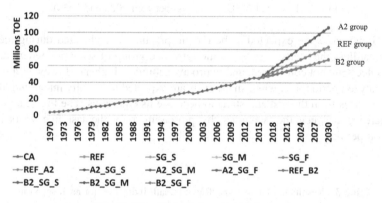

Fig. 1. Historical plus projected demand

Even considering the decrease in hydro generation, the need to import electricity is smaller in the B2 scenarios than it is in the reference scenario due to the demand reduction. The use of smart grids decreases even more the need of importing energy.

The system modeling took into account the characteristics of each generation source. Nuclear, for example, is usually a base source producer rather than a backup, as they have high startup and shut down costs. Contrastingly, solar and wind just generate electricity when have natural conditions.

---

[1] Capacity factor is the ratio between the actual production of a power plant over a period of time and the maximum total capacity in the same period.

[2] Firm energy is the actual energy guaranteed to be available.

According to the model projections, coal, fuel oil and other non-renewable are expected to decrease. In contrast, natural gas is expected to increase, as related emissions are smaller for this source, as well as its maintenance and operation costs, when compared to other thermal concurrent.

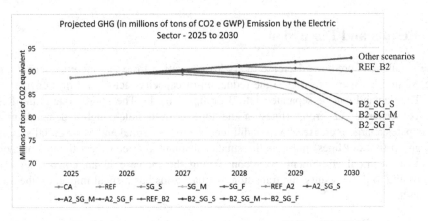

Fig. 2. Projected GHG emissions (between 2025 and 2030)

GHG emissions are expected to be remain practically unchanged during the next years, regardless the scenario (even in the reference group of scenarios where climate change effects are not considered, and hydro availability is higher). However, at the end of the analyzed period, i.e., when smart grids are expected to be fully integrated, the B2 group separates from the rest and starts to decrease its emissions. The Fig. 2 highlights the difference between the scenarios, by presenting only the last five years of the projection period.

Table 3. Results of Sathaye and Ravindranath indicators for each scenario

Scenario	Energy/GDP	CO2/energy
CA	0,321	0,1288
REF	0,343	0,0892
SG_S	0,343	0,0892
SG_M	0,343	0,0892
SG_F	0,343	0,0892
REF_A2	0,295	0,1038
A2_SG_S	0,295	0,1038
A2_SG_M	0,295	0,1038
A2_SG_F	0,295	0,1038
REF_B2	0,296	0,0999
B2_SG_S	0,293	0,0933
B2_SG_M	0,292	0,0918
B2_SG_F	0,290	0,0892

Figure 2 contains five lines: the three lines with lower emissions are the B2 group with smart grids; the fourth line with lower emission corresponds to the scenario B2 without smart grids; the top line is, in fact, six different scenarios overlapping, the reference group and the A2 group, with and without smart grids.

The results presented in Table 3 were obtained through the application of Eq. 1 to each scenario. As none of the scenarios considered any change in the composition of GDP or mix of energy, all changes observed in the table are due to technology and efficiency. Therefore, as presented, only B2 scenarios show an improvement in this indicator as it considers not only smart grids but also other technological changes to increase efficiency. The scenarios that only consider smart grids do not show any difference between the reference scenario and the smart grids scenarios (this difference appears in the energy needs for imports).

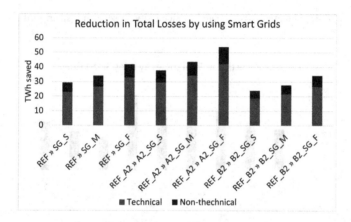

Fig. 3. Technical and non-technical losses reduction

The same is observed in the carbon intensity indicator: only B2 group shows differences between smart grids scenarios and reference cases.

As Brazil is a country with high participation of renewables in its electricity mix, its potential to reduce GHG is considered small. However, it still has potential to mitigate and adapt the sector to climate change.

Figures 3, 4 and 5 present the projected total losses (technical plus non-technical) reduction for each scenario when compared to the reference one. The reduction in technical losses is associated to an increase in efficiency, while the reduction on non-technical losses is associated to increase in control and reduction in thief and fraud. Within each scenarios group, the losses decrease (reduction increases) with higher smart grids penetration.

The Fig. 4 shows the monetization of the actual losses and projected values for 2030 for all scenarios.

Again, within each group of scenarios, the costs decrease as the penetration of smart grids increase.

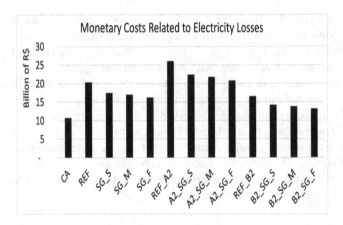

Fig. 4. Monetary costs related to electricity losses using smart grids

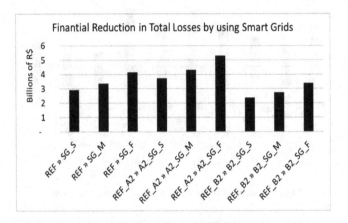

Fig. 5. Total financial saved by using smart grids

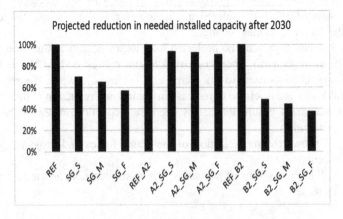

Fig. 6. Projected reduction in needed installed capacity after 2030

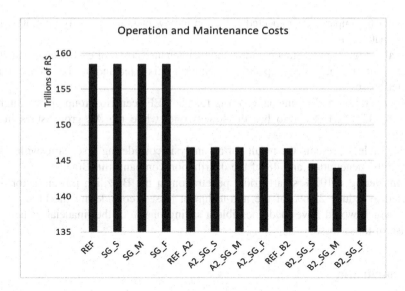

Fig. 7. Projected OM costs in 2030

Figure 5 presents how much could be saved by implementing smart grids, comparing the expected losses from each smart grid scenario with its own (climate) reference scenario.

As expected, as faster the implementation, the higher would be the total amount of money saved by electricity distributions companies by using smart grids.

Table 4. Projections for non-technical losses reduction considered 100% penetration in these 48 distribution companies market

Actual losses	Losses after smart grid	Number of distributors	Actual losses (Thousands of R$)[a,b]	Losses after smart grid (Thousands of R$)[a,b]	Actual losses (TWh)[a]	Losses after smart grid (TWh)[a]
Until 3,5%	Stay the same	30	1.118.050	1.118.050	4,916	4,916
3,5%; 5%	3,5%	5	172.501	154.264	1,653	1,448
5%; 6%	4,5%	2	89.847	71.586	0,670	0,565
6%; 8%	5,5%	2	59.429	44.579	1,407	1,061
8%; 10%	7%	0	–	–	–	–
10%; 15%	9%	3	93.276	69.990	2,560	1,951
Bigger than 15%	10%	6	194.546	113.061	7,843	4,801
Total		48	1.727.649	1.571.531	19,049	14,741
Difference			−9,04%		−22,62%	
Total safe			156.118		4,308	

[a]Base year 2015
[b]Sum of the individual losses of each company in this group

Figure 6 compares the expected reduction in new installed capacity in 2030 within each climate groups.

Finally, Fig. 7 shows the operation and maintenance costs for each scenario in 2030 according to the installed capacity, generation mix, demand to be attended and expected losses.

The operation and maintenance costs for the REF scenario group are the highest ones, the B2_SG_F scenario has the lowest one, while the A2 group stays in the middle.

The Table 4 presents the results of an analysis considering a scenario with 100% smart grids' penetration and 48 of the distribution companies in Brazil1.

Considering a 100% smart grids' penetration in the Brazilian power sector, it is expected to reduce 22,6% of the non-technical losses (considering the base year of 2015), which would have made possible a saving of 9% of the financial value spent because of those losses.

5   Conclusion

This paper presents an analysis of the potential of smart grids' technologies to mitigate climate change and increase the resilience of the power sector. To achieve that, three groups of climate scenarios were considered: Reference, A2 and B2, each one containing three levels of smart grids' penetration, named as Slow, Moderate and Fast plus a reference scenario. Finally, an integrated model of the electric system was used to simulate the future generation for each scenario, the emissions and the impact of smart grids.

By considering the projected firm energy and capacity factor for the A2 and B2 scenarios, a model was created to simulate the future generation with the climate change impacts in temperature and precipitation. After that, twelve scenarios were constructed to simulate the demand, generation and GHG emission in the power sector using the LEAP model.

Results show that smart grids can help save energy by reducing losses and, in the B2 group, can also mitigate the sector reducing GHG emission as this scenario also considers the use of more efficient equipment and a more conscious society. A2 scenario presents a situation where the consumption is higher due to increase of number of hot days, which will increase the use of air conditioner in residential and commercial sectors. Additionally, the B2 scenario, despite of a decrease in hydro generation, still shows the smallest GHG emission, even when compared to the reference scenario, which does not consider the climate change effects and will have a higher hydro participation in 2030.

The smart grids also proved its potential to reduce the operational and maintenance costs and the investments in new power plants after 2030.

The best mitigation and adaptation results came from the B2 group of scenarios, which represented not only the changes in climate but also the changes in society (awareness about environmental issues, use of equipment that is more efficient, sustainable growth, global solutions …).

Another efficiency factor came from the losses reducing potential. Smart grids can offer about 20% reduction in total losses in some of the analyzed scenarios.

Finally, smart grid would also help to reduce electricity tariffs by increasing the ability of the grid to take the most of Brazilian natural characteristics that gives the country great winds, insolation and water, preparing the grid to work with intermittent sources.

Acknowledgments. This work was supported in part by ERASMUS MUNDUS through BE MUNDUS Project. The authors, also, would like to thank the Stockholm Environment Institute and energy community for the permission to use the LEAP model and the support during the research.

References

Baer, W., McDonald, C.: A return to the past? Brazil's privatization of public utilities: the case of the electric power sector. Q. Rev. Econ. Financ. **38**(3), 503–523 (1998). https://doi.org/10.1016/S1062-9769(99)80130-6

De Carvalho, A.L., Menezes, R.S.C., Nóbrega, R.S., Pinto, A.D.S., Ometto, J.P.H.B., von Randow, C., Giarolla, A.: Impact of climate changes on potential sugarcane yield in Pernambuco, northeastern region of Brazil. Renew. Energy **78**, 26–34 (2015). https://doi.org/10.1016/j.renene.2014.12.023

Centro da Memória da Eletricidade no Brasil: Memória da Eletricidade (n.d.). http://memoriadaeletricidade.com.br/. Accessed 5 Aug 2015

Clastres, C.: Smart grids: another step towards competition, energy security and climate change objectives. Energy Policy **39**(9), 5399–5408 (2011). https://doi.org/10.1016/j.enpol.2011.05.024

Duarte, D.P., Maia, F.C., Neto, A.B., Cesar, L.S., Kagan, N., Gouvêa, M.R., Labronici, J., Guimaraes, D.S., Bonini, A., da Silva, J.F.R.: Brazilian smart grid roadmap - an innovative methodology for proposition and evaluation of smart grid functionalities for highly heterogeneous distribution grids (2015)

EPE: PNE 2050 - Estudo sobre a Demanda (2016). http://informesanuales.xm.com.co/2013/SitePages/operacion/3-1-Demanda-de-energia-nacional.aspx

EPE/MME: Balanço energético nacional 2016: Ano base 2015 (2016)

EPE - Empresa de Pesquisa Energética: Plano Nacional de Energia 2030, Rio de Janeiro (2007)

Ferraz, C.: Os Desafios das Renováveis no Brasil. Grupo de Econmia da Energia - GEE (2015). http://www.gee.ie.ufrj.br

GEE: Blog Infopetro (2015). https://infopetro.wordpress.com/. Accessed 4 June 2015

Hamilton, B., Summy, M.: Benefits of the smart grid: part of a long-term economic strategy. IEEE Power Energy Mag. **9**(1), 102–104 (2011). https://doi.org/10.1109/mpe.2010.939468

IBGE: Instituto Brasileiro de Geografia e Estatística (2015). http://www.ibge.gov.br/home/. Accessed Jan 2015

IEA (International Energy Agency): Technology Roadmap (2011)

Krol, M., Jaeger, A., Bronstert, A., Güntner, A.: Integrated modelling of climate, water, soil, agricultural and socio-economic processes: a general introduction of the methodology and some exemplary results from the semi-arid north-east of Brazil. J. Hydrol. **328**, 417–431 (2006). https://doi.org/10.1016/j.jhydrol.2005.12.021

LEAP and Energy Community: Energy Community (n.d.). https://www.energycommunity.org. Accessed 1 Jan 2017 de Lorenzo, H.C.: O setor elétrico Brasileiro: Passado e Futuro. Perspectivas **24–25**, 147–170 (2002)

Losekann, L.: Agenda 2015–2018 do Setor Elétrico no Brasil. Grupo de Econmia da Energia – GEE (2015a). http://www.gee.ie.ufrj.br/

Losekann, L.: Energia Agora 13 - A Crise das Geradoras e a MP 688. Grupo de Econmia da Energia - GEE (2015b). http://www.gee.ie.ufrj.br/

Lucena, A.F.P., Clarke, L., Schaeffer, R., Szklo, A., Rochedo, P.R.R., Daenzer, K., Gurgel, A., Kitous, A., Kober, T.: Climate policy scenarios in Brazil: a multi-model comparison for energy. Energy Econ. (2015). https://doi.org/10.1016/j.eneco.2015.02.005

Lucena, A.F.P., Szklo, A.S., Schaeffer, R., Dutra, R.M.: The vulnerability of wind power to climate change in Brazil. Renew. Energy **35**(5), 904–912 (2010). https://doi.org/10.1016/j. renene.2009.10.022

Lucena, A.F.P.: Proposta metodológica para avaliação da vulnerabilidade às mudanças climáticas globais no setor hidroelétrico. Universidade Federal do Rio de Janeiro (2010)

Lucena, A.F.P., Schaeffer, R., Szklo, A.: Mudanças climáticas e eventos extremos no Brasil: A vulnerabilidade do sistema de energia eléctrica à mudança climática no Brasil. Fundação Brasileira para o Desenvolvimento Sustentável (2012)

Nelson, D.R., Adger, W.N., Brown, K.: Adaptation to environmental change: contributions of a resilience framework. Annu. Rev. Environ. Resourc. **32**, 395–419 (2007)

Perazzoli, M., Pinheiro, A., Kaufmann, V.: Assessing the impact of climate change scenarios on water resources in southern Brazil. Hydrol. Sci. J. **58**(1), 77–87 (2013). https://doi.org/10.1080/02626667.2012.742195

Pereira, E.B., Martins, F.R., Pes, M.P., da Cruz Segundo, E.I., Lyra, A.D.A.: The impacts of global climate changes on the wind power density in Brazil. Renew. Energy **49**, 107–110 (2013). https://doi.org/10.1016/j.renene.2012.01.053

Queiroz, R.: A Crise Política e a Crise do Setor Elétrico. Grupo de Econmia da Energia – GEE (2015). http://www.gee.ie.ufrj.br

Sathaye, J.A., Ravindranath, N.H.: Climate change mitigation in the energy and forestry sectors of developing countries. Annu. Rev. Energy Environ. **23**(1), 387–437 (1998). https://doi.org/10.1146/annurev.energy.23.1.387

Schaeffer, R., Szklo, A.S., de Lucena, A.F.P., de Souza, R.R.: Mudanças Climáticas e Segurança Energética no Brasil, Rio de Janeiro (2009)

Soito, J.L., Freitas, M.V.: Amazon and the expansion of hydropower in brazil: vulnerability, impacts and possibilities for adaptation to global climate change. Renew. Sustain. Energy Rev. **15**(6), 3165–3177 (2011). https://doi.org/10.1016/j.rser.2011.04.006

Working Group on Smart Grids: Smart Grid (2010). https://doi.org/10.1007/978-3-642-22179-8_17

High-Frequency Transformer Isolated AC-DC Converter for Resilient Low Voltage DC Residential Grids

Nelson Santos[1(✉)], J. Fernando Silva[2], and Vasco Soares[1]

[1] Instituto Superior de Engenharia de Lisboa, INESC-ID,
R. Conselheiro Emídio Navarro, 1959-007 Lisbon, Portugal
{nsantos,vesoares}@deea.isel.ipl.pt
[2] Instituto Superior Técnico, Universidade de Lisboa, INESC-ID,
DEEC, AC Energia, Av. Rovisco Pais, 1049-001 Lisbon, Portugal
fernando.alves@tecnico.ulisboa.pt

Abstract. In a global and growing society, it is necessary to rethink strategies in order to minimize the environmental impact resulting from the progressive increase of energy consumption and the misuse of energy resources. Energy power system converters must be considered as a global way to spare most of the wasted energy. Today, in the power distribution infrastructure, including modern residential buildings, most equipment have power supplies with imbedded AC-DC power converters which may have overall losses as high as 25% regarding the equipment output. Therefore, common DC buses for residential applications are being studied to increase equipment efficiency. This paper presents and designs an AC-DC isolated converter that uses a full-bridge matrix topology with high-frequency isolation transformer and non-electrolytic capacitors to integrate into the future residential buildings DC bus, presenting a reliable alternative to AC power. Non-linear control techniques (sliding mode control and backstepping control) are employed to guarantee stability and disturbance robustness to the output DC low voltage, while enforcing sinusoidal input AC current and power factor correction. Control strategies are described and simulation results are presented and discussed.

Keywords: AC-DC converter · High-frequency transformer
Sliding mode control · Backstepping control · DC residential grids
Smart-grids

1 Introduction

In the last years there is a big motivation to electrical energy generation in a decentralized way, using clean energy resources, like solar and wind. The local generation and consumption would avoid transport losses, improving the affordability of electrical networks. However, the inter-connection of sources, like DC (solar) and AC (wind), with different voltage levels and waveforms, implies their adaptation through electronic power converters and their use in micro-grids has to follow well defined characteristics.

© IFIP International Federation for Information Processing 2018
Published by Springer International Publishing AG 2018. All Rights Reserved
L. M. Camarinha-Matos et al. (Eds.): DoCEIS 2018, IFIP AICT 521, pp. 147–155, 2018.
https://doi.org/10.1007/978-3-319-78574-5_14

Main electric networks will continue to use AC voltages for compatibility reasons and easy voltage level transformation. However, nowadays most electric loads need DC power obtained via AC-DC switching converter, between the AC input voltage 230 V and the DC output voltage from 5 V to 400 V. Therefore, most actual loads such as lighting systems, small appliances and mobile devices could advantageously be supplied from a common DC network, eliminating the need of the embedded AC-DC converter and sparing its losses.

With the recent progress in the semiconductor technologies a single high efficiency central AC-DC converter could be advantageous in eliminating several less reliable and less efficient AC-DC embedded converters. Even if the DC voltage needs to be customized, a DC-DC converter is simpler and presents higher efficiency than an AC-DC converter. Considering also the High Voltage Direct Current (HVDC) technology DC buses are currently an alternative to AC grids, with better use of resources, both in transmission, distribution and in the final consumer place. These advantages will be emphasized by the integration of electric vehicles, as they require a large amount power to handle the fast charging of batteries. The fast charging process may benefit from the DC grid due to the simplicity and efficiency of DC-DC converters compared to the typical AC-DC. Moreover, in DC networks there is no reactive power compensation, no need to synchronize generators for the network frequency, no skin effect and easier power failure mitigation when combined with energy storage systems [1]. On the other hand, DC networks are not directly connectable to electric machines, such as transformers and motors demanding the use of switching converters (at least diode bridge rectifiers for doubly fed induction motors driven with 3 phase inverters). Additionally, it is necessary to adapt the handling equipment including equipment and people protections, while needing an efficient isolated AC-DC converter to link with the mains network.

A characteristic of a DC micro-grid is its voltage level. This level must be selected taking into account the type and the power required by connecting loads. Recently, consensus exists in benefits of use the level around 400 Vdc to cover different types of loads, a value advocated by associations of equipment producers, which may possibly be adopted in the new regulations for the residential and commercial DC distribution networks. For example, the Emerge Alliance's [2], standards recommends 24 Vdc for occupied Space standard (low power appliances, lighting, typically in bedrooms and living rooms, guaranteed safety and efficiently at lower voltage) and 380 Vdc for Data and Telecom Center standard (high power loads, kitchen appliances, air conditioning systems for residential air conditioning and electric vehicles). Values in the order of 48 Vdc, 120 Vdc and 230 Vdc have also been considered [3].

Also within the low voltage level, the USB 3.1 standard must be considered [4], with 3 levels of voltage 5 V, 12 V and 20 V and a maximum current of 5 A, being able to feed loads up to 100 W at the voltage of 20 V. There is a growing trend of using devices with USB connector type-C, such as mobile phones, tablets, laptops and other small devices like heaters, blenders and monitors, lighting, already available in the current market. Socially it is a type of system so well accepted that USB wall-sockets are being installed in private houses and offices buildings.

In this framework, the research question is how to implement the concept of AC-DC power converters with galvanic insulation, but without the use of intermediate electrolytic capacitor DC buses, to provide energy to a DC micro-grid.

This paper propose the use of a single-phase AC-DC power converter, using high frequency transformer isolation, to interconnect the AC power distribution grid network at unity power factor and to supply a DC micro-grid for residential or commercial complex use, with the target of reducing losses, cost and footprint, ensuring required quality parameters for suppliers and consumers.

2 Contribution to Resilient Systems

In an increasingly competitive society focused on improving systems efficiency [5], optimizing stability, security and robustness, smart grids play an important role in the community to help achieve these goals. The infrastructure improvement can happen as long as stakeholders have an open mind to change the way energy is used. To ensure the best operating conditions for producers and consumers, human consumption behaviors must be supervised in order to collect all available information to make the best management choice in a way that can decide and adapt all stakeholders for get energy efficiency and a resilient system. In general, smart grids might quickly adapt to new conditions, which may not be initially foreseen, and thus require a natural and sustainable human action.

In this paper we present an AC-DC converter, without electrolytic capacitors, as a solution or an integral part in the conversion of energy for a smart micro grid DC network, introducing a safe and resilient alternative to the existing AC network. This converter can be applied to network distribution or micro-networks in residential complexes, buildings, individual homes and commercial facilities.

3 AC-DC Power Converter Topology

3.1 Circuit Description

The proposed circuit is a high-frequency transformer isolated AC-DC Converter, as depicted in Fig. 1. Active matrix full-bridge with bidirectional semiconductor technology (SiC MOSFET) with LC filter is used, combining proper control strategy to enforce almost unity power factor and maintaining the source current with mostly the fundamental frequency component. Each S_k switch ($k \in \{1,2,3,4\}$) uses two anti-series connected MOSFET semiconductors to handle bipolar voltages from AC source. At load side this AC-DC converter is composed by a diode full-bride rectifier and LC filter to reduce unwanted harmonics of the high frequency diode rectification of the matrix active bridge and the oscillatory nature of AC single phase power.

The differential equations of the proposed circuit are written in (1), with variables and devices shown in Fig. 1.

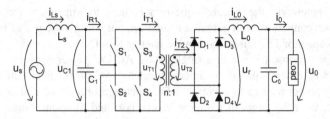

Fig. 1. Proposed high-frequency transformer isolated AC-DC converter.

$$
\frac{d}{dt}\begin{bmatrix} u_0 \\ u_{C_1} \\ i_{L_s} \\ i_{L_0} \end{bmatrix} = \begin{bmatrix} 0 & 0 & 0 & 1/C_0 \\ 0 & 0 & 1/C_1 & 0 \\ 0 & -1/L_s & 0 & 0 \\ -1/L_0 & 0 & 0 & 0 \end{bmatrix}\begin{bmatrix} u_0 \\ u_{C_1} \\ i_{L_s} \\ i_{L_0} \end{bmatrix} + \begin{bmatrix} 0 & 0 & 0 & -1/C_0 \\ 0 & 0 & -1/C_1 & 0 \\ 0 & 1/L_s & 0 & 0 \\ 1/L_0 & 0 & 0 & 0 \end{bmatrix}\begin{bmatrix} u_r \\ u_s \\ i_{r_1} \\ i_0 \end{bmatrix} \quad (1)
$$

The switches states of S_1, S_2, S_3 and S_4 are represented in (2).

$$
\delta(t) = \begin{cases} +1 \rightarrow (S_1, S_4 \, ON) \\ 0 \rightarrow (S_1, S_2 \, ON \, or \, S_3, S_4 \, ON) \\ -1 \rightarrow (S_2, S_3 \, ON) \end{cases} \quad (2)
$$

3.2   Control

Sliding Mode Internal Voltage Control (u_{C_1}). Considering that, $u_{T_1} = \delta(t)u_{C_1}$ and $i_{r_1} = \delta(t)^2 sgn(u_{C_1})ni_{L_0}$ then, using (1) the phase canonical model for the voltage u_{C_1} is written as (3).

$$
\frac{du_{C_1}}{dt} = \frac{1}{C_1}\left(i_{L_s} - \delta(t)^2 sgn(u_{C_1})i_{L_0}/n\right) \quad (3)
$$

The first time derivative of u_{C_1} contains the control action $\delta(t)$, thus the strong relative degree of u_{C_1} is 1 [6], and a suitable sliding surface [7–9], can be obtained as a linear combination of the control error $e_{u_{C_1}} = u_{C_1}^{ref} - u_{C_1}$ where $u_{C_1}^{ref}$ is the reference value to be tracked by the u_{C_1} voltage. Considering a positive gain k_1, used to bound the semiconductors switching frequency, a suitable sliding surface is $S\left(e_{u_{C_1}}, t\right) = k_1 e_{u_{C_1}} = 0$. The switching strategy is obtained applying the sliding mode stability condition $S\left(e_{u_{C_1}}, t\right)\dot{S}\left(e_{u_{C_1}}, t\right) < 0$. The derivative of $\left(e_{u_{C_1}}, t\right)$, $\dot{S}\left(e_{u_{C_1}}, t\right)$ is:

$$
\dot{S}\left(e_{u_{C_1}}, t\right) = k_1\left(\frac{du_{C_1}^{ref}}{dt} - \frac{i_{L_s}}{C_1} + \frac{\delta(t)^2 sgn(u_{C_1})i_{L_0}/n}{C_1}\right) \quad (4)
$$

The stability condition is written as (5).

$$\begin{cases} \text{if } S(e_{u_{C1}}, t) > 0, \text{then } \dot{S}(e_{u_{C1}}, t) < 0 \rightarrow \dfrac{du_{C1}^{ref}}{dt} - \dfrac{i_{L_s}}{C_1} + \dfrac{\delta(t)^2 sgn(u_{C_1}) i_{L_0}/n}{C_1} < 0 \\[3mm] \text{if } S(e_{u_{C1}}, t) < 0, \text{then } \dot{S}(e_{u_{C1}}, t) > 0 \rightarrow \dfrac{du_{C1}^{ref}}{dt} - \dfrac{i_{L_s}}{C_1} + \dfrac{\delta(t)^2 sgn(u_{C_1}) i_{L_0}/n}{C_1} > 0 \end{cases} \quad (5)$$

From (5), the sliding mode reaching condition can be obtained as $\dfrac{\delta(t)^2 sgn(u_{C_1}) i_{L_0}/n}{C_1} >$ $MAX\left(\dfrac{i_{L_s}}{C_1} - \dfrac{du_{C1}^{ref}}{dt}\right)$, Supposing that i_{L_0}/n is high enough to satisfy this reaching condition and the $sgn(u_{C_1}) = sgn(i_{L_s})$, on most situations, the control strategy is written as (6). To impose finite switching frequency, the ripple Δu_{C_1} voltage must be considered and k_1 is selected to ensure stability and fast response.

$$\begin{cases} \text{if } S > 0, \text{then } \dot{S} < 0 \rightarrow \begin{cases} u_{C_1} > 0 \rightarrow \delta(t) = 0 \\ u_{C_1} < 0 \rightarrow \delta(t) = \pm 1 \end{cases} \\[4mm] \text{if } S < 0, \text{then } \dot{S} > 0 \rightarrow \begin{cases} u_{C_1} > 0 \rightarrow \delta(t) = \pm 1 \\ u_{C_1} < 0 \rightarrow \delta(t) = 0 \end{cases} \end{cases} \quad (6)$$

To avoid magnetic saturation of the transformer core an auxiliary control signal, takes advantage of the redundancy presented in (6) to ensure transformer primary voltages with zero average value.

Backstepping Current Control (i_{L_s}). Considering the state variable u_{C_1} whose dynamics can be enforced and using the sub-system (1) the i_{L_s} dynamics can be written as $\frac{di_{L_s}}{dt} = (u_s - u_{C_1})/L_s$. Defining the control objective $e_{i_{L_s}} = 0$, the tracking error is $e_{i_{L_1}} = i_{L_s}^{ref} - i_{L_s}$, and a positive definite Lyapunov function candidate can be selected as $V_1(i_{L_s}, t) = e_{i_{L_s}}^2/2$. From Lyapunov 2nd method of stability, the time derivative of $V_1(e_{i_{L_s}}, t)$ must be negative definite for all (i_{L_s}, t), according to $V_1 \dot{V}_1 < 0$.

$$\frac{dV_1}{dt} = e_{i_{L_s}} \frac{de_{i_{L_s}}}{dt} < 0 \Rightarrow e_{i_{L_s}} \frac{de_{i_{L_s}}}{dt} = -k_2 e_{i_{L_s}}^2 \Rightarrow \frac{de_{i_{L_s}}}{dt} = -k_2 e_{i_{L_s}} \quad (7)$$

Define a positive constant k_2, selected to impose the time constant of the tracking error converging to zero asymptotically. Solving (7) for $u_{C_1}^{ref}$, the virtual control law [10] is (8).

$$u_{C1}^{ref} = u_s - k_2 L_s e_{i_{L_s}} - \frac{di_{L_s}^{ref}}{dt} L_s \quad (8)$$

The current $i_{L_s}^{ref}$ must be sinusoidal for low THD and in phase with the voltage u_s for near unity power factor, with the amplitude adjusted according to the load consumption. Using a controller, choosing the k_2 to ensure fast response and stability condition, it is intended to control the output voltage u_0 using the input RMS value of input current i_{L_s}. Assuming the efficiency parameter η, the source power P_s and output power filter P_{L_0}, the input-output power is $P_s\eta = P_{L_0}$ being $U_{s_{rms}}I_{L_{s_{rms}}}\eta = u_0 i_{L_0}$. Using the sub-system (1) Eq. (9) can be written assuming the load output power is P_0.

$$C_0 u_0 \frac{du_0}{dt} = u_0(i_{L_0} - i_0) \Rightarrow \frac{du_0^2}{dt} = \frac{2}{C_0}\left(U_{s_{rms}}I_{L_{s_{rms}}}\eta - P_0\right) \tag{9}$$

Integral Backstepping Voltage Control (u_0). Considering the state variable $I_{L_{s_{rms}}}$ whose dynamics can be enforced. Consider differential Eq. (9) and define the control objective $e_{u_0^2} = 0$. The tracking error being $e_{u_0^2} = u_0^{ref^2} - u_0^2$, add the integral of the tracking error $e_1 = \int e_{u_0^2} dt$ to ensure the convergence, in steady-state, due to the non-modelled disturbances and parameter uncertainty [11] (as an example the efficiency of system). Use a positive definite Lyapunov function V_2 that weights (with k_4) the integral to be added to a new Lyapunov function (10).

$$V_2 = \frac{1}{2}k_4 e_1^2 + \frac{\left(e_{u_0^2}\right)^2}{2} \tag{10}$$

The time derivative of V_2 must be negative definite ($\dot{V}_2 = -k_3 e_{u_0^2}$, $k_3 > 0$), thus:

$$\frac{dV_2}{dt} = k_4 e_1 \frac{de_1}{dt} + e_{u_0^2}\frac{de_{u_0^2}}{dt} = -k_3\left(e_{u_0^2}\right)^2 \tag{11}$$

Using the differential Eqs. (9) and (11), it results (12).

$$k_4 e_1 e_{u_0^2} + e_{u_0^2}\left(\frac{du_0^{ref^2}}{dt} - \frac{2}{C_0}\left(U_{s_{rms}}I_{L_{s_{rms}}}\eta - P_0\right)\right) = -k_3\left(e_{u_0^2}\right)^2 \tag{12}$$

Solving (12) for $I_{L_{s_{rms}}}$, the virtual control law is (13).

$$I_{L_{s_{rms}}} = \frac{C_0}{2\eta U_{s_{rms}}}\left(k_4 e_1 + k_3 e_{u_0^2} + \frac{2P_0}{C_0}\right) \tag{13}$$

Making $i_{L_s}^{ref} = \sqrt{2}I_{L_{s_{rms}}}\sin(\omega t)$ concludes the external loop control design, the current being synchronized to main power AC grid with angular frequency ω. Gains k_3 and k_4 have to be fine-tuned in order to obtain a sufficiently slow response to avoid line current distortion.

3.3   Input and Output LC Filter Design

According to the desired maximum switching frequency f_s for the converter, the L_s and C_1 filter input parameters are calculated by considering the voltage ripple Δu_{C_1}, the current ripple Δi_{L_s} and the maximum cycle factor D for maximum power condition

$D = u_0 n \sqrt{2}/\eta U_s$, $C_1 = i_{L_s}D/2f_s\Delta u_{C_1}$ and $L_s = \Delta u_{C_1}D/2f_s\Delta i_{L_s}$, [12]. The output filter L_0 and C_0 is chosen to attenuate the harmonic of order two of AC the amplitude voltage component U_{0_A}, due to rectifier bridge. Establishing the target ripples for current Δi_{L_0} and voltage Δu_{C_0}, according to the desired limitation, it can written $L_0 = U_{0_A}/\omega\Delta i_{L_0}$ and $C_0 = \Delta i_{L_0}/2\omega\Delta u_{C_0}$.

4 Simulation Results

The circuit presented in Fig. 1 was simulated using MATLAB/Simulink considering variations of 25% of load nominal power and the simulation contains the semiconductor models, including the effects of switching and losses. For the simulation parameters it is considered $P_0 = 3\,$kW, $u_0 = 48\,$V, $n = 2$, $\eta = 95\%$, $f_s = 50\,$kHz, $\Delta u_{C_1} = 10\%$, $\Delta i_{L_s} = 2\%$, $\Delta u_{C_0} = 8\%$ and $\Delta i_{L_0} = 10\%$. Figure 2(a) presents the current source i_{L_s}, the voltage u_{C_1}, voltage at the transformer u_{T_1} and the voltage u_0.

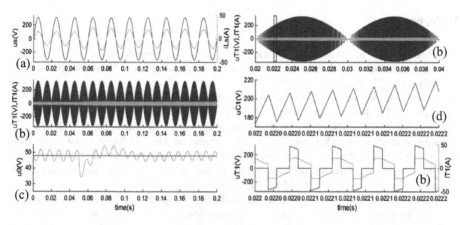

Fig. 2. Temporal evolution; (a) voltage u_s and Current i_{L_s} power source; (b) Transformer voltage u_{T_1} and current i_{T_1}; (c) Output voltage u_0; (d) Voltage u_{C_1}

The current presents maximum THD of 1.8% at nominal load thus improving the power factor, verified in Fig. 2(a). The voltage applied at the transformer primary side (Fig. 2(b)) presents symmetry and zero average value avoiding core saturation and the current presents a rectangular shape with variable frequency and duty cycle, due the control technique, reaching a maximum of 50 kHz. Regarding the output voltage in Fig. 2(c), it presents ripple below 8%. After a load variation of +25% is applied at t=50 ms, the voltage droops 12 V with a recovery time (at 10% of voltage reference) about 60 ms (3 cycles of frequency of main power supply). The Fig. 2(d) represents the voltage ripple at the capacitor C_1 having 10% regarding the source amplitude voltage u_s.

5 Conclusion

This study demonstrates that high-frequency transformer-isolated AC-DC converters can be integrated into future residential buildings with DC bus. Choosing a 48 V DC bus can be a reasonable trade-off regarding human safety and efficiency. This AC-DC isolated converter uses a full-bridge topology with high-frequency isolation transformer working around 40 kHz to improve efficiency and reducing size, and to increase resiliency in energy systems due to the saved energy. This converter combined with energy storage batteries, can further improve resiliency, while reducing the ripple of the output voltage and increasing the response speed to fast request dynamic loads on the DC bus. The non-linear control techniques demonstrate stability and disturbance rejection maintaining sinusoidal input AC current and power factor correction.

Acknowledgments. This work supported by Portuguese national funds through FCT - Fundação para a Ciência e Tecnologia, under Project UID/CEC/50021/2013.

References

1. Chauhan, R., Rajpurohit, B., Hebner, R., Singh, S., Gonzalez-Longatt, F.: Voltage standardization of DC distribution system for residential buildings. J. Clean Energy Technol. **4**(3), 167–172 (2016)
2. Public Overview of the Emerge Alliance Data/Telecom Center Standard "Emerge Alliance Advances DC Power to the Desktop", EMerge Alliance, San Ramon, CA, USA (2012)
3. Priyadharshini, G., Nandhini, N.R., Shunmugapriya, S., Ramaprabha, R.: Design and simulation of smart sockets for domestic DC distribution. In: 2017 International Conference on Power and Embedded Drive Control (ICPEDC), Chennai, pp. 426–429 (2017)
4. Shivakumar, A., Normark, B., Welsch, M.: Household DC networks: state of the art and future prospects. InsightE. Rapid Response Energy Brief (2015)
5. Electricity Distribution in the European Union: An Opportunity for Energy Efficiency in Europe. An IEEE European Public Policy Initiative Position Statement. Adopted 27 March 2017. www.ieee.org/about/ieee_europe/eppi_dc_electricitiy_distribution_june_2017.pdf. Accessed 1 Nov 2017
6. Silva, J.F., Pinto, S.F.: Linear and nonlinear control of switching power converters. In: Power Electronics Handbook, Fourth edn., pp. 1141–1220 (2017)
7. Fernando Silva, J.: Sliding mode control of voltage sourced boost-type reversible rectifiers. In: Proceedings of IEEE/ISIE 1997 Conference (ISBN 0-7803-3937-1), Guimarães, Portugal, Julho, vol. 2, pp. 329–334 (1997)
8. Martins, J.F., Pires, A.J., Fernando Silva, J.: A novel and simple current controller for three-phase PWM power inverters. IEEE Trans. Ind. Electr. **45**(5), 802–805 (1998). Special Section on PWM Current Regulation
9. Santos, N., Silva, J.F.A., Santana, J.: Sliding mode control of unified power quality conditioner for 3 phase 4 wire systems. In: Camarinha-Matos, L.M., Barrento, N.S., Mendonça, R. (eds.) DoCEIS 2014. IAICT, vol. 423, pp. 443–450. Springer, Heidelberg (2014). https://doi.org/10.1007/978-3-642-54734-8_49

10. Martin, A.D., Cano, J.M., Silva, F.A., Vazquez, J.R.: Back-stepping control of smart-grid connected distributed photovoltaic power supplies for telecom equipment. IEEE Trans. Energy Convers. **30**(4), 1496–1504 (2015)
11. Tan, Y., Chang, J., Tan, H.: Advanced motion control scheme with integrator backstepping: design and analysis. In: Power Electronics Specialists Conference, IEEE PESC (2000)
12. Silva, J.F.: Electrónica Industrial – Semicondutores e Conversores de Potência. Fundação Calouste Gulbenkian, 2ª edição (2013)

Analysis of Domestic Prosumer Influence on a Smartgrid

João Carvalhuço[1], R. Pereira[1,2(✉)] (iD), and P. M. Fonte[1,2] (iD)

[1] Instituto Superior de Engenharia de Lisboa, Instituto Politécnico de Lisboa,
R. Conselheiro Emídio Navarro 1, 1959-007 Lisbon, Portugal
a39394@alunos.isel.pt, {rpereira,pfonte}@deea.isel.ipl.pt
[2] Low Carbon Energy Conversion (LCEC),
R. Conselheiro Emídio Navarro 1, 1959-007 Lisbon, Portugal

Abstract. The shift from a centralized production to a distributed generation based on renewable energy production has been increasing, resulting in prosumer arising as new stakeholder of the electrical system. In this paper, a prosumer that owns photovoltaic solar panels and small wind generator, is considered and renewable energy production forecast is given by artificial neural networks (ANN). An energy management model is developed considering a battery storage system. Several case studies are analyzed considering seasonality factors.

Keywords: Smartgrid · Prosumer · Energy management · Storage

1 Introduction

In the electric system five stakeholders can be considered, such as, consumer, producer, prosumer, network operator and regulator. The prosumer is the most recent stakeholder, which derives from a passive consumer who only buys energy from the power grid, for a consumer that also is able to generate electricity from renewable resources. An advantage of using a production unit is the dependency reduction from energy market and also the energy invoice decreasing. When a storage system is incorporated, a better energy management is achieved. Several EMSs that include both domestic energy production and storage system, can be found in literature.

In [1], a review of existing models for energy management are discussed. These models are based on linear programming, artificial intelligence or SCADA systems, which can be applied connected to the grid, standalone or in smartgrids. In [2], the energy management system (EMS) aim is to minimize energy consumption from the power grid and maximize the energy consumption from batteries and renewable energy. An EMS was developed in [3] in order to minimize the power grid operation cost, considering the existence of wind and biomass energy production, as well as an electric vehicle used in a V2G perspective. In [4], standalone EMS is considered. Solar and wind generation associated with a battery storage system is analyzed. Linear programming based in HOMER software is used in [5], applied on an EMS connected to the power grid where a non-constant energy production is considered. This software allows to determine

© IFIP International Federation for Information Processing 2018
Published by Springer International Publishing AG 2018. All Rights Reserved
L. M. Camarinha-Matos et al. (Eds.): DoCEIS 2018, IFIP AICT 521, pp. 156–163, 2018.
https://doi.org/10.1007/978-3-319-78574-5_15

which management strategies led to maximum economic savings. An EMS based in artificial intelligence is referred in [6]. In this paper an EMS based in linear programming, developed to manage production, consumption and a battery storage system, is addressed.

This paper is organized as follows: in Sect. 2 the relationship between the developed work and resilient systems is addressed; in Sect. 3 the implemented EMS including storage mathematical model is described; in Sect. 4 the developed model is applied to case studies, considering energy buying and selling, accordingly to production and consumption values. In these scenarios seasonality is considered and a comparison between a summer month and winter month is addressed. Finally, in Sect. 5 some conclusions about the developed work are drawn.

2 Relationship to Resilient Systems

Resilient systems can be considered as systems that have the adaptation ability to quick changes, absorbing those changes and disturbances, being simultaneously able to mitigate their impact on systems normal functioning. Resilience allow systems to rapidly recover from stress situations [7].

The EMS developed in this paper is inserted in a smartgrid context. A smartgrid enables power delivering using modern information technologies, which supports bidirectionality in communications and also in power flow.

Considering that reliability can be defined as the power system ability to supply a load under any circumstances, then the smartgrid contributes for a more reliable power system [8]. This happens mainly due to the incorporation of renewable energy resources into the power system, which allows to improve the response to a consumption increase and also to reduce outage issues [9].

The main benefits of distributed generation are the reliability improvement, the increasing of renewable sources penetration, environmental benefits, transmission loss reduction and also cost reduction [8]. The increase of distributed generation allows to deferral the implementation of new power plants, needed to satisfy the continuous growth of load demand, and simultaneously allow to decrease the dependence of central production units in the power grid. As mentioned before, this fact not only contributes for diminishing the losses in the power grid, but also contributes for a more resilient power grid, for example by strengthening the grid against different hazards.

Public policies in several countries are trying to avoid climate changes and to reduce the consequences of it. These policies triggered more efficient, less polluting systems and consequently enhanced efficiency, power grid resiliency and security [9]. Resilience can also be achieved by the prosumer if he can be independent of regulatory changes, such as the approach presented in [10]. Knowing in advantage the availability of renewable energy sources, namely through forecasts, allows to support the design and application of storage system, providing financial reliance to prosumers [10]. Using forecast techniques such as ANN, helps to support an adequate EMS because the system efficiency is enhanced, by improving storage management in coordination with decisions on selling or buying energy that leads to a cost reduction and simultaneously guarantee that consumers load is satisfied. In addition, the energy storage system helps to accommodate the intermittency effects that derive from the renewable generation, reducing instability and reduce outage probability in the EMS.

This paper shows that when an energy storage system is included in the energy system, the energy invoice decreases and a more profitable domestic energy management can be reached. In the developed energy management model, the energy selling or buying decisions are made accordingly to current energy prices. Knowing in advance the expected renewable production, the energy prices and load, contributes to augment system resiliency, using the energy storage system when is more advantageous, economically or technically.

3   Energy Management System Mathematical Model

The energy management system mathematical model considered is based on [1], and includes energy selling to the power grid and renewable energy production. In the developed mathematical model the time variable, t, corresponds to 1 h. For the months under study, 31-days are considered, where t assume values between 1 and 744 h. The objective function is given by (1).

$$min\ F = min\left(C_{buy} - R_{sell}\right).$$

(1)

Where C_{buy} is energy buying price, and R_{sell} is energy sale revenue, both in €, and given by (2–3).

$$C_{buy} = \sum_0^t M_{buy,t} P_{grid,t}, \text{ if } P_{grid,t} \geq 0.$$

(2)

Where $M_{buy,t}$ is the buy energy price at t in €/kWh and $P_{grid,t}$ is the demanded power or sold power to the grid at t, in kWh.

$$R_{sell} = \sum_0^t M_{sell,t} P_{grid,t}, \text{ if } P_{grid,t} < 0.$$

(3)

Where M_{cellt} is the selling energy price at t in €/kWh. The objective function (1) is subjected to the energy balance Eq. (4) and operation restriction Eqs. (5–8).

$$P_{load,t} = P_{solar,t} + P_{wind,t} \pm P_{grid,t} \pm P_{battery,t}.$$

(4)

Where $P_{load,t}$ is the power consumption at t in kWh, $P_{solar,t}$ is the photovoltaic solar panels power generation at t in kWh, $P_{wind,t}$ is the wind turbine power generation at t in kWh and $P_{battery,t}$ is the battery power flow at t, in kWh. Expression (5) gives the minimum and maximum battery capacity.

$$0 \leq E_{battery,t} \leq E_{battery}^{max}.$$

(5)

Where $E_{battery,t}$ is the batteries stored energy at t in kWh, and $E_{battery}^{max}$ is maximum battery stored energy at t in kWh. It is also considered that battery initial capacity is equal to battery final capacity and that battery storage capacity is given by (6).

$$E_{battery,t} = \sum_{0}^{t} \eta_b \cdot \alpha_b \cdot P_{battery,t}. \tag{6}$$

Where η_b is the converter and batteries efficiency and α_b indicates the batteries state. Which is +1 for the charging state and −1 for the discharging state.

$$P_{battery}^{min} \leq P_{battery,t} \leq P_{battery}^{max}. \tag{7}$$

Where $P_{battery}^{min}$ and $P_{battery}^{max}$ are, respectively, the minimum and maximum battery power flow. Expression (8) ensures that sold energy to the power grid derives from renewable energy sources.

$$P_{grid,t} \geq -(P_{solar,t} + P_{wind,t}). \tag{8}$$

4   Case Studies

The energy management model was applied to monthly time periods, January and July, in order to include seasonality factor and simultaneously allowing the proper management of an adequate data set. Two scenarios were tested: (1) without storage and; (2) considering a battery storage system. Hourly radiation, temperature and wind speed data were obtained from the Portuguese Monchique meteorological station [11]. Is assumed that the prosumer is subjected to the Portuguese Tri-hour tariff and the energy sale price is 90% of the price indicated by the OMIE [12]. Selling energy prices at January and July were 6,437 c€ and 4,374 c€, respectively and buying prices at peak, full and valley were, respectively, 22,470 c€, 17,68 c€ and 10,230 c€. Is considered that prosumer owns an Ennera Windera S wind generator, five LG NeON 2 photovoltaic solar panels, and the *Sonnem Batterie eco* 8.2/10, battery storage system. These equipment's technical characteristics are detailed in [13–15], respectively. It is also assumed that the storage system is discharged at the beginning and at the end of the time period under study. The case studies were focused in two different months, January and July, in order to analyse the influence of each considered renewable production. The optimization algorithm was implemented on GAMS software considering BARON solver and was run to the entire month (744 h).

In Figs. 1 and 2 the consumed and generated energy are depicted, where P_{load} is the consumed power and $P_{gen.}$ is the sum of photovoltaic and wind power generation. These figures only show 72 h, corresponding to 3 significant days of January and July. The choice of 3 significant days for each month helps to avoid an overloaded and rough graphical analysis of 744 h.

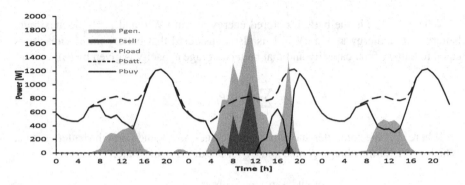

Fig. 1. Scenario 1- power diagram for January.

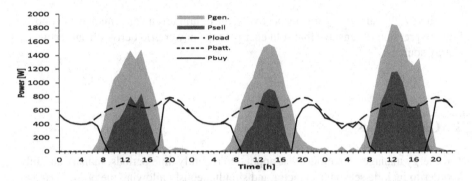

Fig. 2. Scenario 1- power diagram for July.

Where P_{buy} is bought energy and P_{sell} is sold energy. Because in this case study no storage system is considered, the stored energy, $P_{batt.}$ is null. It is also considered that energy is bought only to satisfy the load and sold when a production surplus is verified.

A comparison between the two considered months is shown in Table 1. Total energy consumed in each month is shown, as well as the generated by each renewable power source.

Table 1. Accumulated energy data for January and July

	January	July
E_{load} [kWh]	591,4	444,6
$E_{generated}$ [kWh]	333,0	371,0
E_{solar} [kWh]	96,1	311,1
E_{wind} [kWh]	236,9	60,0

January is characterised by higher energy consumption than in July, probably due to heating systems consumption, and naturally the photovoltaic production is higher in July, whereas the wind production present higher values in January. July energy consumption is lower, but photovoltaic production assumes greater importance when compared with the one observed in January. In Fig. 1, it is clear that in January, although during a few hours, the production is higher than consumption. In the remaining hours is noticeable a lack of renewable production.

Under a purely technical point of view, since consumer is connected to the power grid, is possible to sell the excess of energy or buy it when the production is not enough. However, when results are analysed under an economic perspective, considering the energy cost over time, the energy selling period may not be coincident with higher selling revenues. The same problem may occur during the energy buying periods, namely if they are coincident with full or peak periods.

Results of the EMS implementation, considering energy storage, are shown in Figs. 3 and 4.

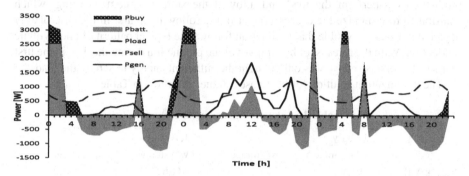

Fig. 3. Scenario 2- power diagram for January.

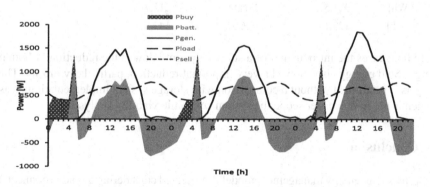

Fig. 4. Scenario 2- power diagram for July.

From Figs. 3 and 4 analysis, results that energy is only bought from the power grid when the production is not sufficient to satisfy the load or if it is necessary to charge batteries. The surplus produced energy can be stored or sold to the power grid. This is the main added value of the proposed EMS. The percentage of sold energy decrease, being the energy stored in periods where produced energy surplus occurs and lower selling price is verified. That stored energy can be used in periods of high consumption and high energy bought costs. For instance, in Fig. 3 is clear the high quantity of bought energy at first hours of each day, even when the load is low. Because the storage system was discharged and energy buying price was low, the bought energy was used to completely charge batteries and feed the load, once the generation is null. Another detail can be noticed between 3 h and 6 h. Even with storage system fully charged, it was preferable to buy energy to satisfy the load, because is more profitable to delay batteries discharge to a period, in which, although there is some energy production, is also coincident to higher energy buying prices. In this case, storage allowed shifting the energy, not to a period with low production but to a period with higher energy buying prices. At 15 h the same proceeding was adopted. The EMS decided to storage energy, rather than use the existing production to satisfy the load, mainly because in following hours, the difference between production and the load is significantly bigger than the one observed before. Also, analyzing the second day shown in Fig. 3, it is verified that the maximum production out-perform the load and allowed the storage system charging, which contributed to minimize the energy bought in the following hours. In the case of July depicted in Fig. 4, is visible that production follows the typical pattern of photovoltaic production. With the forecast of high photovoltaic production during the day, in the first hours of the first two days was only bought the sufficient energy to satisfy the load.

The optimization solution for both cases studies is shown in Table 2.

Table 2. Solution with and without storage

	January		July	
	w/o storage	w/ storage	w/o storage	w/ storage
E_{load} [kWh]	591,4		444,6	
$E_{generated}$ [kWh]	333,0		371,0	
E_{buy} [kWh]	381,5	331,0	215,7	77,1
E_{sold} [kWh]	123,1	72,6	142,0	3,4

In both cases the introduction of a storage system allowed the reduction of bought energy. Sold energy also showed a very important reduction, particularly in July. This analysis is performed without a comparison with storage system implementation costs, as such is not in this paper scope the system economic viability analysis.

5   Conclusions

In this paper, an energy management model is presented considering a system connected to the power grid with a battery storage system. Also, solar and wind generation are considered in the referred developed energy management model. Based on obtained results can be observed that the energy management model used has achieved the desired energy bill reduction for different scenarios. The obtained values are strongly dependent on solar radiation and wind speed availability. When comparing January with July it is verified that consumption is higher during January and that photovoltaic production is higher in July. The introduction of an energy storage system allows energy bill reduction, either by storing surplus energy production or by buying energy in off peak hours, where the energy has the lowest value, in order to use that stored energy in peak periods where energy has a higher cost. Future work developments include: comparison analysis of battery charging only from renewable production and also from a combination between renewable production and energy bought from power grid; considering annual time horizon analysis; economic study of different scenarios, including storage system implementation cost; and return of investment analysis.

References

1. Olatomiwa, L., Mekhilef, S., Ismail, M.S., Moghavvemi, M.: Energy management strategies in hybrid renewable energy systems: a review. Renew. Sustain. Energy Rev. **62**, 821–835 (2016)
2. Pascual, J., Barricarte, J., Sanchis, P., Marroyo, L.: Energy management strategy for a renewable-based residential microgrid with generation and demand forecasting. Appl. Energy **158**, 12–25 (2015)
3. Battistelli, C., Baringo, L., Conejo, A.J.: Optimal energy management of small electric energy systems including V2G facilities and renewable energy sources. Electr. Power Syst. Res. **92**, 50–59 (2012)
4. Ismail, M.S., Moghavvemi, M., Mahlia, T.M.I.: Techno-economic analysis of an optimized photovoltaic and diesel generator hybrid power system for remote houses in a tropical climate. Energy Convers. Manage. **69**, 163–173 (2013)
5. Dalton, G.J., Lockington, D.A., Baldock, T.E.: Feasibility analysis of renewable energy supply options for a grid-connected large hotel. Renew. Energy **34**(4), 955–964 (2009)
6. Jiang, Q., Xue, M., Geng, G.: Energy management of microgrid. IEEE Trans. Power Syst. **28**(3), 3380–3389 (2013)
7. Banica, A., Istrate, M.: Towards a resilient energy system in Eastern Romania – from fossil fuels to renewable sources. Ann. Univ. Oradea **2**, 148–156 (2015)
8. Misra, S., Krishna, P.V., Saritha, V., Obaidat, M.S.: Learning automata as a utility for power management in smart grids. IEEE Commun. Mag. **51**(1), 98–104 (2013)
9. Martín-Martínez, F., Sánchez-Miralles, A., Rivier, M.: Prosumers' optimal DER investments and DR usage for thermal and electrical loads in isolated microgrids. Electr. Power Syst. Res. **140**, 473–484 (2016)
10. Sanduleac, M., Ciornei, I., Albu, M., Toma, L., Sturzeanu, M., Martins, J.F.: Resilient prosumer scenario in a changing regulatory environment—the UniRCon solution. Energies **10**(12), 1–22 (2017)
11. Sistema nacional de informação de recursos hídricos. http://snirh.apambiente.pt
12. OMIE. http://www.omel.es/files/flash/ResultadosMercado.swf
13. Ennera, W.S.: http://www.ennera.com/es/windera-s#descargas
14. LG. http://www.lg.com/us/business/solar-panel/all-products/lg-LG400N2W-A5
15. SonnenBatterie. https://www.sonnenbatterie.de/en-au/sonnenbatterie

Sensing Systems

Fair Resource Assignment at Sensor Clouds Under the Sensing as a Service Paradigm

Joel Guerreiro, Luís Rodrigues, and Noélia Correia[(✉)]

CEOT, University of Algarve, 8005-139 Faro, Portugal
{jdguerreiro, lrodrig, ncorreia}@ualg.pt

Abstract. The Sensing as a Service (Se-aaS) is a Cloud-based service model for sensors/data to be shared, allowing for a multi-client access to sensor resources, and multi-supplier deployment of sensors. This way, everyone can benefit from the Internet of Things (IoT) ecosystem. Client applications can have components binded to mashups stored at the Cloud, where each mashup is a workflow wiring together virtual Things. Property requirements are specified for each mashup element. For an efficient use of resources, multiple mashup elements (from different applications) can be materialized into the same physical Thing, if the requested functionality is the same and requirements/constraints are not competing. That is, clusters of mashup elements are built for materialization. The problem of choosing for the best clustering and materialization assignment, having costs as a basis of decision, is addressed. The problem is mathematically formalized and the fair resource assignment results are compared against the unfair ones.

Keywords: Internet of Things · Cloud · Sensing as a Service
Fairness

1 Introduction

To prevent the Internet of Things (IoT) from becoming just a collection of Things, unable to be discovered, a move towards the Web of Things (WoT) is required. When real-world objects become part of the World Wide Web, they become accessible to a large pool of application developers and mashups combining physical Things with virtual Web resources can be created. The WoT is, therefore, envisaged as the key for an efficient resource discovery, access and management [1]. Naturally, security is a critical issue in IoT/WoT deployments.

The more Things become discoverable and accessible, the more it makes sense to rely on Cloud infrastructures for storage and processing. The Sensing as a Service (Se-aaS) is a Cloud-based service model for sensors/data to be shared, allowing for a multi-client access to sensor resources, and multi-supplier deployment of sensors [2]. This way, everyone can benefit from the IoT ecosystem. When incorporating Se-aaS platforms at the application architecture, software components/sections can be binded to mashups stored at the Cloud. In such scenario, a mashup is a workflow linking elements (virtual Things), each specifying a functionality and a set of property requirements (done by the client application). The exact physical materialization of

© IFIP International Federation for Information Processing 2018
Published by Springer International Publishing AG 2018. All Rights Reserved

L. M. Camarinha-Matos et al. (Eds.): DoCEIS 2018, IFIP AICT 521, pp. 167–174, 2018.
https://doi.org/10.1007/978-3-319-78574-5_16

each element (mashup element to physical device assignment) is to be done at the Cloud, and basically depends on the requirements of mashup elements and registered physical devices. For resources to be used efficiently, multiple mashup elements (from different client applications) can be materialized into the same physical Thing, if the requested functionality is the same and their requirements/constraints are not competing. That is, clusters of mashup elements for materialization are built at the Cloud.

This article addresses resource allocation at sensor Clouds under the Se-aaS paradigm, i.e., the problem of choosing for the best cluster to physical Thing assignment, having costs as a basis of decision. A materialization cost reflects the gap between the requirements of mashup elements (at a cluster) and physical Thing properties. The fair resource assignment problem (minimize highest cluster-Thing assignment cost) is mathematically formalized and results are compared against the unfair one (minimize overall cost). By minimizing the highest assignment cost, physical things with characteristics (e.g. camera resolution, localization accuracy) above what is requested, can be left for future requests. Since these can fit more requirement ranges, and due to the unpredictable nature of requests dynamically arriving, more clients can be served.

The remainder of this article is organized as follows. In Sect. 2, the resilience benefits of sensor Clouds, and importance of carefull resource assignments for resilience, are presented. In Sect. 3, work related with the Se-aaS paradigm is discussed. Section 4 analyses the resource assignment approach, and results are discussed in Sect. 5. Section 6 concludes the article.

2   Contribution to Resilient Systems

Besides allowing everyone to benefit from the IoT ecosystem, the Se-aaS service model allows highly-available, or resilient, applications to be developed. Basically, resilient applications require planning at both software development and application architecture levels, and Se-aaS platforms can serve as a basis for the last. Besides ensuring robust storage and scalability, such service model allows physical Things to be dynamically allocated to clients/applications because users remain unaware of physical devices involved in the process. That is, the client ends up having no deployment and maintenance costs, while having an on-demand fault tolerant service because clients/applications can always use other available physical Things.

The Se-aaS service model supports the development of multi-sensing applications with mashups combining physical Things with virtual Web resources. In such scenarios, a carefull assignment of sensors to client applications becomes important not only to face current and future requests, but also for applications to quickly recover when devices are experiencing problems (i.e., ensure available backup devices). One way to achieve this is for multiple mashup elements (from different client applications) to be materialized into the same physical Thing, if the requested functionality is the same and their requirements/constraints are not competing. This way more devices are free for future requests or for backup. Different clusters will have different materialization costs (sum of gaps between physical properties and requirements).

A criterion that can be used, when deciding for cluster to physical Thing assignments, is the minimization of the overall cost (all materializations). This leads, however, to asymmetries, i.e., some assignments may have a low cost while others may have a high cost. In this article we analyse the impact of using a fair criterion. More specifically, the goal is to minimize the highest cluster-Thing assignment cost. Results show that the produced assignments reduce high property gaps (note that there are multiple properties and, therefore, the cost of a cluster includes multiple physical property to requirements gaps). Physical Things with features that are above what is requested can, therefore, be kept idle as they can be needed for future requests and/or backup. This makes applications more resilient.

3 Related Work

Se-aaS was initially introduced at [3], where Cloud based sensing services using mobile phones (crowd sensing) are discussed. In the context of smart cities, Se-aaS is discussed both in [2, 4]. The first addresses technological, economical and social perspectives, while the last proposes the abstraction of physical objects through semantics, so that devices can be integrated by neglecting their underlying architecture. In [5, 6], the semantic selection of sensors is also addressed.

The integration of Wireless Sensor Networks (WSNs) with the Cloud is addressed in [7–9], where data storage and/or device assignment to tasks is the focus.

In this article, and contrarily to previous works, the focus in on multi-supplier and multi-client Sensor Clouds where client applications can request for available devices and build mashups. The focus is not on crowd sensing, making data from mobile phones available to multiple clients, but instead on how applications can share devices registered at the Cloud and build mashups, also not considered in [7–9].

4 Resource Assignment Approach

4.1 Assumptions and Motivation

In the following sections the overall set of physical devices is denoted by D, and each $d \in D$ is assumed to have a functionality and one or more properties (sensor owners voluntarily register/deregister physical Things to/from the Cloud). Client applications are connected to one or more mashups, each with one or more nodes/elements. The overall set of mashup nodes is denoted by N, and each $n \in N$ specifies a functionality requirement and one or more property constraints. As devices are registered, and client applications create mashups, the Cloud is able to:

- maintain a graph $G(N, L)$, where N includes all mashup elements and L denotes a set of links. A link between n_i and $n_j \in N$ exists if: (*i*) nodes have similar functionality requirements; (*ii*) property requirements are not incompatible.
- update, for each $n \in \mathcal{N}$ a cost vector $C_n = \left\{ c_n^{d_1}, c_n^{d_2}, \ldots, c_n^{|D|} \right\}$ where $c_n^{d_i}$ is the cost of using device d_i for the materialization of the maximum clique, in graph $G(N, L)$, including n. Such clique is largest set of compatible mashup elements. This way, high clustering of mashup elements is ensured.

Assuming that gaps, between a property requirement and device property supply, are normalized using $\{\Delta_1, \ldots, \Delta_5\}$, where Δ_1 is the lowest cost and Δ_5 is the highest (moderate and extreme levels), a cost $c_n^{d_i}$ will be the sum of all property requirement to device property gaps. If more than one node at the clique has a requirement for a certain property, then the lowest one is chosen (e.g., assuming requests for 12.1 MP and 24.2 MP camera resolutions, and a physical Thing providing 48.4 MP, then the 24.2 to 48.4 MP gap is the one to be considered; the other request is considered to be fulfilled).

Therefore, several materialization possibilities, with different costs, are provided as input information. The problem is which materializations to chose. Here the mini- mization of the highest cluster-Thing assignment cost is analysed. The goal is to see if (contrarily to an unfair approach where the overall cost is minimized) this leads to assignments with lower property gaps, allowing physical Things with features that are above what is requested to be left for future requests and/or backup. This makes applications more resilient.

4.2   Mathematical Formalization of Problem

The following information is assumed to be known:

N Set of all mashup nodes/elements.

D Set of physical devices.

$D(n)$ Set of physical devices that can be used for the materialization of node $n \in N$.

$C(n, d)$ Set of mashup nodes (clique) able to join $n \in N$ if $d \in D(n)$ is used for materialized.

$\Delta(n, d)$ Cost of $C(n, d)$ materializing

Since multiple components contribute to a materialization cost (i.e., there are multiple properties and, therefore, gaps between physical properties and requirements), and to fairly compare the clusters, $\Delta(n, d)$ is divided by the number of summed up property gaps. This way, the highest cluster-Thing assignment cost (used at the objective function) will more easily reduce.

The variables will be:

λ^{max} Highest clique materialization cost from selected materializations.

δ_n^d One if device $d \in D$ was selected for the materialization of $n \in N$, zero otherwise

Where Δ^{max} is the highest possible $\Delta(n, d)$ cost. Basically, the primary goal is to minimize the upper bound.

- Objective function:

$$OF^{fair:} Minimize\ \lambda^{max} + \frac{\sum_{n \in N} \sum_{d \in D(n)} \Delta(n, d) \times \delta_n^d}{|N| \times |D| \times \Delta^{max}} \tag{1}$$

Subject to:

- Upper bound limitation:

$$\sum_{d \in D(n)} \Delta(n, d) \times \delta_n^d \leq \lambda^{max}, \forall n \in N \tag{2}$$

- Single materialization for a mashup node:

$$\sum_{d \in D(n)} \delta_n^d = 1, \forall n \in N \tag{3}$$

- Materialization of cliques:

$$\delta_{n'}^d \geq \delta_n^d, \forall n \in N, \forall d \in D(n), \forall n' \in C(n, d) \tag{4}$$

- Allocation of device to a single clique materialization:

$$\sum_{n' \in N \setminus C(n,d)} \delta_{n'}^d \leq \left(1 - \delta_n^d\right) \times |N|, \forall n \in N, \forall d \in D(n) \tag{5}$$

- Non-negative variables:

$$\delta_n^d \in \{0, 1\}; \lambda^{max} \geq 0. \tag{6}$$

5   Analysis of Results

5.1   Scenario Setup

To evaluate the allocation of resources, a pool of 10 functionalities were generated for the Cloud, each with its own pool of 10 properties. Based on these, a population of physical Things and population of mashup elements were generated as follows:

- Each physical Thing has a randomly selected functionality, each functionality including 100% of the properties from corresponding pool.
- The mashup element functionality requirement is also randomly selected from the pool of functionalities, together with 50% of its properties. Each pair $n_i, n_j \in N$ sharing the same functionality requirement is assumed to be compatible with probability $\delta = 0.5$ or $\delta = 0.75$ (creation of cliques).
- The gap between a property requirement and device property supply is randomly selected from $\{\Delta_1 = 1, \Delta_2 = 2, \ldots, \Delta_5 = 5\}$(moderate and extreme levels).

The population of mashup elements is 50 and teste were done for 80, 90, 100 and 110 physical things. The CPLEX[1] optimizer was used to solve this problem.

5.2 Fair vs Unfair Approach

To evaluate the impact of the fair approach, a second objective function is also implemented. This second objective function has no fairness into consideration (no minimization of the upper bound is performed) and chooses to reduce the overall cost. It is defined as follows:

$$OF^{unfair}: Minimize \sum_{n \in N} \sum_{d \in D(n)} \Delta(n, d) \times \delta_n^d. \tag{7}$$

5.3 Discussion

Plots in Figs. 1 and 2 show the cumulative frequency values obtains for the fair and unfair objective functions for 90 and 110 physical Things, respectively. The results obtained for 100 physical Things were similar (no difference between fair and unfair approaches exists), and for this reason are not shown. From plot at Fig. 1, it is possible to observe that the fair approach reaches a cumulative frequency of 50 at clique materialization cost of 10, while the unfair reaches a cumulative frequency of 50 at clique materialization cost of 12. That is, the fair approach avoids 11 and 12 costs. Since these cost reflect the gap between the requirements of mashup elements and physical Things, lower gaps allow physical Things with features that are above what is requested to be left for future requests and/or backup. This makes applications more resilient. Regarding plot at Fig. 2, a similar behavior, although not so foreshadowed, can be observed.

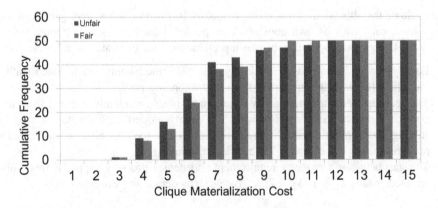

Fig. 1. Number of physical things: 90.

---

[1] IBM ILOG CPLEX Optimizer.

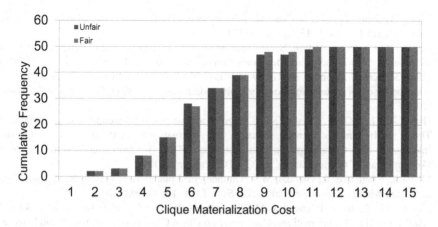

Fig. 2. Number of physical things: 100

Both fair and unfair approaches have as a basis cost vectors $C(n, n) \in N$, that give the materialization costs based on maximum cliques. This reduces the search space (and, therefore, execution time) but eliminates potential solutions including cliques of smaller size. When increasing the population of cliques, better results can be obtained.

6   Conclusions and Further Work

This article addresses resource allocation at sensor Clouds under the Se-aaS paradigm. The basis of decision it the materialization cost, which reflects the gap between the requirements and physical Thing properties. A fair resource assignment, when compared against the unfair one, minimizes the highest gap/cost in many scenarios. This allows physical Things with features that are above what is requested to be left for future requests and/or backup, meaning that applications have a lower probability of not finding available devices (devices fulfilling the requested property requirements). As future work, algorithms to find populations of cliques with more potential, while avoiding increasing too much the search space, should be developed.

Acknowledgment. This work was supported by FCT (Foundation for Science and Technology) from Portugal within CEOT (Center for Electronic, Optoelectronic and Telecommunications) and UID/MULTI/00631/2013 project.

References

1. Guinard, D., Trifa, V.: Building the Web of Things. Manning Publications, Shelter Island (2016)
2. Perera, C., Zaslavsky, A., Christen, P., Georgakopoulos, D.: Sensing as a service model for smart cities supported by Internet of Things. Trans. Emerg. Telecommun. Technol. **25**(1), 81–93 (2014)

3. Sheng, X., Tang, J., Xiao, X., Xue, G.: Sensing as a service: challenges, solutions and future directions. IEEE Sens. J. **13**(10), 3733–3741 (2013)
4. Petrolo, R., Loscrì, V., Mitton, N.: Towards a smart city based on cloud of things, a survey on the smart city vision and paradigms. Trans. Emerg. Telecommun. Technol. **28**(1), 1–12 (2015)
5. Misra, S., Bera, S., Mondal, A., Tirkey, R., Chao, H.-C., Chattopadhyay, S.: Optimal gateway selection in sensor-cloud framework for health monitoring. IET Wirel. Sens. Syst. **4**(2), 61–68 (2014)
6. Hsu, Y.-C., Lin, C.-H., Chen, W.-T.: Design of a sensing service architecture for Internet of Things with semantic sensor selection. In: International Conference on Ubiquitous Intelligence and Computing, on Autonomic and Trusted Computing, and on Scalable Computing and Communications, UTC-ATC-ScalCom (2014)
7. Misra, S., Chatterjee, S., Obaidat, M.S.: On theoretical modeling of sensor cloud: a paradigm shift from wireless sensor network. IEEE Syst. J. **11**, 1084–1093 (2014)
8. Zhu, C., Li, X., Ji, H., Leung, V.C.M.: Towards integration of wireless sensor networks and cloud computing. In: International Conference on Cloud Computing Technology and Science, CloudCom (2015)
9. Dinesh Kumar, L.P., Shakena Grace, S., Krishnan, A., Sumalatha, M.R.: Data filtering in wireless sensor networks using neural networks for storage in cloud. In: International Conference on Recent Trends In Information Technology, ICRTIT (2012)

Elephant Herding Optimization Algorithm for Wireless Sensor Network Localization Problem

Ivana Strumberger[1(✉)], Marko Beko[2,3], Milan Tuba[4] (iD), Miroslav Minovic[5], and Nebojsa Bacanin[1]

[1] Faculty of Informatics and Computing, Singidunum University,
Danijelova 32, 11000 Belgrade, Serbia
{istrumberger,nbacanin}@singidunum.ac.rs
[2] COPELABS/ULHT, Lisbon, Portugal
[3] CTS/UNINOVA, Caparica, Portugal
mbeko@uninova.pt
[4] Department of Technical Sciences, State University of Novi Pazar,
Vuka Karadzica bb, 36300 Novi Pazar, Serbia
tuba@ieee.org
[5] Faculty of Organizational Sciences, State University of Belgrade,
Jove Ilica 154, 11000 Belgrade, Serbia
miroslav.minovic@mmklab.org

Abstract. This paper presents elephant herding optimization algorithm (EHO) adopted for solving localization problems in wireless sensor networks. EHO is a relatively new swarm intelligence metaheuristic that obtains promising results when dealing with NP hard problems. Node localization problem in wireless sensor networks, that belongs to the group of NP hard optimization, represents one of the most significant challenges in this domain. The goal of node localization is to set geographical co-ordinates for each sensor node with unknown position that is randomly deployed in the monitoring area. Node localization is required to report the origin of events, assist group querying of sensors, routing and network coverage. The implementation of the EHO algorithm for node localization problem was not found in the literature. In the experimental section of this paper, we show comparative analysis with other state-of-the-art algorithms tested on the same problem instance.

Keywords: Elephant herding optimization · Swarm intelligence
Wireless sensor networks · Node localization problem · Metaheuristics

1 Introduction

1.1 Motivation

Recent advances and development of distributed environments such as Peer-to-Peer networks, cloud and grid computing have led to the growth in popularity of Wireless Sensor Networks (WSN). WSN is an emerging paradigm of computing and networking, which can be defined as a network of minuscule, diminutive, in- expensive, and keenly

© IFIP International Federation for Information Processing 2018
Published by Springer International Publishing AG 2018. All Rights Reserved
L. M. Camarinha-Matos et al. (Eds.): DoCEIS 2018, IFIP AICT 521, pp. 175–184, 2018.
https://doi.org/10.1007/978-3-319-78574-5_17

intellective devices called sensor nodes [1]. Sensor nodes, that are placed in different locations within the space that is being observed, exchange data gathered from the monitoring field by using wireless communication channels. Gathered data is sent to the sink node, that either processes the data locally, or sends it to the other networks with more processing power [2].

One of the most fundamental challenges in WSNs is node localization. There are many instances of node localization problem and they belong to the group of NP-hard optimization [2, 3]. NP-hard problems cannot be solved in a reasonable amount of computational time by traditional deterministic techniques and algorithms. In this case, it is better to employ non-deterministic (stochastic) algorithms, like metaheuristics.

Swarm intelligence metaheuristics simulate groups of organisms from the nature such as flock of birds and fish, colony of bees and ants, groups of bats and cuckoo birds, etc. These algorithms are population-based, stochastic and iterative search methods that are based on four self-organization principles: positive feedback, negative feedback, multiple interactions and fluctuations [4].

In this paper, we propose elephant herding optimization (EHO) algorithm adopted for tackling localization problem in WSN. EHO algorithm was proposed by Wang et al. in 2015 for global unconstrained optimization [5].

1.2   Research Question

According to the literature survey [5, 6] and our previous work [9, 11] we concluded that EHO is promising approach for tackling NP-hard problems such as node localization. Therefore, the research question can be formulated as follows:

How to design an efficient (in terms of solutions quality and convergence) EHO algorithm for node localization problem in WSN that will outperform other approaches for same problem formulation?

In order to address the research question, we have formulated the following hypothesis:

EHO algorithm, as promising approach for dealing with NP-hard tasks, can be adapted for solving localization problem and can obtain lower value of localization error than similar approaches tested on the same localization model.

1.3   Related Work

EHO algorithm is relatively new approach that belongs to the group of swarm intelligence metaheuristics, and there have only been few implementations of the EHO found in the literature. In [6, 7], EHO was tested on standard benchmark problems. EHO was also applied on multilevel image thresholding [9], and other practical problems [10, 11].

On the other hand, according to the literature survey, node localization, as one of the most fundamental challenges in WSNs, was tackled by swarm intelligence algorithms in both, basic and modified/hybridized implementations. For example, a localization algorithm based on multi-objective particle swarm optimization (PSO) was presented in [12]. In [13], differential evolution (DE), firefly algorithm (FA) and hybridized FA-DE were presented for tackling localization problems. Also, modified bat algorithm

(BA) [14], cuckoo search (CS) [3] and fireworks algorithm (FWA) [15] were implemented for this kind of problem. In this paper, we used the same model as in [19], and compared our approach with four localization algorithms based on artificial bee colony (ABC) and PSO swarm intelligence metaheuristics [19].

1.4 Contributions

In this paper, we show an implementation of elephant herding optimization (EHO) algorithm adopted for tackling localization problem in WSN. This is the very first implementation of the EHO algorithm for this kind of problem. Our approach obtains better results in terms of localization error than other metaheuristics tested within the same experimental conditions. Even if we compare mean results generated from a set of 30 independent algorithm runs, our approach outperforms other algorithms included in the comparative analysis.

2 Relationship to Resilient Systems

Resilient systems involve tasks of actuation, sensing, and control with a goal of analyzing and describing an environment, and making critical decisions based on the available data in an adaptive or predictive manner. However, in many applications, the information gathered by the sensors is meaningless if not associated to accurate location of where the changes are occurring (e.g., a system might be configured to rapidly respond to changes in sensor data). Therefore, determining accurate location of sensors is a very important task in forming a smart, resilient system.

A sensor network is a good example of such a system. A sensor network is composed of a large number of sensors that collaborate between themselves in order to respond in an active and adaptive manner to the changes in the environment registered by sensors. An important example where target localization helps building a resilient system is ocean/sea monitoring. It has been used for search and rescue operations, detection of tsunamis[1], etc. Clearly, such systems require robust, reliable and fast algorithms for detection, localization, tracking, classification, and activity analysis [16].

Furthermore, accurate localization of people and objects in both outdoor and indoor environments allows new applications in emergency services that can improve safety in everyday life (e.g., assistance for elderly or people with disabilities) [17, 18].

3 Localization Problem in Wireless Sensor Networks

Localization problem is one of the most studied challenges in WSN, because, if the location of sensor nodes is not known, coverage, power and routing will not be optimal. Because of this particular problem, localization is essential in WSN. The location of

---

[1] After the 2004 Indian Ocean tsunami when 228000 people lost their lives, the problem of early detection of tsunamis using underwater sensor networks to detect the generation of a tsunami wave has received much attention [16].

some sensor nodes can be defined by Global Positioning System (GPS) and these nodes are referred as anchor or beacon nodes, while other sensor nodes are distributed randomly in search space. These nodes are referred as unknown nodes or sensor nodes. Due to the battery life of each node, cost, climatic conditions, locations of only few of them are determined by GPS coordinates while for other nodes, the location needs to be estimated by employing localization algorithms [19].

Localization algorithms that are proposed for the localization of sensor nodes in WSN, with anchor and unknown nodes, are conducted in two phase processes. The first phase is referred as ranging phase, where algorithms determine the distance between unknown nodes and neighboring anchor nodes. The second phase estimates the position of the nodes by using various methods for collecting ranging information in the first phase, such as Angle of Arrival (AOA), Time of Arrival (TOA), Time Difference of Arrival (TDOA), Round Trip Time (RTT), Radio signal strength (RSS), etc. [20].

3.1 Problem Statement

The objective of localization problem in WSN that consists of M sensor nodes is to estimate locations of N unknown nodes using location information of M-N anchor nodes, with transmission range R. If a sensor node is within transmission range of three or more anchors, then it is considered to be localized. This is 2D localization problem with total number of $2N$ unknown coordinates.

In this paper, the RSS method is being considered for estimating range between nodes. Whatever ranging method is being employed, imprecise measures can occur. The position estimation of the coordinates of the N unknown nodes can be formulated as an optimization problem, involving the minimization of objective function representing the error in locating the nodes [19]. The objective function for this problem is represented by the sum of squared ranged errors between the N unknown nodes and $M N$ neighboring anchor nodes [19].

Along with the RSS, Trilateration will be considered for tackling localization problem in WSN. The principle of this method is based on known locations of three anchor nodes. The location of the unknown node can be estimated within the transmission range of three anchor nodes.

Each node estimates the distance from i-th anchor as $\hat{d} = d_i + n_i$, where n_i is Gaussian noise, and d_i is actual distance calculated by using the equation:

$$d_i = \sqrt{\left(x - x_i\right)^2 + \left(y - y_i\right)^2},$$ (1)

The objective function that should be minimized is given as mean squared error (MSE) between actual and estimated distances of computed node coordinates and the actual node coordinates:

$$MSE = \frac{1}{M} \sum_{i=1}^{M} (d_i - \hat{d})^2,$$ (2)

where d_i is actual distance and $\hat{d}_i$ is estimated distance (the value d_i obtained from noisy range measurements) and $M \geq 3$ (location of the sensor node needs a minimum of three anchors within transmission range R).

Due to the noisy range measurements in node localization, optimization methods and approaches, such as swarm intelligence metaheuristics, are used in order to estimate adequate distance between nodes.

4 Elephant Herding Optimization Algorithm for WSN Optimization Problem

EHO was inspired by social behavior of elephants in herds [21]. It was proposed by Wang et al. for solving global optimization tasks [5]. Authors have developed a general-purpose heuristic search based on the coexistence of the elephants in clans under the leadership of a matriarch. The matriarch is the oldest chosen female in the clan. Other members of the clan are mostly females and calves, where, upon the full growth, male elephants leave the habitation to live separately. Even though they live independently, male elephants communicate with the others from the clan through low frequency vibrations.

This structural independence and social communication in elephant herding can be portrayed as two different environments [5]: the first environment where all elephants live under the influence of the matriarch, and the second environment, where male elephants live independently but still communicate with the clan. These environments are modeled as updating and separating operators [5]. In EHO algorithm, each solution j in each clan ci is updated by its current position and matriarch ci through updating operator. After that, through the separating operator, the population diversity is enhanced at the later generations of the algorithm execution.

Each individual in the population is represented as an integer number vector with the dimension of $2\,N$, where N denotes the number of unknown sensor nodes. First the population is divided into n clans. The updating operator is modeled by altering position of each solution j in the clan ci by the influence of the matriarch ci which has the best fitness value in generation [5]:

$$x_{new,ci,j} = x_{ci,j} + \alpha \times \left(x_{best,ci} - x_{ci,j}\right) \times r, \tag{3}$$

where $x_{new,ci,j}$ represents the old position of the individual j in the clan ci, and $x_{best,ci}$ is the best solution in the clan ci found so far. Parameter $\alpha \in [0, 1]$ is a scale factor that designates the influence of matriarch ci on $x_{ci,j}$, while $r \in [0, 1]$ is random variable with uniform distribution.

The following expression is employed to update the fittest solution in each clan ci [5]:

$$x_{new,ci,j} = \beta \times x_{center,ci}, \tag{4}$$

where $\beta \in [0, 1]$ denotes the influence factor of the $x_{center,ci}$ on the updated individual. The center of the clan ci for d-th dimension problem can be calculated as [5]:

$$x_{center,ci,d} = \frac{1}{n_{ci}} \times \sum_{j=1}^{d} x_{ci,j,d}, \qquad (5)$$

where $1 \leq d \leq D$ represents the d-th dimension, D is total dimension of the search space, and n_{ci} indicates the number of solutions in clan ci.

The separating operator is applied at each generation of algorithms execution on the worst individual in population [5]:

$$x_{worst,ci} = x_{min} + \left(x_{max} - x_{min} + 1\right) \times rand, \qquad (6)$$

where x_{max} and x_{max} represent the upper and lower bound of the position of the individual, respectively, $x_{worst,ci}$ indicates the individual with the worst fitness in the clan ci, and $rand \in [0, 1]$ is a random number chosen by uniform distribution.

Pseudo-code of EHO algorithm is given in Algorithm 1.

Algorithm 1 Pseudo-code of EHO algorithm

Initialization. Generate individuals; divide population into
n clans; calculate fitness for each individual; set generation
counter $t = 1$ and maximum generation *MaxGen*.
while $t < MaxGen$ **do**
    Sort all solutions according to their fitness
    for all clans ci **do**
        for all solution j in the clan ci **do**
            Update $x_{ci,j}$ and generate $x_{new,ci,j}$ using Eq. 3
            Select and retain better solution between $x_{ci,j}$ and $x_{new,ci,j}$
            Update $x_{best,ci}$ and generate $x_{new,ci,j}$ using Eq. 4
            Select and retain better solution between $x_{best,ci}$ and $x_{new,ci,j}$
        end for
    for all clans ci in the population **do**
        Replace the worst solution in clan ci using Eq. 6
    end for
    Evaluate population and calculate fitness
end while
return the best solution among all clans

In the literature, there are two approaches for tackling WSN localization problem [19]: single-stage and multi-stage. In single-stage localization all un-known nodes are localized in a single stage where unknown nodes with more neighboring anchor nodes will get better location estimation than unknown nodes with the less neighboring anchor nodes.

In our EHO implementation, we have used multi-stage localization where the localization process is performed in multiple stages. In each stage, only unknown sensor nodes with three or more neighboring anchor nodes are being localized. Sensor nodes which are localized in one stage become anchor sensor nodes for the following stage. The localization process is repeated until all unknown nodes are localized.

Pseudo-code for multi-stage localization using EHO algorithm is given in Algorithm 2.

Algorithm 2 Multi-stage localization using EHO algorithm

1: Initialize network topology
2: Find unknown nodes with three or more anchor nodes in their neighborhood
3: Run EHO algorithm to find location of unknown nodes identified in step 2
4: Add localized sensor nodes in current stage to the anchor node list and remove them from unknown node list
5: Repeat steps 2-4 until all sensor nodes are localized

5   Experimental Discussion and Results

For experimental purposes we used mathematical formulation presented in Sub-Sect. 3.1. Parameters of EHO algorithm were adjusted as follows: number of clans $n = 5$, number of solutions in each clan $n_{ci} = 10$, and maximum generations number *MaxGen* = 800, which yields to the total number of 40,000 objective function evaluations. Additionally, we set values for scale factors α and β to 0.5 and 0.1, respectively. We performed simulations with different values of α and β parameter and found optimal values for this particular problem. Same values for n, n_{ci}, α and β parameters were used in [5]. The objective function that is subject of minimization is shown in Eq. (2).

We used similar experimental setup of wireless network topology as in [19]. We deployed 1000 sensor nodes with varying transmission range (from 20 m to 50 m), and anchor node density in 100 * 100 m square network domain. The anchor node density is very small and varies from 2.5% to 10% of total number of nodes. Comparative analysis was performed with particle swarm optimization (PSO), multi-stage PSO (MPSO), artificial bee colony (ABC), and multi-stage ABC (MABC) [19]. All algorithms included in comparative analysis were tested using the same experimental conditions.

The algorithm was executed in 30 independent runs, each starting from a different random number seed. In the results tables we show both, best and mean values obtained from 30 independent algorithm's runs to prove the robustness of EHO for this problem.

The experimental results in the Table 1, show the best values of the MSE (Eq. (2)), obtained by each algorithm included in comparative analysis. The best results for each setup (% of anchor, the transmission range) are presented in bold format.

According to the experimental results presented in Table 1, on average, EHO algorithm obtains better results than all the other approaches included in comparative analysis. In all test instances, EHO performs better than the single-stage PSO, ABC, and the multi-stage ABC (MABC), while in most test cases, EHO generates better solutions than the multi-stage PSO (MPSO) approach.

Table 1. Simulation results for best values obtained for MSE objective function

No. of nodes	Anchor no. (%)	Range	PSO	MSPSO	ABC	MSABC	EHO
1000	2.5	20	28.149	27.667	45.128	40.038	**27.493**
1000	5	20	27.273	27.073	43.386	38.038	**26.923**
1000	7.5	20	26.156	26.035	40.925	36.374	**26.019**
1000	10	20	25.273	25.035	40.079	34.193	**24.978**
1000	2.5	35	27.320	**27.167**	43.990	39.273	27.180
1000	5	35	26.089	25.936	41.759	35.238	**25.903**
1000	7.5	35	25.169	25.005	40.279	33.273	**24.971**
1000	10	35	24.763	**24.573**	38.462	30.283	24.591
1000	2.5	50	26.537	25.893	41.534	35.928	**25.857**
1000	5	50	25.534	**25.382**	40.178	33.229	25.428
1000	7.5	50	24.810	24.673	38.289	32.037	**24.668**
1000	10	50	24.485	24.109	36.972	30.028	**24.097**

In Table 2, we show the mean values that EHO obtains for each experimental setup. From this table, we can see that even when comparing mean values, EHO on average outperforms all other approaches included in the comparative analysis. If we compare the experimental results for best and mean values (Tables 1 and 2, respectively), we can see only slight degradation in results. That means that EHO is robust approach and that it obtains similar results in each algorithm's run.

Table 2. Simulation results for mean values obtained For MSE objective function

No. of nodes	Anchor no. (%)	Range	PSO	MSPSO	ABC	MSABC	EHO
1000	2.5	20	28.149	27.667	45.128	40.038	**27.582**
1000	5	20	27.273	27.073	43.386	38.038	**26.983**
1000	7.5	20	26.156	26.035	40.925	36.374	**26.031**
1000	10	20	25.273	25.035	40.079	34.193	**25.013**
1000	2.5	35	27.320	**27.167**	43.990	39.273	27.180
1000	5	35	26.089	25.936	41.759	35.238	**25.927**
1000	7.5	35	25.169	**25.005**	40.279	33.273	25.019
1000	10	35	24.763	**24.573**	38.462	30.283	24.635
1000	2.5	50	26.537	25.893	41.534	35.928	**25.872**
1000	5	50	25.534	**25.382**	40.178	33.229	25.443
1000	7.5	50	24.810	**24.673**	38.289	32.037	24.681
1000	10	50	24.485	24.109	36.972	30.028	**24.106**

It should be noted that results obtained in problem instances where EHO outperforms all other approaches (values in Tables 1 and 2 marked bold) are the lower bound of achievable MSE performance (Eq. (2)).

Similarly, as in [19], we have observed that the localization error decreases significantly with the increase in transmission range. However, with the increase in percentage of anchor nodes, the localization error does not decrease significantly, and this is the case with all algorithms included in comparative analysis.

6   Conclusion

In this paper, EHO algorithm was adapted for solving the localization problem in WSN. This problem is one of the most fundamental challenges in this domain and its goal is to find the coordinates of unknown nodes randomly deployed in the monitoring field.

To test the robustness of EHO algorithm when dealing with this type of problem, we used similar experimental condition as in [19]. According to the simulation results and comparative analysis with other approaches, EHO proved to be robust and efficient metaheuristics when tackling localization problem in WSN. As part of the future research, EHO algorithm could also be applied to other problems from this domain, such as coverage and energy efficiency in WSNs.

Acknowledgements. This research is supported by the Ministry of Education, Science and Technological Development of Republic of Serbia, Grant No. III-44006. The work of M. Beko was supported in part by Fundação para a Ciência e a Tecnologia under Projects PEst-OE/EEI/ UI0066/2014 (UNINOVA) and Program Investigador FCT (IF/00325/2015).

References

1. Akyildiz, I.F., Vuran, M.: Wireless sensor networks. No. 978-0-470-03601-3, Wiley (2010)
2. Rawat, P., Singh, K.D., Chaouchi, H., Bonnin, J.M.: Wireless sensor networks: a survey on recent developments and potential synergies. J. Supercomput. **68**, 1–48 (2014)
3. Goyal, S., Patterh, M.S.: Wireless sensor network localization based on cuckoo search algorithm. Wireless Pers. Commun. **79**, 223–234 (2014)
4. Bonabeau, E., Dorigo, M., Theraulaz, G.: Swarm Intelligence: From Natural to Artificial Systems. Oxford University Press, Oxford (1999)
5. Wang, G.-G., Deb, S., dos Santos Coelho, L.: Elephant herding optimization. In: Proceedings of the 2015 3rd International Symposium on Computational and Business Intelligence (ISCBI), pp. 1–5, December 2015
6. Wang, G.-G., Deb, S., Gao, X.-Z., dos Santos Coelho, L.: A new metaheuristic optimization algorithm motivated by elephant herding behaviour. Int. J. Bio-Inspired Comput. **8**, 394–409 (2017)
7. Tuba, V., Beko, M., Tuba, M.: Performance of elephant herding optimization algorithm on CEC 2013 real parameter single objective optimization. WSEAS Trans. Syst. **16**, 100–105 (2017)
8. Tuba, E., Stanimirovic, Z.: Elephant herding optimization algorithm for support vector machine parameters tuning. In: Proceedings of the 2017 International Conference on Electronics, Computers and Artificial Intelligence (ECAI), pp. 1–5, June 2017
9. Tuba, E., Alihodzic, A., Tuba, M.: Multilevel image thresholding using elephant herding optimization algorithm. In: Proceedings of 14th International Conference on the Engineering of Modern Electric Systems (EMES), pp. 240–243, June 2017

10. Gupta, S., Singh, V.P., Singh, S.P., Prakash, T., Rathore, N.S.: Elephant herding optimization based PID controller tuning. Int. J. Adv. Technol. Eng. Explor. **3**, 194–198 (2016)
11. Strumberger, I., Bacanin, N., Beko, M., Tomic, S., Tuba, M.: Static drone placement by elephant herding optimization algorithm. In: Proceedings of the 24th Telecommunications Forum (TELFOR), November 2017
12. Sun, Z., Tao, L., Wang, X., Zhou, Z.: Localization algorithm in wireless sensor networks based on multiobjective particle swarm optimization. Int. J. Distrib. Sens. Netw. **2015**, 1–9 (2015)
13. Harikrishnan, R., Kumar, J.S., Ponmalar, P.S.: A comparative analysis of intelligent algorithms for localization in wireless sensor networks. Wireless Pers. Commun. **87**, 1057–1069 (2016)
14. Goyal, S., Patterh, M.S.: Modified bat algorithm for localization of wireless sensor network. Wireless Pers. Commun. **86**, 657–670 (2015)
15. Tuba, E., Tuba, M., Beko, M.: Node localization in ad hoc wireless sensor networks using fireworks algorithm. In: Proceedings of the 5th International Conference on Multimedia Computing and Systems (ICMCS), pp. 223–229, September 2016
16. Comfort, L.K., Boin, A., Demchak, C.C.: Designing Resilience: Preparing for Extreme Events. University of Pittsburgh Press, Pittsburgh (2010)
17. Barsocchi, P., Chessa, S., Furfari, F., Potorti, F.: Evaluating aal solutions through competitive benchmarking: the localization competition. IEEE Pervasive Comput. Mag. **12**, 72–79 (2013)
18. Tomic, S., Beko, M., Dinis, R., Montezuma, P.: Distributed algorithm for target localization in wireless sensor networks using RSS and AoA measurements. Pervasive Mobile Comput. **37**, 63–77 (2017)
19. Lavanya, D., Udgata, Siba K.: Swarm intelligence based localization in wireless sensor networks. In: Sombattheera, C., Agarwal, A., Udgata, S.K., Lavangnananda, K. (eds.) MIWAI 2011. LNCS (LNAI), vol. 7080, pp. 317–328. Springer, Heidelberg (2011). https://doi.org/10.1007/978-3-642-25725-4_28
20. Liu, Y., Yang, Z., Wang, X., Jian, L.: Location, localization, and localizability. J. Comput. Sci. Technol. **25**(2), 274–297 (2010)
21. Sukumar, R.: The Asian Elephant: Ecology and Management. Cambridge University Press, Cambridge Studies in Applied Ecology and Resource Management (1993)

Hybrid RSS/AoA-Based Target Localization and Tracking in Wireless Sensor Networks

Slavisa Tomic[1,5(✉)], Marko Beko[1,2], and Rui Dinis[3,4]

[1] COPELABS, ULHT, Lisbon, Portugal
{slavisa.tomic,beko.marko}@ulusofona.pt
[2] CTS - UNINOVA – Campus FCT/UNL, Caparica, Portugal
[3] IT, Lisbon, Portugal
rdinis@fct.unl.pt
[4] DEE/FCT/UNL, Caparica, Portugal
[5] ISR/IST, LARSyS, UL, Lisbon, Portugal

Abstract. This paper presents a brief overview of recently developed algorithms for target localization and tracking using combined angle of arrival (AoA) and received signal strength (RSS) observations. Due to the non-linearity and non-convexity of the problem at hand, our main focus here is on the estimators that solve the localization problem by transforming it into more suitable frameworks. In other words, these estimators share the main objective in which the original problem is approximated by another (possibly convex) one whose solution can be readily found. Different convex relaxation techniques are described, and the paper presents recent advancement of the techniques in the sense of both computational complexity and localization accuracy.

Keywords: Target localization and tracking · Received signal strength (RSS)
Angle of arrival (AoA) · Wireless sensor network (WSN)

1 Introduction

The aspiration for accurate awareness of the position of a mobile object at any given time has recently awakened a large amount of scientific research. This is due to a firm growth of the range of technologies and devices, together with the need for continuous solutions for position-based services. Apart from localization precision, another prerequisite for evolving solutions is that they are cost-restrained, regarding computational and financial budget. Therefore, deriving different positioning approaches from previously installed technologies, *e.g.*, from various terrestrial radio frequency fonts is of substantial interest in real-world. These approaches are practically unlimited and include concepts based on received signal strength (RSS), angle of arrival (AoA), time of arrival (ToA), or perhaps a mixture of them [1–18].

This work presents latest developments in emerging procedures for localization and tracking, of a target, describing the state-of-the-art methods and recent accomplishments in localization and tracking of a target based on amalgam, RSS/AoA, observations.

© IFIP International Federation for Information Processing 2018
Published by Springer International Publishing AG 2018. All Rights Reserved
L. M. Camarinha-Matos et al. (Eds.): DoCEIS 2018, IFIP AICT 521, pp. 185–201, 2018.
https://doi.org/10.1007/978-3-319-78574-5_18

1.1   Research Question and General Approach

In this work, we describe target localization and tracking problems by using combined RSS/AoA measurements in wireless sensor networks (WSNs). The considered problem is very difficult to solve directly, because it is highly non-linear and non-convex, and no closed-form solution is available. Hence, solving the originally posed localization problem (*e.g.*, by a grid search method) could be computationally burdensome, and iterative methods (*e.g.*, gradient descent algorithm) could get trapped into local optima resulting in large errors.

Furthermore, in practice, WSNs are prone to alterations in topology (*e.g.*, sensor and/or edge failure) and in indoor and highly dense urban environments, not all (if any) links are line-of-sight. To further aggravate the problem, energy resources are often limited by sensor's battery and the quality of the measurements by hardware imperfections and noise.

Because of all these challenges, the development of even the simplest algorithms is not an easy task. Thus, in order to tackle the problem in its most general form, the main research question of this work is:

How to develop an effective (highly precise and computationally abstemious) localization algorithm, robust to network topology and channel variations, flexible to different settings of the localization problem, appropriate for real-time applications?

The answer to the above research question is not straightforward or obvious, and our research started with the following hypothesis:

An effective localization algorithm can be designed by applying convex optimization tools together with statistical modeling in order to tightly approximate the original non-linear and non-convex localization problem into a convex one. Also, by thoughtful development of weighting strategies, the impact of potentially corrupt links can be minimized and the impact of potentially good ones boosted. In addition, by incorporating prior knowledge within an estimator, the performance of an estimator can be enhanced significantly.

1.2   Main Findings and Paper Organization

Our research started by studying low-cost localization systems, such as RSS ones (distributed solutions) [1] (centralized solutions) [2]. The main conclusions in these works are that the RSS solution is highly unpredictable and that even small variations in the channel can lead to unstable performance of an estimator. In order to minimize its drawbacks, our research led us to integrate RSS with AoA measurements [3–18]. It turned out that this measurement combination function very well, and that the drawbacks of each of the individual systems can be efficiently overcome, by their combination. In the case of a moving target, the performance of an estimator can be further enhanced by integrating prior knowledge within the estimator, so that it *learns* the target's movement along the trajectory.

The rest of the paper is organized as follows. Section 2 described the relationship of our research to resilient systems. In Sect. 3, a formal representation of the considered problem is introduced. Sections 4 and 5 present recently developed localization and tracking schemes, respectively. A set of numerical results is presented is Sect. 6, and finally, the main conclusions are summarized in Sect. 7.

2   Relationship to Resilient Systems

This section relates the main research topic of this work to resilient systems. We describe a possible application of a WSN in forming a resilient system and highlight the importance of target localization in such a scenario. It is noteworthy that one of the fundamental applications of WSNs is in preventing natural disasters by data collection in risk-prone communities in order to increase resilience in such communities [19].

WSNs are autonomous in terms of human interaction and owing to low device costs, they find application in various areas, like event detection (fires, earthquakes, tsunamis), monitoring (health care, industrial, agricultural, environmental), surveillance, exploration (underwater, underground, outer space), and energy-efficient routing [1–19] to name a few. A possible application in fire detection/prevention in forests is illustrated in Fig. 1. In this figure, some targets might be dropped out of an airplane and/or can be mobile, and they could be used to measure the temperature in their vicinity. As soon as any of them detects high temperature (fire danger) they can communicate their location, together with a valid warning message to its user (fire-fighters) through a sink.

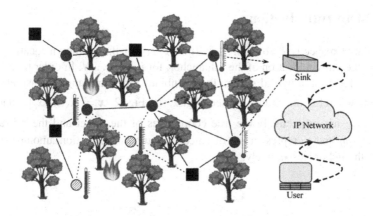

Fig. 1. Application of a WSN in fire prevention/detection.

Nevertheless, having the information about fire danger at hand gains much more relevance if it is closely associated with the location of the sensors which detected such data. Therefore, one of the most important tasks and at the same time one of the biggest challenges is to accurately determine the locations of all sensors. Although it might seem as a simple task, there are plenty of factors that make this task a very challenging one. For instance, one could not use GPS devices in dense forest environments, since these would require at least line-of-sight links between the sensors and at least three GPS satellites. Besides, GPS devices are relatively expensive, and in an application such as fire detection it is likely that a lot of sensors get destroyed, rendering such devices cost-inefficient. Consequently, derivation of diverse positioning schemes from previously installed technologies, *i.e.*, from various terrestrial radio frequency fonts is of substantial interest in practice. This (development of localization techniques that make use of already deployed terrestrial technologies) is exactly the main research topic of this work.

Most proposed schemes require the location of some sensors in the network to be known in advance. These sensors are termed anchors or reference nodes. The unresolved sensors are termed targets, and they estimate their location after sufficient information is gained through the exchange of messages with reference nodes.

The techniques that will be presented here offer a solution to the critical problem of location discovery, which is essential for designing sensor networks that offer promising means for early detection of forest fires. They represent an advance in technical systems that could support the timely monitoring of key indicators of fire risk and the transmission of information to emergency service organizations to enable them to take action and reduce risk for their communities. Essentially they contribute to distributed cognition in practical scenarios, *i.e.*, they are instruments that can contribute to achieving resilient systems in preventing natural disasters.

3   Problem Introduction

Let us denote respectively by $x, a_i \in \mathbb{R}^q (q = 1, 2 \, or \, 3)$ the unknown location of the target and the known location of the i-th anchor, for $i = 1, \ldots, N$. In order to determine the target's location, a hybrid system that combines range and angle measurements is employed, we refer the reader to see Fig. 2. In Fig. 2, $x = \left[x_x, x_y, x_z \right]^T$ and $a_i = \left[a_{ix}, a_{iy}, a_{iz} \right]^T$ are respectively the true coordinates of the target and the i-th anchor, whereas d_i, ϕ_i and α_i represent the distance, azimuth and elevation among the target and the i-th anchor, respectively.

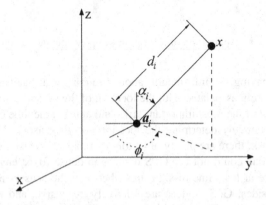

Fig. 2. Illustration of the considered scenario in 3-dimensions: the true locations of the target, x, and an anchor, a_i, and the units of interest (distance, d_i, azimuth, ϕ_i, and elevation, α_i) amongst them.

In this work, we assume that the distance is drawn from the RSS observations only, since ranging founded on RSS does not require any additional hardware [1, 2]. The noise-free RSS amongst the target and the i-th anchor can be modeled as [6]:

$$P_i = P_T \left(\frac{d_i}{d_0}\right)^{\gamma} 10^{-\frac{L_0}{10}}, \text{ for } i = 1, \ldots, N, \tag{1}$$

where $d_i = \|x - a_i\|$ is the distance amongst the i-th anchor and the target, P_T is the target's transmit power, γ is the rate at which the signal power deteriorates with distance (known as the path loss exponent), L_0 is the reference path loss measured at a short distance d_0 (usually $d_0 = 1$ m and $d_0 \le d_i$). Therefore, according to (1), a log-distance RSS model can be derived as:

$$P_i = P_0 - 10\gamma \log_{10} \frac{d_i}{d_0} + n_i, \tag{2}$$

where $P_0(\text{dBm}) = P_T(\text{dBm}) - L_0(\text{dB})$ is the RSS received at d_0, and $n_i \sim \mathcal{N}\left(0, \sigma_{n_i}^2\right)$ represents the log-normal shadowing term, represented here as a zero-mean Gaussian random variable with standard deviation σ_{n_i} (dB).

The AoA observations can be obtained by implementing a directional antenna or antenna arrays, or even video cameras [5] at anchor nodes. Thus, applying simple geometry, azimuth angle and elevation angle observations can be respectively modeled as [20]:

$$\phi_i = \tan^{-1}\left(\frac{x_y - a_{iy}}{x_x - a_{ix}}\right) + m_i, \text{ for } i = 1, \ldots N, \tag{3}$$

and

$$\alpha_i = \cos^{-1}\left(\frac{x_z - a_{iz}}{\|x - a_i\|}\right) + v_i, \text{ for } i = 1, \ldots N, \tag{4}$$

where m_i (rad) and v_i (rad) are respectively the observation inaccuracies of azimuth angle and elevation angle, represented here as $m_i \sim \mathcal{N}\left(\sigma_{m_i}^2\right)$ and $v_i \sim \mathcal{N}\left(\sigma_{v_i}^2\right)$.

According to (2), (3) and (4), we can derive the maximum likelihood (ML) estimator [3] as

$$\hat{x} = \arg \min_x \sum_{i=1}^{3N} \frac{1}{\sigma_i^2} [\theta_i - f_i(x)]^2, \tag{5}$$

where $\theta = \left[P^T, \phi^T, \alpha^T\right]^T$ ($\theta \in \mathbb{R}^{3N}$), with $P = [P_i]^T$, $\phi = [\phi_i]^T$, $\alpha = [\alpha_i]^T$, and

$$f(X) = \begin{bmatrix} \vdots \\ P_0 - 10\gamma\log_{10}\frac{\|x-a_i\|}{d_0} \\ \vdots \\ \tan^{-1}\left(\frac{x_y - a_{iy}}{x_x - a_{ix}}\right) \\ \vdots \\ \cos^{-1}\left(\frac{x_z - a_{iz}}{\|x-a_i\|}\right) \\ \vdots \end{bmatrix}, \sigma = \begin{bmatrix} \vdots \\ \sigma_{n_i} \\ \vdots \\ \sigma_{m_i} \\ \vdots \\ \sigma_{v_i} \\ \vdots \end{bmatrix}.$$

The above problem is non-convex and does not have a closed-form solution. Solving (5) (*e.g.*, by grid search algorithm) could be computationally burdensome and applying iterative methods (*e.g.*, gradient descent algorithm) could produce big inaccuracies, owing to non-convexity of (5); see Fig. 3a. Hence, it will be shown in the following text that the ML problem can be closely approximated with an alternative estimator whose global optimum is readily obtained, as shown in Fig. 3b.

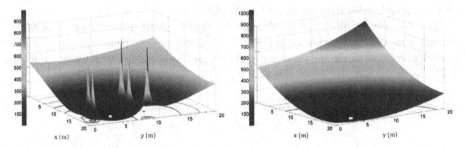

Fig. 3. Illustration of optimization-based principle; the true target location at $[17.35, 4.77]^T$.

Figure 3 illustrates the main idea when solving the localization problem used in the literature. It represents a possible realization of the objective function in (5), Fig. 3a, and a possible realization of its tight approximation that will be developed in the further text. All sensors were randomly placed within a region of 20×20 m^2 size. The true location of the target was set at $[17.35, 4.77]^T$, and $N = 5$ anchors were able to communicate with the target. The rest of the parameters were set as follows: $P_0 = -10$ dBm, $\sigma_{n_i} = 5$ dB, $\sigma_{m_i} = 8°$, $\sigma_{v_i} = 8°$, $\gamma = 3$, and the objective functions were contrived versus x (m) and y (m) match (the step size in the mesh grid was 0.1 m). Figure 3a exhibits that the cost function in (5), using the true locations of the target, has an overall minimum at $[17.5, 4.7]^T$, and some local minima and saddle points around it. Figure 3b shows that the approximated cost function has a global minimum at $[18, 4.2]^T$, and is much smoother than (5). One can also see that the two objective functions have analogous comportment: both monotonically grow and drop in the equal areas. Hence, from Fig. 3, it is obvious that the derived cost function is a tight approximation of the cost function in (5).

4   Hybrid RSS/AoA-Based Target Localization

This section presents various localization estimators that were developed recently.

4.1   SDP Estimator

In [17], the authors showed how to transform the ML estimator in (5) into another, convex, one by applying semidefinite programming (SDP) relaxations. First, from (2), (3) and (4) respectively it is written:

$$\lambda_i \|x - a_i\| \approx \eta d_0, \forall i, \tag{6}$$

$$c_i^T (x - a_i) \approx 0, \forall i, \tag{7}$$

$$k^T (x - a_i) \approx \|x - a_i\| cos(\alpha_i), \forall i, \tag{8}$$

where $\lambda_{ij} = 10^{\frac{P_i}{10\gamma}}$, $\eta = 10^{\frac{P_0}{10\gamma}}$, $c_i = [-\sin(\phi_i), \cos(\phi_i), 0]^T$, $k = [0, 0, 1]^T$.

Rendering the least squares (LS) criterion, from (6), (7) and (8), an estimate of the target's location is obtained by solving the following problem:

$$\hat{x} = \arg\min_x \sum_{i=1}^N (\lambda_i \|x - a_i\| - \eta d_0)^2 + \sum_{i=1}^N (c_i^T (x - a_i))^2$$
$$+ \sum_{i=1}^N (k^T (x - a_i) - \|x - a_i\| cos(\alpha_i))^2. \tag{9}$$

The estimator in (9) is clearly non-convex and does not have a closed-form solution. To transform it into a convex estimator, we first define auxiliary variables $r_i = \|x - a_i\|$, $z = [z_i]$ ($z \in R^{3N}$), where $z_i = \lambda_i \|x - a_i\| - \eta d_0$, for $i = 1, \ldots, N$, $z_i = c_i^T (x - a_i)$, for $i = N+1, \ldots, 2N$, and $z_i = k^T (x - a_i) - \|x - a_i\| \cos(\alpha_i)$, for $i = 2N+1, \ldots, 3N$.

Then, by introducing an epigraph variable, e, and applying semidefinite cone constraint relaxation, one gets:

$$\underset{x, r, z, e}{\text{minimize}} \, e$$

subject to

$$\|x - a_i\| \leq r_i, \text{for } i = 1, \ldots, N,$$
$$z_i = \lambda_i \|x - a_i\| - \eta d_0, \text{for } i = 1, \ldots, N,$$
$$z_i = c_i^T (x - a_i), \text{for } i = N+1, \ldots, 2N, \tag{10}$$
$$z_i = k^T (x - a_i) - \|x - a_i\| \cos(\alpha_i), \text{for } i = 2N+1, \ldots, 3N,$$
$$\left\| \begin{bmatrix} I_{3N} & z \\ z^T & e \end{bmatrix} \right\| \succeq 0_{3N+1}.$$

The problem in (10) is an SDP estimator, and can be efficiently resolved by the CVX package for specifying and solving convex programs. Notice that the constraint was rewritten $\|z^2\| \leq e$ into a semidefinite cone constraint form by applying the Schur complement [17]. In the further text, the estimator in (10) is denoted as "SDP".

4.2   SOCP Estimator

Note that writing the constraint $\|z\|^2 \leq e$ into a semidefinite cone constraint increases the computational complexity of an algorithm, i.e., its execution time. This constraint, as well as the LS estimator in (9) can be rewritten as a second-order cone programming (SOCP), which significantly reduces the computational complexity [16]. To this end, first, introduce auxiliary variables $r_i = \|\mathbf{x} - \mathbf{a}_i\|$, $z = [z_i]$, with $z_i = \lambda_i \|\mathbf{x} - \mathbf{a}_i\| - \eta d_0$, $\mathbf{g} = [g_i]$ with $g_i = \mathbf{c}_i^T(\mathbf{x} - \mathbf{a}_i)$, and $\mathbf{h} = [h_i]$ with $h_i = \mathbf{k}^T(\mathbf{x} - \mathbf{a}_i) - \|\mathbf{x} - \mathbf{a}_i\| \cos(\alpha_i)$. Furthermore, introduce epigraph variables e_1, e_2 and e_3. This yields:

$$\underset{x,r,z,e}{\text{minimize}}\ e$$

subject to

$$\|\mathbf{x} - \mathbf{a}_i\| \leq r_i, \text{for } i = 1, \ldots, N,$$
$$z_i = \lambda_i \|\mathbf{x} - \mathbf{a}_i\| - \eta d_0, \text{for } i = 1, \ldots, N,$$
$$g_i = \mathbf{c}_i^T(\mathbf{x} - \mathbf{a}_i), \text{for } i = 1, \ldots, N, \tag{11}$$
$$h_i = \mathbf{k}^T(\mathbf{x} - \mathbf{a}_i) - \|\mathbf{x} - \mathbf{a}_i\| \cos(\alpha_i), \text{for } i = 1, \ldots, N,$$
$$\left\|\begin{bmatrix} 2z \\ e_1 - 1 \end{bmatrix}\right\| \leq e_1 + 1, \left\|\begin{bmatrix} 2g \\ e_2 - 1 \end{bmatrix}\right\| \leq e_2 + 1, \left\|\begin{bmatrix} 2h \\ e_3 - 1 \end{bmatrix}\right\| \leq e_3 + 1.$$

The estimator in (11) is an SOCP estimator, and is readily resolved by CVX package. The main difference between this estimator and the one given in (10) is that SOCP relaxation technique was applied in (11), whereas SDP relaxation technique was used in (10) to *convexify* the derived non-convex estimators. In the further text, (11) is referred to as "SOCP".

4.3   SR-WLS Estimator

Although the computational complexity is significantly decreased by the derived SOCP estimator in (11), it is still relatively high, and can be further reduced. Hence, a linear estimator for solving the ML problem was proposed in [3], solved by means of a bisection procedure.

By squaring (6), it can be written:

$$\lambda_i^2 \|\mathbf{x} - \mathbf{a}_i\|^2 \approx \eta^2 d_0^2, \forall i. \tag{12}$$

In order to give more importance to *nearby* links, introduce weights, $w = [\sqrt{w_i}]$, where each w_i is defined as $w_i = 1 - \dfrac{\widehat{d_i}}{\sum_{i=1}^{N} \widehat{d_i}}$, with $\widehat{d_i} = d_0 10^{\frac{P_0 - P_i}{10\gamma}}$ denoting the ML assessment of the distance acquired from (2).

Then, replacing $\|x - a_i\|$ with $\widehat{d_i}$ in (8) and using weights, a weighted LS (WLS) problem can be derived rendering to (12), (7) and modified (8) yields:

$$\widehat{x} = \arg\min_{x} \sum_{i=1}^{N} w_i \left(\lambda_i^2 \|x - a_i\|^2 - \eta^2 d_0^2 \right)^2 + \sum_{i=1}^{N} w_i \left(c_i^T (x - a_i) \right)^2$$
$$+ \sum_{i=1}^{N} w_i \left(k^T (x - a_i) - \widehat{d_i} \cos(\alpha_i) \right)^2. \tag{13}$$

Like (9), the LS estimator in (13) is non-convex. However, (13) can be expressed as a quadratic programming estimator whose exact resolution can be obtained readily [3]. Applying the replacement $y = \left[x^T, \|x\|^2 \right]^T$, the estimator in (16) can be posed as:

$$\underset{y}{\text{minimize}} \| W(Ay - b) \|^2 : y^T D y + 2 p^T y = 0, \tag{14}$$

where $W = I_3 \otimes \text{diag}(w)$, with $\otimes$ denoting the Kronecker product, and

$$A = \begin{bmatrix} \vdots & \vdots \\ -2\lambda_i^2 a_i^T & \lambda_i^2 \\ \vdots & \vdots \\ c_i^T & 0 \\ \vdots & \vdots \\ k^T & 0 \\ \vdots & \vdots \end{bmatrix}, b = \begin{bmatrix} \vdots \\ \eta^2 d_0^2 \\ \vdots \\ c_i^T a_i \\ \vdots \\ k^T a_i + \widehat{d_i} \cos \alpha_i \\ \vdots \end{bmatrix}, D = \begin{bmatrix} I_q & 0_{q \times 1} \\ 0_{1 \times q} & 0 \end{bmatrix}, p = \begin{bmatrix} 0_{q \times 1} \\ -\frac{1}{2} \end{bmatrix},$$

and I_U and $0_{U \times U}$ respectively denote the identity and the zero matrix of size U.

Both the objective and the constraint in (14) are quadratic. This type of problems are identified as generalized trust region sub-problems (GTRS) [3, 9]. Even though still non-convex in general, an interval in which the objective function is monotonically decreasing can be easily computed and thus, it can be solved exactly by a bisection procedure [3]. The estimator in (14) is denoted by "SR-WLS" in the remaining text.

4.4 WLS Estimator

The estimator given by (14) has linear computational complexity in the number of anchors, but it is executed iteratively, which raises its execution time. To further reduce the computational complexity, the authors in [6] presented a different WLS estimator that does not require iterative execution, and has closed-form solution.

By switching from Cartesian to spherical coordinates, vector $x - a_i$ can be expressed as $x - a_i = r_i u_i$, for $i = 1, .., N$. The unit vector, u_i, can be obtained by employing the available AoA information as $u_i = [\cos(\phi_i)\sin(\alpha_i), \sin(\phi_i)\sin(\alpha_i), \cos(\alpha_i)]^T$. Apply the described conversion in (6) and (8), and multiply with $u_i^T u_i$, to get respectively:

$$\lambda_i u_i^T r_i u_i \approx \eta d_0 \Leftrightarrow \lambda_i u_i^T(x - a_i) \approx \eta d_0 \tag{15}$$

$$k^T r_i u_i \approx u_i^T r_i u_i \cos(\alpha_i) \Leftrightarrow (\cos(\alpha_i)u_i - k)^T(x - a_i) \approx 0. \tag{16}$$

According to the WLS criterion and by taking advantage of (7), (15), (16), the following estimator is obtained:

$$\hat{x} = \arg\min_x \sum_{i=1}^{N} w_i\left(\lambda_i u_i^T(x - a_i) - \eta d_0\right)^2 + \sum_{i=1}^{N} w_i\left(c_i^T(x - a_i)\right)^2$$

$$+ \sum_{i=1}^{N} w_i\left((\cos(\alpha_i)u_i - k)^T(x - a_i)\right)^2,$$

which, rewritten in an equal matrix formulation is represented as

$$\underset{x}{\text{minimize}} \left\| W(\tilde{A}x - \tilde{b}) \right\|^2, \tag{17}$$

with

$$\tilde{A} = \begin{bmatrix} \vdots \\ \lambda_i u_i^T \\ \vdots \\ c_i^T \\ \vdots \\ (\cos(\alpha_i)u_i - k)^T \\ \vdots \end{bmatrix} \quad \text{and} \quad \tilde{b} = \begin{bmatrix} \vdots \\ \lambda_i u_i^T a_i - \eta d_0 \\ \vdots \\ c_i^T a_i^T \\ \vdots \\ (\cos(\alpha_i)u_i - k)^T a_i \\ \vdots \end{bmatrix}.$$

The closed-form solution to (17) is readily given by $\hat{x} = (\tilde{A}W^T W\tilde{A})^{-1}(\tilde{A}^T W^T \tilde{b})$. We will refer to (17) as "WLS" in the remaining text.

5 Hybrid RSS/AoA-Based Target Tracking

Consider now that the target is moving within the WSN and that we want to estimate its location at every time instant, t. For simplicity and without loss of generality, let us consider the 2-dimensional space, and denote by x_t the target's true location at time moment t. We assume a nearly constant velocity motion model (e.g., disturbed merely by wind gust) and denote by $\theta_t = \left[x_t^T, v_t^T \right]^T$ the true target state at instant t (described by its location and velocity, v_t, in x- and y-directions). The target movement can then be represented as [7, 10]

$$\theta_t = S\theta_{t-1} + r_t, \tag{18}$$

where S is the state transition matrix and $r_t \sim \mathcal{N}(0, \mathbf{Q})$ is the state process noise, i.e.,

$$S = \begin{bmatrix} 1 & 0 & \varDelta & 0 \\ 0 & 1 & 0 & \varDelta \\ 0 & 0 & 1 & 0 \\ 0 & 0 & 0 & 1 \end{bmatrix}, \mathbf{Q} = \xi \begin{bmatrix} \varDelta^3/3 & 0 & \varDelta^2/2 & 0 \\ 0 & \varDelta^3/3 & 0 & \varDelta^2/2 \\ \varDelta^2/2 & 0 & \varDelta & 0 \\ 0 & \varDelta^2/2 & 0 & \varDelta \end{bmatrix}$$

with ξ and $\varDelta$ denoting the state process noise intensity and the sampling interval between two consecutive time steps.

A state estimate, $\widehat{\theta}_{t|t}$, of θ_t can be obtained from $p(\theta_t | z_{1:t})$ by employing the maximum a posteriori (MAP) criteria [7], i.e., by maximizing the marginal posterior probability distribution function (PDF)

$$\widehat{\theta}_{t|t} = \arg\max_{\theta_t} p(\theta_t | z_{1:t}) \approx \arg\max_{\theta_t} p(z_t | \theta_t) p(\theta_t | z_{1:t-1}). \tag{19}$$

The problem in (19) resembles the ML estimator, apart from the existence of the prior PDF. This problem is greatly non-convex and its analytical resolution cannot be attained in general. Therefore, it needs to be approximated (19) by another estimator whose solution is readily available.

In Sect. 4.4, it was already revealed how to tightly approximate the ML estimator by a linear one, whose closed-form resolution is obtained efficiently. Thus, one only needs to deal with the prior knowledge part in (19). It can be assumed that $p(\theta_{t-1} | z_{1:t-1}) \sim \left(\widehat{\theta}_{t-1|t-1}, \widehat{\Sigma}_{t-1|t-1} \right)$ [7]. Then, according to the prediction step, typical for Bayesian methods, one obtains

$$p(\theta_t | z_{1:t-1}) \approx \frac{1}{k} \exp\left\{ -\frac{1}{2} \left(\theta_t - \widehat{\theta}_{t|t-1} \right)^T \widehat{\Sigma}_{t|t-1}^{-1} \left(\theta_t - \widehat{\theta}_{t|t-1} \right) \right\},$$

where k is a constant, and $\widehat{\theta}_{t|t-1}$ and $\widehat{\Sigma}_{t|t-1}$ represent respectively the mean and covariance of the one-step predicted state, acquired via (18) as

$$\widehat{\theta}_{t|t-1} = S\widehat{\theta}_{t-1|t-1},$$
$$\widehat{\Sigma}_{t|t-1} = S\widehat{\Sigma}_{t-1|t-1}S^T + \mathbf{Q}. \tag{20}$$

Therefore, (19) can be written as

$$\widehat{\boldsymbol{\theta}}_{t|t} = \arg\min_{\boldsymbol{\theta}_t} (z_t - \boldsymbol{h}(x_t))^T \boldsymbol{C}^{-1} (z_t - \boldsymbol{h}(x_t))$$

$$+ \left(\boldsymbol{\theta}_t - \widehat{\boldsymbol{\theta}}_{t|t-1}\right)^T \widehat{\boldsymbol{\Sigma}}_{t|t-1}^{-1} \left(\boldsymbol{\theta}_t - \widehat{\boldsymbol{\theta}}_{t|t-1}\right). \tag{21}$$

Similarly as in (8), the problem in (12) can be rewritten as

$$\widehat{\boldsymbol{\theta}}_{t|t} = \arg\min_{\boldsymbol{\theta}_t} \| \boldsymbol{W}(\boldsymbol{H}\boldsymbol{\theta}_t - \boldsymbol{\varphi}) \|^2, \tag{22}$$

where $\boldsymbol{H} = \left[\tilde{\boldsymbol{A}}; \widehat{\boldsymbol{\Sigma}}_{t|t-1}^{-\frac{1}{2}}\right]$, $\boldsymbol{\varphi} = \left[\tilde{\boldsymbol{b}}; \widehat{\boldsymbol{\Sigma}}_{t|t-1}^{-\frac{1}{2}} \widehat{\boldsymbol{\theta}}_{t|t-1}\right]$. The solution to (22) is given by $\widehat{\boldsymbol{\theta}}_{t|t} = \left(\boldsymbol{H}^T\boldsymbol{W}^T\boldsymbol{W}\boldsymbol{H}\right)^{-1}\left(\boldsymbol{H}^T\boldsymbol{W}^T\boldsymbol{\varphi}\right)$, and we denote (22) as "MAP" in the further text.

6 Performance Assessment

In this section, a set of numerical results is presented to assess the performance of the presented approaches as well as some state-of-the-art ones. All of the considered approaches were solved via MATLAB. The main performance metric used to assess the performance of the algorithms is the root mean square error (RMSE), defined as $\text{RMSE} = \sqrt{\sum_{i=1}^{M_c} \frac{\|x_i - \widehat{x}_i\|^2}{M_c}}$, where $\widehat{x}_i$ represents the estimate of the true target location, x_i, in the i-th Monte Carlo, M_c, run at any time instant.

In order to give a better overview of the presented approaches and compare their performance in terms of computational complexity, we present the computational complexity results in Table 1 below. In the table, we used K to denote the maximum number of iterations in the bisection procedure.

Table 1. Overview of the considered methods

Method	Overview	Complexity
SDP	The SDP localization algorithm in (10)	$\mathcal{O}(N^{4.5})$
SOCP	The SOCP localization algorithm in (11)	$\mathcal{O}(N^{3.5})$
SR-WLS	The bisection-based localization algorithm in (14)	$\mathcal{O}(KN)$
WLS	The WLS localization algorithm in (17)	$\mathcal{O}(N)$
MAP	The MAP tracking algorithm in (22)	$\mathcal{O}(N)$

We divide this section into two parts, in which we present the numerical results for target localization and target tracking, respectively.

6.1 Target Localization

In this section, the performance of the presented estimators is compared with the state-of-the-art approaches for hybrid RSS-AoA localization, namely with the least squares (LS) in [20] and the weighted linear least squares (WLLS) in [21]. It is worth mentioning that both LS and WLLS assume perfect knowledge of the target's transmit power, while the algorithms presented in Sects. 4.1, 4.2, 4.3 and 4.4 treat this parameter as unknown and estimate it alongside the target location. Furthermore, the WLLS assumes perfect knowledge of the noise power to enhance its performance. Nevertheless, this assumption might not stand in practice; hence, we also include the results for WLLS when a systematic error in estimating the noise power is present. Finally, the Cramer-Rao lower bound (CRLB) is also included in the figures as a theoretical lower bound.

Unless stated otherwise in the figures below, the values of the parameters are fixed as $N = 4$, $\sigma_{n_i} = 6$ dB, $\sigma_{m_i} = 10°$, $\sigma_{v_i} = 10°$, $\gamma_i \sim \mathcal{U}[2.2, 2.8]$, $\gamma = 2.5$, $B = 15$ m, $P_0 = -10$ dBm, $d_0 = 1$ m, $M_c = 50000$.

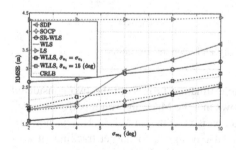

Fig. 4. RMSE versus N assessment.

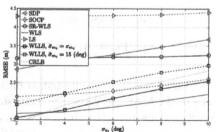

Fig. 5. RMSE versus σ_{n_i} (dB) assessment.

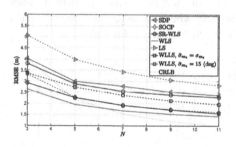

Fig. 6. RMSE vs σ_{m_i} (degrees) assessment.

Fig. 7. RMSE vs σ_{v_i} (degrees) assessment

Figures 4, 5, 6 and 7 illustrate RMSE versus N, σ_{n_i} (dB), σ_{m_i} (degrees) and σ_{v_i} (degrees), performance comparison, respectively. In these figures, one can observe natural behavior of all algorithms, depending on the scenario. For example, all algorithms improve their performance as N is increased, and all of them suffer a deterioration as the noise power is increased, as expected. Also, one can clearly see the progress in terms of the estimation accuracy between the presented algorithms. Not only that the WLS is the least computationally complex one, it is also the most accurate one in general. It can also be seen that almost all algorithms perform better than the CRLB, which means that they are biased. Nevertheless, this conclusion is not unusual for RSS-based estimators and is in concordance with the results in [2].

6.2 Target Tracking

In this section, the performance of the presented estimator in Sect. 4 is compared with the state-of-the-art approaches for hybrid RSS-AoA tracking, namely with the Kalman filter (KF) in [22], the extended Kalman filter (EKF), the unscented Kalman filter (UKF) and the particle filter (PF) in [10]. Moreover, the performance of the WLS estimator presented in Sect. 4.4 is also included in order to show the benefit of employing the Bayesian approach in contrast to the *classical* one (which disregards the prior knowledge and only exploits the measurements).

We consider a couple of essentially diverse target trajectories shown in Fig. 8: one in which the target moves in a *constant* direction and takes sharp maneuvers after certain period of time (Fig. 8a) and a different one, where the target continuously alters its path, but with not so extreme turns as in the first scenario (Fig. 8b).

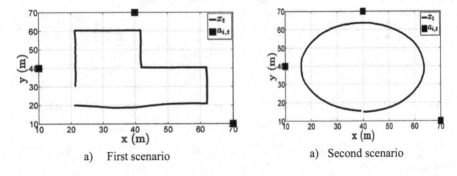

a)   First scenario   a)   Second scenario

Fig. 8. True target's trajectory and anchors' locations.

In Fig. 9, we show an illustration of the estimation process in the two considered scenarios for MAP and WLS. The figure shows that the MAP algorithm performs much smoother along the whole trajectory in comparison with the WLS. This is not surprising, since the MAP algorithm is specifically designed for target tracking (it uses prior knowledge to improve its performance), while the WLS is a localization algorithm which only utilizes observations and disregards the prior knowledge. Thus, we refer to the WLS as a *naive* tracking approach. The figures also show that both algorithms suffer a small deterioration in the proximity of any of the anchors. This fact can be explained by the fact that when the target gets close to an anchor, the weights introduced in both algorithms give a lot of importance to that specific measurement, making all others negligible. Finally, the Fig. 9 (left) exhibits loss in performance of the MAP algorithm whenever the target makes a sharp maneuver. Nevertheless, this is

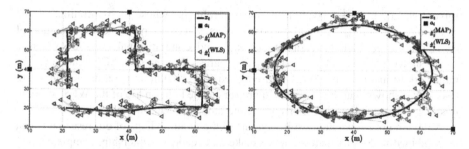

Fig. 9. Illustration of the estimation process in the first (figure on the left) and the second (figure on the right) considered scenario.

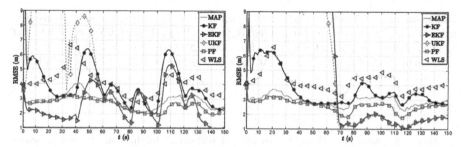

Fig. 10. RMSE versus t (s) comparison in the first (figure on the left) and the second (figure on the right) considered scenarios for $N = 3, \sigma_{n_i} = 9\,\mathrm{dB}, \sigma_{m_i} = 4°, \sigma_{v_i} = 4°, P_0 = -10\,\mathrm{dBm},$ $\gamma_i \sim \mathcal{U}[2.7, 3.3], \gamma = 3, \xi = 2.5 \times 10^{-3} \frac{\mathrm{m}^2}{\mathrm{s}^3}, M_c = 1000.$

anticipated because whenever the target takes this abrupt turn, the prior knowledge accumulated until that moment is lost. It is worth noting that these deteriorations are relatively mild, and that the MAP algorithm recuperates quickly from them.

In Fig. 10, we present the RMSE versus t (s) performance comparison of all considered tracking algorithms. It is worth noting that all considered algorithms were initialized in the same manner, *i.e.*, by solving the WLS in $t = 0$. The figure shows a relatively stable performance of the MAP algorithm in both considered trajectories, in contrast to almost all other approaches. One can see that only the PF algorithm has more reliable performance than the MAP one. However, the PF algorithm is based on a large amount of particles, which considerably raise its computational complexity.

7 Conclusions

This work presented a brief overview of recently developed RSS/AoA-based localization and tracking algorithms. A common objective of the described algorithms was to estimate the unknown location of the target by solving a tight approximation of the original problem that represent an excellent framework even under inopportune network configuration and strong measurement noise. A strong emphasis was made on convex relaxations and derivation of convex problems, whose global minima can be readily obtained through general purpose solvers. The work presented an evolution of the algorithms in both localization accuracy and computational complexity sense, having started with an SDP estimator, and progressed through an SOCP and a GTRS framework to just a simple WLS whose closed-form resolution is efficiently obtained. Also, it was shown that the Bayesian approach can be built on top of the WLS solution, by simply incorporating the prior knowledge alongside with the observations.

Although not shown here explicitly, one of the main findings of our research is that the combined RSS/AoA measurements make an excellent partnership, complementing each other's drawbacks and emphasizing their individual strong points. As far as the future work is concerned, it will involve other possible measurement integration (*e.g.*, RSS/ToA) and target localization in adverse (indoor) environments.

Acknowledgments. This work was partially supported by Fundação para a Ciência e a Tecnologia under Programa Investigador FCT (Grant IF/00325/2015), Project UID/EEA/50008/2013, Project PEst-OE/EEI/UI0066/2014 and Project UID/EEA/50009/2013. The research of Slavisa Tomic was supported by Fundação para a Ciência e a Tecnologia under the grant SFRH/BD/91126/2012.

References

1. Tomic, S., Beko, M., Dinis, R.: Distributed RSS-based localization in wireless sensor networks based on second-order cone programming. Sensors **14**(10), 18410–18432 (2014)
2. Tomic, S., Beko, M., Dinis, R.: RSS-based localization in wireless sensor networks using convex relaxation: noncooperative and cooperative schemes. IEEE Trans. Veh. Technol. **64**(5), 2037–2050 (2015)
3. Tomic, S., Beko, M., Dinis, R.: 3-D target localization in wireless sensor network using RSS and AoA measurement. IEEE Trans. Veh. Technol. **66**(4), 3197–3210 (2017)
4. Tomic, S., Beko, M., Dinis, R.: Distributed RSS-AoA based localization with unknown transmit powers. IEEE Wirel. Commun. Lett. **5**(4), 392–395 (2016)
5. Tomic, S., Beko, M., Dinis, R., Montezuma, P.: Distributed algorithm for target localization in wireless sensor networks using RSS and AoA measurements. Pervasive Mob. Comput. **37**, 63–77 (2016)
6. Tomic, S., Beko, M., Dinis, R., Montezuma, P.: A closed-form solution for RSS/AoA target localization by spherical coordinates conversion. IEEE Wirel. Commun. Lett. **5**(6), 680–683 (2016)
7. Tomic, S., Beko, M., Dinis, R., Gomes, J.P.: Target tracking with sensor navigation using coupled RSS and AoA measurements. Sensors **17**(7), 1–26 (2017)
8. Tomic, S., Beko, M., Dinis, R., Montezuma, P.: A robust bisection-based estimator for TOA-based target localization in NLOS environments. IEEE Commun. Lett. **21**(11), 2488–2491 (2017)
9. Tomic, S., Beko, M.: A bisection-based approach for exact target localization in NLOS environments. Sign. Process. **143**, 328–335 (2018)
10. Tomic, S., Beko, M., Dinis, R., Tuba, M., Bacanin, N.: Bayesian methodology for target tracking using RSS and AoA measurements. Phys. Commun. **25**(Part 1), 158–166 (2017)
11. Tomic, S., Beko, M.: Exact robust solution to TW-ToA-based target localization problem with clock imperfections. IEEE Sign. Process. Lett. (2017, submitted)

12. Beko, M., Tomic, S., Dinis, R., Montezuma, P.: Método de geolocalização 3-D em redes de sensores sem fio não cooperativas. PT, no. 108735 (2015, pending). Published in the Industrial Property Bulletin no. 21/2017 (2017)
13. Beko, M., Tomic, S., Dinis, R., Montezuma, P.: Método para localização tridimensional de nós alvo numa rede de sensores sem fio baseado em medições de potência recebida e angulos de chegada do sinal recebido. PT, no. 108963 (2015, pending) (2017)
14. Beko, M., Tomic, S., Dinis, R., Montezuma, P.: Method for RSS/AoA target 3-D localization in wireless networks. USA, no. 15287880 (2016, pending) (2017)
15. Tomic, S., Beko, M., Dinis, R., Tuba, M., Bacanin, N.: RSS-AoA-Based Target Localization and Tracking in Wireless Sensor Networks, 1st edn. River Publishers, Gistrup (2017)
16. Tomic, S., Marikj, M., Beko, M., Dinis, R., Tuba, M: Hybrid RSS/AoA-based localization of target nodes in a 3-D wireless sensor network. In: Yurish, S.Y., Malayeri, A.D. (eds.) Sensors & Signals. Book Series, pp. 71–85. IFSA Publishing (2015). ISBN 978-84-608-2320-9
17. Tomic, S., Beko, M., Dinis, R., Tuba, M.: Target localization in cooperative wireless sensor networks using measurement fusion. In: Yurish, S.Y. (ed.) Advances in Sensors: Reviews, vol. 3 Sensors, Transducers, Signal Conditioning and Wireless Sensors Networks. Book Series, vol. 3, pp. 329–344. IFSA Publishing (2016). ISBN 978-84-608-7705-9
18. Tomic, S., Beko, M., Dinis, R., Tuba, M.: Distributed algorithm for multiple target localization in wireless sensor networks using combined measurements. In: Yurish, S.Y. (ed.) Advances in Sensors: Reviews, vol. 4 Sensors and Applications in Measuring and Automation Control Systems. Book Series, vol. 4, pp. 263–275. IFSA Publishing (2017). ISBN 978-84-617-7596-5
19. Ling, H., Znati, T., Comfort, L.K.: Designing resilient systems: integrating science, technology, and policy in international risk reduction. In: Comfort, L.K., Boin, A., Demchak, C.C. (eds.) Designing Resilience: Preparing for Extreme Events, pp. 244–271. University of Pittsburgh Press, Pittsburgh (2010). ISBN 978-0-8229-7370-6
20. Yu, K.: 3-D localization error analysis in wireless networks. IEEE Trans. Wirel. Commun. 6(10), 3473–3481 (2007)
21. Khan, M.W., Kemp, A.H., Salman, N., Mihaylova, L.S.: Localisation of sensor nodes with hybrid measurements in wireless sensor networks. Sensors 16(7), 1–16 (2016)
22. Khan, M.W., Salman, N., Ali, A., Khan, A.M., Kemp, A.H.: A comparative study of target tracking with Kalman filter, extended Kalman filter and particle filter using received signal strength measurements. In: IEEE ICET, Peshawar, Pakistan, pp. 1–6 (2015)

Electrical Systems

Experimental Set-up for an IoT Power Supply with an 130 nm SC DC-DC Converter

Ricardo Madeira[1,2(✉)], Nuno Correia[2], João P. Oliveira[1,2], and Nuno Paulino[1,2]

[1] Department of Electrical Engineering (DEE),
Faculty of Sciences and Technology (FCT NOVA), Caparica, Portugal
r.madeira@campus.fct.unl.pt, jpao@fct.unl.pt
[2] Centre for Technologies and Systems (CTS) – UNINOVA, Caparica, Portugal
{nuno.correia,nunop}@uninova.pt

Abstract. This paper presents an experimental set-up for performing preliminary tests for energy power systems, such as the ones found in the IoT systems. It is composed by an energy harvesting device that charges a supercapacitor and a Switched Capacitor (SC) DC-DC converter, implemented in 130 nm bulk CMOS technology, to convert the voltage variation of the supercapacitor into a stable power supply for the IoT system. The experimental results show that the SC DC-DC converter achieves a maximum energy efficiency of 75%, and the overall system achieves a maximum energy efficiency of 47%.

Keywords: IoT system · Supercapacitor · Energy harvesting
Power Management Unit · Switched Capacitor DC-DC Converters

1 Introduction

The deployment of a large number of IoT systems requires these systems to have a lifetime as high as possible to become a viable solution in terms of cost [1]. These systems must be autonomous and thus, usually, they make use of one or more energy harvesters sources to charge an energy storing device [2, 3]. Using supercapacitors, instead of the traditional lithium batteries, allows to extend the lifetime of the system because of the millions of cycles endured by these devices [4]. Another crucial part is to ensure that the Power Management Unit (PMU) is also reliable, in its durability and also in its capability of providing a constant power supply and protect the system from overvoltage and overcurrent stresses [5]. Switched Capacitor (SC) DC-DC converters have gathered interest in recent years to achieve monolithic integration of PMUs [6, 7]. Unlike inductive DC-DC converters, the SC DC-DC converters only use transistors and capacitors that are available in CMOS technology, requiring no extra external components. This results in a production cost reduction, by reducing the bill of materials and the System on Chip (SoC) area. Before deploying IoT systems in the field, it is important to perform a preliminary field test to validate and adjust the theoretical design [8], and thus achieve an optimal solution that minimizes the implementation costs. This paper presents an IoT Power supply set-up for preliminary tests before the deployment

© IFIP International Federation for Information Processing 2018
Published by Springer International Publishing AG 2018. All Rights Reserved
L. M. Camarinha-Matos et al. (Eds.): DoCEIS 2018, IFIP AICT 521, pp. 205–213, 2018.
https://doi.org/10.1007/978-3-319-78574-5_19

of the final system. It makes use of a Switched Capacitor (SC) DC-DC converter implemented in 130 nm bulk CMOS technology, to power the IoT system.

2 Relationship to Resilient Systems

A Power Management Unit (PMU) designed for IoT systems must be as reliable as possible to extend the system durability so that it becomes a viable solution in terms of cost – due to the large number of systems that are deployed. Furthermore, to protect both the IoT system and the energy storage device, the PMU must be a resilient system as it must adapt and react to keep its normal behavior to upcoming events, which most of the time are unpredictable. For example: the voltage level variation of the super-capacitor; the inrush current from the system when it turns ON and control the power down when it turns OFF; block the reverse current when the supercapacitor voltage is lower than the output voltage; have hysteresis in the comparisons to avoid multiple transitions; have under voltage lock-out (UVLO) mechanisms to turn the system OFF if the voltage on the supercapacitor drops below a threshold value; and so on. Hence the developed SC DC-DC converter, has real-time mechanisms that try to withstand some of these problems while keeping the normal operation.

3 IoT Power Supply Architecture

The simplified schematic of the IoT power supply architecture is shown in Fig. 1. It is composed by an energy harvester, in this case a Photovoltaic (PV) cell, that charges a supercapacitor. Because the IoT node requires a stable power supply, it is necessary to convert the exponential voltage variation from the supercapacitor into a stable output voltage using an SC DC-DC converter [9] implemented in 130 nm bulk CMOS technology. The converter has an input voltage range of 1.2 V to 2.3 V, and can delivery up to 1 mW of output power. Three different SC circuit topologies (1/2, 2/3, and 1/1) were merged into a single multi-ratio SC DC-DC converter, composed by 2 flying capacitors and 9 switches, to keep the voltage conversion efficiency high throughout the whole input voltage range. These switches are turned ON/OFF according to the clock signal and the topology (state) in which the converter is oper-ating. The clock signal is generated by an Asynchronous State Machine (ASM) with a maximum frequency of operation of 10 MHz. The active state is selected by a second ASM. The input voltage ranges of each state are: 2.3 V to 1.9 V for the 1/2 state, 1.9 V to 1.5 V for the 2/3, and 1.5 V to 1.2 V for the 1/1 state. These voltages are generated using a resistive ladder (with hysteresis) from an external reference voltage of 0.45 V. A double-tail dynamic comparator is used to compare these voltage limits with the supercapacitor voltage. The output regulation is done by sensing the output voltage and compare it with the external reference voltage using a comparator. This comparator enables (disables) the clock if the voltage is lower (higher) than 0.9 V - Single-bound hysteretic controller. The converter has also a start-up circuit that charges the output node directly from the supercapacitor (using an internal CMOS switch) when the voltage on the supercapacitor is below 1.2 V.

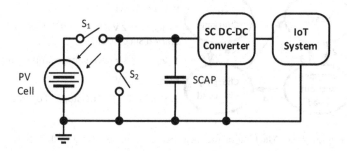

Fig. 1. IoT power supply node architecture.

4   Experimental Evaluation of the Deployment Conditions

The measurement of the system operation (voltages and currents) is carried out using a programmable system-on-chip (PSoC) from Cypress (CY8C5888LTI-LP097). The PSoC is used to both measure and control the harvester connection to the superca-pacitor and to the bank of output loads of the DC-DC converter. Figure 2 shows the state diagram implemented in the PSoC. It initiates on the *Start* state that goes directly to the *Reset* state. In this, the PV cell is disconnected from the supercapacitor, and the supercapacitor is discharged to ground. Only when the supercapacitor is completely discharged (0 V) the state machine changes to the *Charging without Load* state. In this state the supercapacitor is charged by the PV cell without any load at the output of the converter. When the voltage on the supercapacitor reaches 1.2 V, the converter should be operating. Thus, if the voltage at the output of the DC-DC is higher than 0.7 V, it means that the DC-DC is working and so the state machine changes to the *Charging with load state*; if the output voltage of the converter is lower than 0.7 V, it means that the DC-DC is not working properly and the state machine goes back to the *Reset* state, and repeats the process. When the output voltage reaches 0.9 V the converter is working properly, and a load can be connected. The state machine only leaves this state in two conditions: when the supercapacitor voltage reaches the maximum voltage allowed by the DC-DC converter (2.3 V), which disconnects the PV cell – *Stop Charging* state. It goes back to the previous state when the voltage of the superca-pacitor is below 2.3 V. The other possible transition is to go back to the *Reset* state when the voltage on the DC-DC is lower than 0.7, that is, when the DC-DC has stopped its operation. This can happen when the voltage of the supercapacitor is lower than 1.2 V, or if a sudden failure had occurred.

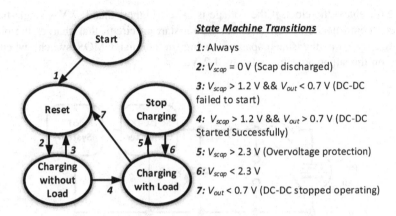

Fig. 2. State Machine of the measurements implemented in the PSoC.

While keeping the system running, through the state machine, the PSoC also collects the electrical signals from the system. To do so a Delta-Sigma ($\Sigma\Delta$) ADC is connected to a multiplexer that sense the following signals:

- I_{scap} – the current that charges the supercapacitor from the PV cell;
- V_{scap} – the voltage at the supercapacitor;
- $I_{in_{DC-DC}}$ – the input current of the DC-DC converter, provided by the supercapacitor/PV cell;
- $V_{in_{DC-DC}}$ – the voltage at the input of the converter;
- $I_{out_{DC-DC}}$ – the output current of the DC-DC converter;
- $V_{out_{DC-DC}}$ – the output voltage at the DC-DC converter;
- V_{RESET} – the reset signal generated by the DC-DC converter;
- $V_{S12}, V_{S23}, V_{S11}$ – the voltage signals of the three states of operation (1/2, 2/3, and 1/1).

Note that all the currents are sensed by measuring the voltage across a $10\,\Omega$ resistor in series. The PSoC is continuously sensing each one of the electrical signals.

The power at any given point in time is given by the multiplication between the voltages and currents (1), (2). Hence, the voltage conversion efficiency (η_{VCE}) is given by the ratio between the output and input power (2). The η_{VCE} of the converter varies from state to state, and so for a given input voltage and output power the η_{VCE} value can vary significantly. In terms of the system, the instantaneous efficiency does not provide an accurate measure of the overall system efficiency. So instead of looking at the voltage conversion efficiency in terms of the ratio between instantaneous power values, we look at the efficiency in terms of the amount of energy consumed or provided. The energy is calculated by integrating the power values over time. There are three important energy values that are recorded:

- E_{SCAP} – the energy stored in the supercapacitor, provided by the PV cell;
- $E_{USED_{DC-DC}}$ – the energy absorbed by the DC-DC converter, from the supercapacitor;
- E_{LOAD} – the energy dissipated by the output load (IoT system).

These energy values are then used to determine the energy efficiency of the converter ($\eta_{Energy_{DC-DC}}$) and of the overall system ($\eta_{Energy_{SYSTEM}}$) (3), (4). The last Eq. (4) is very important as it gives the minimum amount of energy necessary for the operation of any given system. Thus, by carry out preliminary tests with the IoT system in a real scenario, the energy efficiency of the overall system can be determined and then be used to determine the minimum capacitance value of the supercapacitor value [8], or either to determine the minimum PV cell size that must be used.

$$P_{charging} = V_{scap} I_{scap} \tag{1}$$

$$\eta_{VCE} = \frac{P_{out_{DC-DC}}}{P_{in_{DC-DC}}} = \frac{V_{out_{DC-DC}} I_{out_{DC-DC}}}{V_{in_{DC-DC}} I_{in_{DC-DC}}} \tag{2}$$

$$\eta_{Energy_{DC-DC}} = \frac{E_{LOAD}}{E_{USED_{DC-DC}}} \tag{3}$$

$$\eta_{Energy_{SYSTEM}} = \frac{E_{LOAD}}{E_{SCAP}} \tag{4}$$

$$E_{Vscap} = \frac{1}{2} C_{scap} V_{scap_{max}}^2 \tag{5}$$

To this end, the experimental set-up (Fig. 3) is set so that it is close to a real condition of deployment. It is composed by two monocrystalline Solar Cells (SLMD121H04L) connected to the supercapacitor bank through a switch (TPS22860). In the SC DC-DC converter test board there are four supercapacitors of different types and values that are manually selected – 10 F (Maxwell - BCAP0010 P270 T01), 1 F (Eaton - HV0810-2R7105-R), 0.474 F (Kemet - FC0V474ZFTBR24), and 0.1 F (Kemet - FTW0H104ZF). To simulate an IoT system, three different resistance values are used to generate three output power levels – 1 mW, 0.5 mW, and 0.1 mW. These are selected using three load switches (TPS22860) controlled by the PSoC. All the data collected by the PSoC is sent via UART to a graphical interface developed using the Processing environment (Fig. 4). The interface displays in plots the above voltages and currents sensed by the PSoC. With those signals it determines the electric power and energy overt time. It also allows to act on the PSoC (Start/Stop) and generates a csv file with all the collected data.

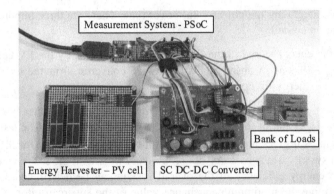

Fig. 3. System set-up.

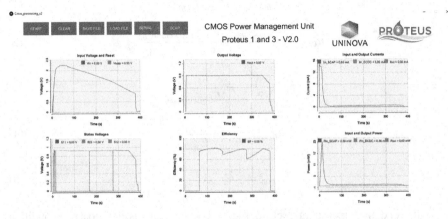

Fig. 4. Graphical Logger interface developed using the Processing environment.

5 Experimental Results

The performed experiments were carried out as follows: the supercapacitor was charged to the maximum value allowed by the DC-DC converter (2.3 V) without any load connected at the output. When the supercapacitor voltage reaches 2.3 V, the PV cell is disconnected (by turning OFF the switch) and, at the same time, one of the loads is connected. The supercapacitor then slowly discharges until the output voltage of the converter is lower than 10% of the nominal output voltage ($0.9 \times 10\% = 0.81$ V). At that voltage, the supercapacitor is discharged to ground, and a new cycle with a different load begins. There are in total three cycles, one per each load (1 mW, 0.5 mW, and 0.1 mW), for each supercapacitor. Table 1 shows the measurement results of the already mentioned signals (3), (4); the theoretical equation of the energy stored in a capacitor (5) to compare with the measured energy value of the supercapacitor (E_{SCAP}). And lastly, the voltage left in the supercapacitor ($V_{SCAP_{LEFT}}$) after the converter stopped.

Table 1. Experimental results

Scap value (F)	P_{OUT} (mW)	$V_{SCAP_{MAX}}$ (V)	$V_{SCAP_{LEFT}}$ (V)	$E_{SCAP_{Theoretical}}$ (mJ)	E_{SCAP} (mJ)	$E_{USED_{DC-DC}}$ (mJ)	E_{LOAD} (mJ)	$\eta_{Energy_{DC-DC}}$ (%)	$\eta_{Energy_{SYSTEM}}$ (%)
0.1	1	2.3	1.121	264.5	318.5	185.4	139.6	75.29	**43.83**
	0.5		0.977	264.5	327.8	198.5	148.7	74.88	**45.35**
	0.1		0.852	264.5	362.3	174.9	119.5	68.35	**32.99**
0.474	1	2.3	1.115	1253.7	993.4	552.7	413.3	74.78	**41.60**
	0.5		0.974	1253.7	1005.1	611.7	458.7	74.98	**45.64**
	0.1		0.852	1253.7	1000.5	558.6	381.7	68.33	**38.15**
1	1	2.3	0.823	2645.0	3251.4	1776.9	1330.7	74.89	**40.93**
	0.5		0.861	2645.0	2980.8	1874.8	1411.3	75.28	**47.35**
	0.1		0.852	2645.0	3124.5	1448.9	982.44	68.81	**21.96**
10	1	2.3	0.823	26450.0	37336.0	20797.0	15440.0	74.24	**32.65**
	0.5		0.824	26450.0	40249.0	21229.0	16006.0	75.40	**39.77**
	0.1		0.852	26450.0	37074.0	17.274	12.189	70.57	**32.88**

6 Conclusions

The values of the theoretical energy stored in the supercapacitor $(E_{SCAP_{Theoretical}})$ differ significantly from the measured energy values (E_{SCAP}). For the 0.1 F (Kemet - FTW0H104ZF) the $E_{SCAP} > E_{SCAP_{Theoretical}}$ by a maximum of 37%; the 0.474 F (Kemet - FC0V474ZFTBR24) the $E_{SCAP} < E_{SCAP_{Theoretical}}$ by a minimum of −20.76%; and for the 1 F (Eaton - HV0810-2R7105-R) and 10 F (Maxwell - BCAP0010 P270 T01), the $E_{SCAP} > E_{SCAP_{Theoretical}}$ with a maximum of 18.7% and 34.3%, respectively. This can be explained by the variation in the capacitance of the supercapacitors, which according to the datasheets can go as high as 20% for some supercapacitors. Hence, the importance of previously measure the real energy stored in the supercapacitor.

As for the energy efficiency of the SC DC-DC converter, $\eta_{Energy_{DC-DC}}$, it reaches values of approximately 75% for output power values higher than 0.5 mW. As the output power decreases, $\eta_{Energy_{DC-DC}}$ also decreases, reaching values of approximately 68%. This is caused by the power consumption of the internal circuits that does not scale with the output power level. Furthermore, the $\eta_{Energy_{DC-DC}}$ values are kept constant independent of the type and value of the supercapacitor. The overall energy efficiency of the system, $\eta_{Energy_{SYSTEM}}$, shows that in the best case the system uses 47% of the energy stored in the supercapacitor. The main reason for this value is the input voltage limitation of the SC DC-DC Converter, that is unable to work when the supercapacitor voltage is below 1.12 V to 0.823 V, depending on the output power level. Moreover, the $\eta_{Energy_{SYSTEM}}$ decreases substantially for lower output values (0.1 mW) and higher supercapacitor values (1 F, 10 F). This is because both condition result in a longer measuring time, which allows for the energy lost due the leakage current in the supercapacitor to become more important in the overall energy balance. Knowing the $\eta_{Energy_{SYSTEM}}$, it is possible to determine the required energy to be stored on the supercapacitor for a specific E_{LOAD}, consumed by any IoT system. Hence the value of the supercapacitor can be roughly sized using the theoretical Eq. (5), or by performing the charging process and measure the amount of energy stored.

Acknowledgments. This work was supported by the Portuguese Foundation for Science and Technology under a Ph.D. Grant (SFRH/BD/115543/2016), and by PROTEUS project funded by the European Union's H2020 Programme under grant agreement no 644852.

References

1. Omairi, A., Ismail, Z.H., Danapalasingam, K.A., Ibrahim, M.: Power harvesting in wireless sensor networks and its adaptation with maximum power point tracking: current technology and future directions. IEEE Internet Things J. **4**, 2104–2115 (2017)
2. Mondal, S., Paily, R.: On-chip photovoltaic power harvesting system with low-overhead adaptive MPPT for IoT nodes. IEEE Internet Things J. **4**, 1624–1633 (2017)
3. Habibzadeh, M., Hassanalieragh, M., Ishikawa, A., et al.: Hybrid solar-wind energy harvesting for embedded applications: supercapacitor-based system architectures and design tradeoffs. IEEE Circuits Syst. Mag. **17**, 29–63 (2017)

4. Weddell, A.S., Merrett, G.V., Kazmierski, T.J., Al-Hashimi, B.M.: Accurate supercapacitor modeling for energy harvesting wireless sensor nodes. IEEE Trans. Circuits Syst. II Express Briefs **58**, 911–915 (2011)
5. Carreon-Bautista, S., Huang, L., Sanchez-Sinencio, E.: An autonomous energy harvesting power management unit with digital regulation for IoT applications. IEEE J. Solid-State Circuits **51**, 1457–1474 (2016)
6. Mondal, S., Paily, R.: Efficient solar power management system for self-powered IoT node. IEEE Trans. Circuits Syst. I Regul. Pap. **64**, 2359–2369 (2017)
7. Sarafianos, A., Steyaert, M.: Fully integrated wide input voltage range capacitive DC-DC converters: the folding Dickson converter. IEEE J. Solid-State Circuits **50**, 1560–1570 (2015)
8. Kim, S., Chou, P.H.: Size and topology optimization for supercapacitor-based Sub-Watt energy harvesters. IEEE Trans. Power Electron. **28**, 2068–2080 (2013)
9. Madeira, R., Paulino, N.: Analysis and implementation of a power management unit with a multiratio switched capacitor DC-DC converter for a supercapacitor power supply. Int. J. Circuit Theory Appl. **44**, 2018–2034 (2016)

Check for updates

Wireless Battery Charger for EV with Circular or Planar Coils: Comparison

L. Romba[1,2], E. N. Baikova[1,3], C. Borges[1,2], R. Melicio[4,5(✉)], and S. S. Valtchev[1,2]

[1] UNINOVA-CTS, Caparica, Portugal
[2] FCT of Universidade Nova de Lisboa, Caparica, Portugal
[3] EST Setúbal, Instituto Politécnico de Setúbal, Setúbal, Portugal
[4] Departamento de Física, Escola de Ciências e Tecnologia, ICT,
Universidade de Évora, Évora, Portugal
ruimelicio@gmail.com
[5] IDMEC, Instituto Superior Técnico, Universidade de Lisboa, Lisbon, Portugal

Abstract. This paper presents the experimental results obtained in the wireless energy transfer (WET) system prototype based on coils: circular or planar. With these experimental results we can choose the tuning settings to improve the efficiency of power transmission of the WET systems. In WET for electric vehicle batteries charging, the coil shape and the range between the coils are the most important issues of those systems.

Keywords: Wireless energy transfer · Magnetic coupling · Planar coil
Circular coil · V2G · V2H · V2B

1 Introduction

The WET systems are associated with innovative technologies to create new possibilities: to charge batteries of different devices, to eliminate the cables, the plugs and sockets, to increase the reliability, to guarantee a maintenance-free operation of the critical systems. The WET is especially used in biomedical implants, electric vehicles, and robotic systems [1]. The WET technologies are classified in two categories: non-radiative and radiative. The non-radiative category is used at short and mid-range distances. The small devices, like cell phone, require a small amount of power to charge their batteries and their charging systems operate at a short distance, a few millimeters, between the transmitter and receiver. The charging system for electric vehicles is an example of mid-range distance. Both examples are defined as near field or non-radiative technologies. The power in this case, is transferred by magnetic field using inductive coupling between coils. The long-range or far-field energy transfer is a radiative technology. The power is transferred by electromagnetic waves like microwaves or laser beams. An example of that far-field technology is the solar power satellite [2].

The near field is usually considered to the one at a distance up to $(1/2\pi)\lambda$, that is 0.159λ (λ is the wavelength of the frequency generated by the source) [3–5]. The essence of the mid-range WET is the inductive coupling through a strong magnetic coupling (SMC). The inductive coupling process allows for an electric energy transfer using only

© IFIP International Federation for Information Processing 2018
Published by Springer International Publishing AG 2018. All Rights Reserved
L. M. Camarinha-Matos et al. (Eds.): DoCEIS 2018, IFIP AICT 521, pp. 214–223, 2018.
https://doi.org/10.1007/978-3-319-78574-5_20

the magnetic field. Originally, the strong magnetic coupling technology was proposed in 2007 by Marin Soljacic. This pioneer technology has broken the traditional WET model for inductive power transfer (IPT) process for which the efficiency strictly depends on the coupling coefficient k between the coils. The energy transmission distances were extended from a few millimetres to distances of some meters. The possibility to stretch the distance represented a significant advance in WET systems [4–6]. The strong magnetic coupling diagram is shown in Fig. 1.

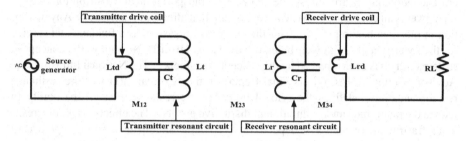

Fig. 1. Strong magnetic coupling diagram.

In Fig. 1 the electric structure used in this technology contains the transmitter drive coil and the transmitter resonant circuit L_t, C_t at the transmitter side, a receiver resonator L_r, C_r and a load coil (receiver drive coil) at the receiver side. The utilization of two excitation coils involves two additional mutual coupling coefficients M_{12}, M_{34}. The mutual coupling between the two resonators is referenced by M_{23}. Those two extra coefficients introduce extra freedom in spreading the transfer distance. However, the overall transfer efficiency will not exceed 50% because of the independence matching requirement [7]. It is important to differentiate two basic concepts, namely: (1) the maximum power transfer principle; and (2) the maximum energy efficiency principle. The impedance matching method adopted in many WET projects is based on the maximum power transfer theorem. The maximum power transfer principle requires impedance matching between the source and the load. The maximum power theorem applies to a situation in which the source impedance is fixed. When the maximum power transfer occurs at impedance matching, in WET systems, the maximum energy efficiency under the maximum power transfer approach never exceed 50% [8].

A coil size problem has been observed in systems of simultaneous charge of batteries in portable devices, e.g. mobile phones, laptops, and electric vehicles. The use of planar coils has minimized the negative impacts of the use of circular coils for certain applications. In this paper, the magnetic field created by the two types of circular and planar coils is compared. The tuning control process of the transmitter and the receiver resonant circuits, with these two types of coils, is also investigated.

2   Technological Innovation for Resilient Systems

The last decades have been challenging due to the constant uncertainties motivated by the unpredictability of catastrophic climatic events, terrorist attacks, as examples. Those events have caused social unrest in which almost everything is uncertain [9]. The recent natural disasters have alerted the societies to the interdependence and functionality of critical structures. Failures in interdependent structures can cause cascade effects on critical structures. The critical infrastructures are highly interconnected and mutually dependent in complex ways. What happens to one infrastructure can directly and indirectly affect other infrastructures, with impact in large geographic regions. The most important challenges are to understand, identify and analyze the interdependencies of critical national infrastructures. The production, transportation and distribution of electric power is an example of a basic infrastructure that affects our daily lives. Any disruption in this area has implications for the entire social system as a fundamental service for the security and social well-being of any nation [10, 12]. Nowadays the concept of sustainability has been replaced by the concept of resilience. The National Infrastructure Advisory Council (NAC) in its Final Report on the "Critical infrastructure resilience recommendations" defines the critical infrastructure resilience as: "...the ability to reduce the magnitude and/or duration of disruptive events". The effectiveness of a resilient infrastructure or enterprise depends on its ability to anticipate, absorb, adapt to, and/or rapidly recover from a potentially disruptive event."

The energy industry is changing: renewables and decentralized energy sources, the need for continuous grid optimization to name only a few. The future smart grids ambient and cyber-physical systems have to have a layered of cyber infrastructures accessing resilient power application that are able to give security and reliability, having the ability to act in order to maintain and correct infrastructure components without affecting the service [11]. A new approach in terms of decentralized energy sources can be achieved through purely electric vehicles and plug-in hybrid vehicles. Depending on the model, each electric vehicle can store approximately between 5 kWh–85 kWh of energy. This energy can be used not only for propulsion but also for supplying other consumers like homes (V2H), or buildings (V2B) and even for supply ancillary services to distribution grid operators, through concept (V2G). These vehicles can charge the batteries overnight and provide the stored energy during the day when and where needed, to the grid, for a building or for a home from a terminal in their parking lot. The importance of this new concept is most evident in the case of a disruption of the electrical power system, when the EV can support essential services for some time [13].

3   Modeling

3.1   Circular Coil

The mutual inductance between the two coils is determined experimentally by two measurements of the self-inductance. During one of the measurements a current flow in the same direction in the two coils and in the other, it flows in opposite directions. If L and L' are the two values of self-inductance determined experimentally and M is their mutual inductance [14] given by:

$$M = \frac{L - L'}{4}$$

(1)

The sum method for determining the inductance of a solenoid assumes that the solenoid is composed of n equal circular coaxial circular loops spaced uniformly at a distance corresponding to the pitch g of the winding. Thus, the total coil length is ng. The current is assumed to get from one ring to the next by means of connections of negligible inductance; that is, the helicity of the actual winding is neglected, the current being assumed to flow in planes perpendicular to the axis of the coil. The inductance of the coil [15] is then given by:

$$L = nL_1 + 2 \sum_1^{n-1} M_p \qquad (2)$$

Where L_1 is the inductance of a loop, and M_p is the mutual inductance of two turns separated by a distance pg. Using the accurate formula for the inductance of a circular loop of a mean radius a having a cross section of radius ρ [15] is given by:

$$nL_1 = 4\pi na \left[\left(1 - \frac{\rho^2}{8a^2}\right) log_e \frac{8a}{\rho} + \frac{\rho^2}{24a^2} - \frac{7}{4} \right] \qquad (3)$$

The mutual inductance M_p is obtained by the Maxwell series formula for the mutual inductance of coaxial circular filaments near together [15] is given by:

$$M_p = 4\pi a \left[\left(1 + \frac{3}{16} \frac{p^2 g_2}{a^2} + \dots \right) log_e \frac{8a}{pg} - \left(2 + \frac{p^2 g^2}{16a^2} + \dots \right) \right] \qquad (4)$$

3.2 Planar Coil

Though it may seem implicitly clear, but it will be stated here that in this work the planar coil is the one that is wound in a spiral way, so its thickness is only one wire diameter. The circular winding will be named the one that has all the turns at the same diameter, making the winding thicker and concentrated at almost the same radius. Considering the first type of winding, the spiral shaped coil exhibits parameters that depend on the outer diameter D_0, the number of turns N, the pitch between the turns p, and the diameter of the conductor used W, as it is shown in Fig. 2. In that figure, the calculation result for the inductance L, the total resistance R, the distributed capacitance C, and the quality

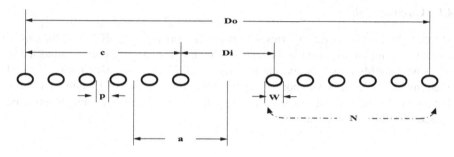

Fig. 2. Cross-sectional view of planar coil [16].

factor Q, are presented. The inner diameter D_i, the total wire length l, the distance a, and the thickness of the winding c, are defined in (5) to (8) from [16]. All the dimensions are shown in meters.

$$D_i = D_o - 2N(W + p) \tag{5}$$

$$\ell = \frac{1}{2}N\pi(D_o + D_i) \tag{6}$$

$$a = \frac{1}{4}(D_o + D_i) \tag{7}$$

$$c = \frac{1}{2}(D_o - D_i) \tag{8}$$

The self-inductance for planar coils is given by:

$$L(H) = \frac{N^2[D_o - N(W - p)]^2}{16D_o + 28N(W + p)} \times \frac{39,37}{10^6} \tag{9}$$

Where L is derived from a modification of Wheeler's formula for a single-layer helical coil, while accounting for the conversion from inches to meters 39.37 in/m and μH to H (10^{-6}). In (9) is valid for most geometries except: (1) when the coil has very few turns, (2) when the pitch is very large relative to the wire diameter ($p \gg W$), and (3) when $p \gg W$ and $c/a < 0.3$ [16].

4   Case Study

The WET at mid-range distances is carried through the air and based on the magnetic field. Therefore, the study of the magnetic field created by different types of coils is very important [17, 18]. In this case study the comparison of the magnetic field originated by a planar coil and by a circular coil is made. In addition, a tuning control system is presented based on the magnetic core reactor (MCR), which allows tuning of these coils in resonant circuits.

4.1   Circular Coil

The circular coil connected in series with magnetic core reactor (MCR) and the capacitance (resonant circuit) used in the tests is shown in Fig. 3. This coil is connected in series with the MCR and the fixe capacitance $C_t = 0.23 \mu F$ (resonant circuit). The coil is assembled by 35 turns with normal copper cable with a section of 1.5 mm². The outer diameter is 240 mm. The average thickness of the coil is 35 mm. The cable length used to assembly this coil is 29 m.

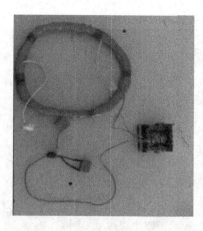

Fig. 3. Circular coil connected in series with MCR and the capacitance (resonant circuit).

The MCR tuning capacity was tested to the frequency range between 15 kHz and 30 kHz. The different tuning points when the MCR control DC current ranges from 0.0 A up to 3.0 A is shown in Fig. 4 for the above two frequencies. The frequency adjustment range is between the two points of maximum impedance.

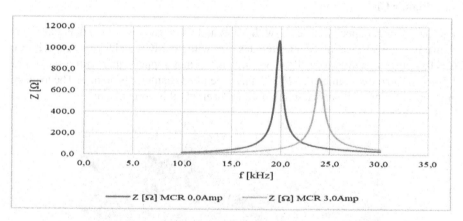

Fig. 4. Tuning points by varying the MCR DC control current for circular coil.

The magnetic field created by this coil (alone) was simulated when it is driven by a current of 15 A with the frequency of 17 kHz. The simulation results for circular coil: magnetic field distribution is shown in Fig. 5.

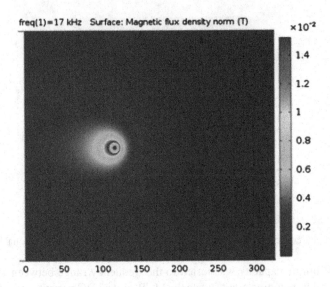

Fig. 5. Circular coil: magnetic field created for the frequency of 17 kHz.

4.2 Planar Coil

The planar coil connected in series with MCR and the capacitance (resonant circuit) is shown in Fig. 6. This coil is connected in series with the MCR and the fixe capacitance $C_t = 0.3$ μF. The coil is formed by 35 turns of normal copper cable with a section of 1.5 mm². The outer diameter is 240 mm and the inner diameter is 50 mm. The average pitch coil is 2 mm. The cable length used to assembly this coil is 18 m.

Fig. 6. Planar coil connected in series with MCR and the capacitance (resonant circuit).

The MCR tuning capacity was tested to the frequency range between 20 kHz and 60 kHz. The tuning points when the MCR control DC current ranges from 0.0 A to 3.0 A is shown in Fig. 7. The main advantage of using the MCR is that the inductance L can be continuously varied. The frequency adjustment range can varying continuously in between the two points of the maximum impedance as indicated in the Fig. 7.

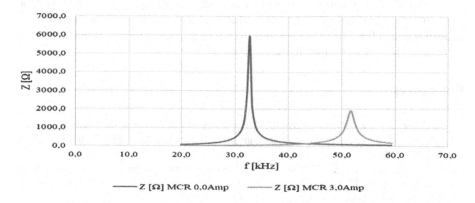

Fig. 7. Tuning points by varying the MCR DC control current for planar coil.

The magnetic field created by this coil (alone) was simulated when it is driven by a current of 15 A with the frequency of 17 kHz. The simulation results for planar coil: magnetic field distribution created for the frequency of 17 kHz is shown in Fig. 8.

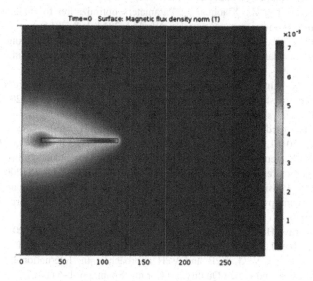

Fig. 8. Planar coil: magnetic field created for the frequency of 17 kHz.

5   Conclusions

The magnetic field created by the circular coil is more intense than that created by the planar coil. However, the spreading of the magnetic field created by the planar coil is larger by comparison of Figs. 5 and 8. This implies a greater efficiency in power transmission. Also note that although both coils have the same number of turns, the circular coil wire length was significantly higher yield which enables a greater magnetic field. The values of the impedances at resonance are larger in planar coil compared with the values of the circular coil, although obtained at different frequencies but within the close frequency range, few *kHz*. Also, the frequency tuning range is greater in the planar coil is compared with that obtained in the circular coil, Figs. 4 and 7.

References

1. Moradewicz, A.J., Kazmmierkowski, M.P.: High efficiency contactless energy transfer system with power electronic resonant converter. Bull. Polish Acad. Sci. Techn. Sci. **57**(4), 375–381 (2009)
2. Qiu, C., Chau, K.Y., Liu, C., Chan, C.C.: Overview of wireless power transfer for electric vehicle charging. In: ESV27 International Battery, Hybrid and Fuel Cell Electric Vehicle Symposium, Barcelona, Spain, pp. 1–9 (2013)
3. Umenei, A.E.: Understanding Low Frequency Non-Radiative Power Transfer. Fulton Innovation, LLC (2011)
4. Kurs, A., Karalis, A., Moffat, R., Joannopoulos, J.D., Fisher, P., Soljacic, M.: Wireless power transfer via strongly coupled magnetic resonances. Science **317**(5834), 83–86 (2007)
5. Karalis, A., Joannopoulos, J.D., Soljacic, M.: Efficient wireless non-radiative mid-range energy transfer. Ann. Phys. **323**(1), 34–38 (2008)
6. Changsheng, L., He, Z., Xiaohua, J.: Paramaters optimization for magnetic resonance coupling wireless power transmission. Sci. World J. **2014**, 1–8 (2014)
7. Xiao, L., Ping, W., Dusit, N., Dong, I.K., Zhu, H.: Wireless charging technologies: fundamentals, standards and network applications. IEEE Commun. Surv. Tutor. **18**(2), 1413–1452 (2016)
8. Hui, S.Y.R., Wenxing, Z., Lee, C.K.: A critical review of recent progress in mid-range wireless power transfer. IEEE Trans. Power Electron. **29**(9), 4500–4511 (2014)
9. Laugé, A., Hernantes, J., Sarriegi, J.M.: The role of critical infrastructures' interdependencies on the impacts caused by natural disasters. In: Luiijf, E., Hartel, P. (eds.) CRITIS 2013. LNCS, vol. 8328, pp. 50–61. Springer, Cham (2013). https://doi.org/10.1007/978-3-319-03964-0_5
10. Steven, M.R., Peerenboom, J.P., Terrence, K.K.: Identifying, understanding and analyzing critical infrastructures independencies. IEEE Control Syst. **21**(6), 11–25 (2001)
11. Gomes, I.L.R., Pousinho, H.M.I., Melicio, R., Mendes, V.M.F.: Bidding and optimization strategies for wind-pv systems in electricity markets assisted by CPS. Energy Procedia **106**, 111–121 (2016)
12. Wang, J., Gharavi, H.: Power grid resilience. Proc. IEEE **105**(7), 1199–1201 (2017)
13. Garcia-Villalobos, J., Zamora, I., San Martin, J.I., Junquera, I., Eguia, P.: Delivering energy from PEV batteries; V2G, V2B and V2H approaches. In: International Conference on Renewable Energy and Power Quality, La Coruña, Spain, pp. 1–6 (2015)
14. Rosa, E.B.: Calculation of the self-inductance of a single layer coils. Bull. Bureau Standards **2**(2), 161–187 (1906)
15. Grover, F.W.: A comparison of the formulas for the calculation of the inductance of coils and spiral wound with wire of large cross sections. Bureau Standards J. Res. **3**, 163–190 (1929)
16. Benjamin, H.W., Brody, J.M., Gunbok, L., Joshua, R.S.: Optimal coil size ratios for wireless power transfer applications. In: IEEE International Symposium on Circuits and Systems, Melbourne, Australia, pp. 2045–2048 (2014)

17. Baikova, E.N., Romba, L., Valtchev, S.S., Melicio, R., Pires, V.F., Krusteva, A., Gigov, G.: Electromagnetic field generated by a wireless energy transfer system: comparison of simulation to measurement. J. Electromag. Waves Appl. **32**(5), 1–18 (2017). https://doi.org/10.1080/09205071.2017.1399832

18. Romba, L.F., Valtchev, S.S., Melicio, R.: Improving magnetic coupling for battery charging through 3D magnetic flux. In: IEEE 17th International Conference on Power Electronics and Motion Control, Varna, Bulgaria, pp. 291–297 (2016)

An Outline of Fault-Tolerant Control System for Electric Vehicles Operating in a Platoon

António Lopes[(⊠)] and Rui Esteves Araújo

INESC TEC, Faculty of Engineering,
University of Porto, 4200-391 Porto, Portugal
{antonio.lopes, raraujo}@fe.up.pt

Abstract. High level vehicle automation systems are currently being studied to attenuate highway traffic and energy consumption by applying the concept of platooning, which has gained increased attention due to progresses in the next generation of mobile communication (5G). The introduction of more complex automation systems originates, however, fault entry points that hinders the system safety and resilience. This paper presents an initial control architecture for the electric vehicles platoon from a fault-tolerant control perspective. To achieve a fault-tolerant platooning structure an over-actuated electric vehicle topology is proposed which may allow the implementation of different redundancies. Furthermore, some of the major challenges in the platooning network control system (NCS) are presented and the techniques to overcome these issues are explored.

Keywords: Platoon · Fault-tolerant control · Automated highway system

1 Introduction

In today's modern society, control techniques have been introduced to obtain a more reliable and safe operation of modern-day systems. One of these systems that has been taking advantage of its utilization is the ground vehicle system. Vehicle control systems should meet several goals such us safety of the vehicle and ride comfort. For these purposes, systems such as: parking assist, lane keep assist and adaptive cruise control have spread through the automobile industry and have earned the approval of users around the world.

The concept of platooning assumes that a group of vehicles can be closely tracked to follow a leader vehicle that may or may not be driven by a human driver. This concept is not new, it was actually proposed in the 90's by a consortium named California Partners for Advanced Transportation Technology (PATH). The basic idea behind vehicle platooning is that the platoon allows a more efficient use of energy by reducing aerodynamic drag and can contribute to the reduction of the evident traffic jam problem that is even more notorious nowadays [1, 2].

In the present work, the developments in the ground electric vehicle system that may enable a fault-tolerant platoon solution are explored. The technological innovations that withstand this concept are carefully discussed as well as the advantages and challenges that arise from their implementation.

© IFIP International Federation for Information Processing 2018
Published by Springer International Publishing AG 2018. All Rights Reserved
L. M. Camarinha-Matos et al. (Eds.): DoCEIS 2018, IFIP AICT 521, pp. 224–231, 2018.
https://doi.org/10.1007/978-3-319-78574-5_21

The document is structured as follows: the state of the art and the research questions are presented in Sect. 1, the relation to resilient systems is presented in Sect. 2, the research contribution will be explored in Sect. 3, the solution is debated in Sect. 4 and the conclusions are presented in Sect. 5.

1.1 State of the Art

The adoption of both connected vehicle technologies and cloud computing is changing the very nature of how automotive industry operates. An example of this is the vehicle platooning, which has recently regained relevance thanks to the developments in mobile communication technology (5G). The fast-speed access provided by wireless communication directed the focus to the Car-to-Car communication which allows the exchange of information between vehicles and infrastructures to avoid traffic or road congestion and improve road safety [3, 4]. On the other and, the IEEE entity amended the 802.11 standard with 802.11p in vehicle environments and dedicated short-range communications(DSRC)[1]. Both technologies, 5G and DSRC, are pointed as potential solutions for vehicle communication.

These developments have led to the investigation of control systems that would provide a suitable solution for the platooning system. For example, the solution presented in [5] assumes that all vehicles are equipped with car-to-car (C2C) communications, where a distributed model predictive control (MPC) method is proposed for heterogeneous vehicle platoon under unidirectional topologies. The platoon nodes are dynamically decoupled but constrained by spatial geometry and an open-loop optimal controller is defined for each one of the vehicles of the platoon relying only on the information of the neighbors. The proposed cost function penalizes the error between the actual spatial trajectory and the predicted solution. A decentralized approach is proposed, which allows a more suitable implementation in real applications and allows for an easier scalability of the solution [5].

One of the key concepts that arises from the implementation of this kind of platooning solution is the concept of String Stability [6–8]. In an automatic vehicle-following a control system it is imperative to consider the disturbance attenuation along the vehicle string, i.e. the string stability of the system [7]. When considering a string of vehicles, the propagation of the system response along a cascaded system must be stable, i.e. disturbance in one node of the system should not be amplified throughout the platoon [7]. In [9] a Lyapunov stability notion is employed to the string stability concept, focusing on the initial condition perturbations. This approach however lacks practical relevance due to the drawback of ignoring the effect of the perturbations of the initial conditions as well as the effects of external disturbances on the system [7]. The string stability of the system can also be assessed from the inspection of the system eigenvalues, nonetheless, the stability properties of the finite-length strings might not

---

[1] IEEE Standard for Information Technology–Telecommunications and information exchange between systems–Local and metropolitan area networks–Specific requirements–Part 11: Wireless LAN Medium Access Control (MAC) and Physical Layer (PHY) specifications Amendment 6: Wireless Access in Vehicular Environments.

converge in an infinite-length string [7]. Another popular approach is the performance-oriented approach in which the string stability is characterized by the amplification, in upstream direction, of the distance, velocity or acceleration error [7].

In [10] the concept of networked control system (NCS) is presented as a system in which the control loop is closed via a communication channel. The proposed H∞ optimal controller achieves good performance even in the presence of intermittent losses or delays of the information communicated between the chain of vehicles [10].

In [11] a longitudinal platoon control for a limited number of vehicles is developed. The control objectives are formulated as one set of linear matrix inequalities (LMI) which are solved by the controller [11]. The C2C communication increases "vehicle awareness" and improves the ability to anticipate potential disturbances on the platoon. Nonetheless, the control loop may be affected by communication delays which may lead to a deterioration of the controller performance [10, 11]. The behavior and stability of the platoon can also be highly affected by faults or failures.

The reliability of hardware present in an Autonomous vehicle is evaluated in [12]. According to this study, the communication does not presents a high reliability and its performance largely depends on the power, signal processing, inter-vehicle distance and environment interferences [12]. Actuator and sensor faults are described as more reliable and its mitigation can be achieved through a suitable control design [12]. The normal AHS control hierarchy can be extended to achieve a fault-tolerant architecture. The solution is achieved by partitioning the system into more manageable sub-systems and proposes a formalize solution to each of them [13]. The fault-tolerant architecture in [13] is constituted by the hierarchical levels: *sensor hierarchy, capability monitor* and *performance monitor*. Each one of the hierarchy levels are divided into two layers, the top layer called the *supervisor*, that receives the capability information and switches between strategies accordingly, and the lower layer called the *regulator*, responsible for implementing the chosen strategy [13]. Collision situations are expected to be completely mitigated to achieve a safe and reliable AHS [14]. Other hazardous situations are also described in [14] such as cut-in situations, when other vehicles merge into the platoon, and driving off the road to avoid collision.

These issues are currently being pursued and investigated to potentiate the implementation of a higher automation into electric vehicles in order to obtain a synergy between the superior dynamic performance achieved by advanced motion controllers for over-actuated electric vehicles and the benefits that arise from a platoon formation.

1.2 Research Questions

In this section, it will be presented the research problem and it clearly defines the objectives of the proposed research. To achieve such a description, a set of questions, that are carefully formulated to steer the research into a clear direction, are defined. These questions aim to answer some gaps in the scientific knowledge. Even more, there will be an effort to study emerging technologies that may improve the performance of the aforementioned systems.

The main purpose is to investigate the architecture of the controller so that the string stability requirement is explicitly incorporate in the design specification, while also considering fail-safety with respect to different faults and failure modes. In the light of the state of the art, the following research questions are raised: What kind of assumption and prior information do we need to formulate a fail-safe controller for a string stable platoon system? How should we exploit the advanced motion systems to increase performance, safety and coordination of platoons?

The previous questions are going to define the main objectives of the investigation and are the guidelines used to steer the discussion.

2   Relationship to Resilient Systems

The interest in vehicle platooning has receiving and increasing amount of attention due to the growing problems in traffic congestion, pollution and energy efficiency. Furthermore, this system would also reinforce the automated vehicle paradigm change in a social level as it would offer an opportunity to attack the population ageing problem by offering a safe and comfortable solution for this particular group.

Nevertheless, this automated system would require serious developments in different fields and a complete synergy between those systems. The increase in complexity of the system will inevitably raise the number of different elements of the system which can provide an entry point for system faults or even complete failure (see Fig. 1). The presence of a fault, in any underlying element, will modify the dynamic of the system and may hinder the nominal operation of the control system. For this reason, the control design must be able to accommodate faults or even complete failure, ensuring the performance of the system in the nominal region as well as in a degraded region of operation.

The aforementioned reasons provide the motivation behind fault-tolerant control design. To this end an over-actuated electric vehicle topology is proposed to provide a physical redundancy necessary towards a fault-tolerant motion control solution.

In a higher-level, the introduction of a communication layer allows a greater knowledge of the adjacent environment providing "awareness" to the system. The ability of communicating with other members of the platoon (car-2-car) through mobile

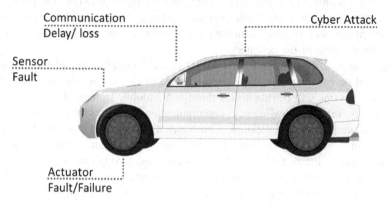

Fig. 1. Potential fault entry points in the vehicle system.

communication improves the string stability properties of the system. Nonetheless, communications delay and package-loss, present in every communication, introduce a complex challenge as the closed-loop of the control system is significantly affected by the communication channel's properties. The resulting networked controlled system can also present other challenges, such as cyber-attacks or information breach, which must be addressed to achieve a resilient control system.

3 Problem Definition and Conceptualization

The previous analysis shows that, in order to achieve an automated intelligent vehicle system, some procedures must be taken in consideration into the design of the control system. Evidently, the increasing complexity will affect the system if no action is taken to ensure the reliability and the overall performance of the system.

In this section, it will be proposed a fault-tolerant platooning architecture. The limits of the problem are clearly stated and the main contributions to achieve a suitable solution are exposed.

3.1 Assumptions

As a reference for our study, we consider a road environment where the traction system of the ground vehicle undergoes a paradigm shift from an internal combustion engine to a purely electric motor solution. The torque response and the efficiency of the electric motor have proved to be an asset that ensures a more dynamic system response. Furthermore, multi-motor topologies have become more popular introducing the concept of actuation redundancy to the system and allowing advanced stability control solutions that potentiate the increase in degrees of freedom of the system. The four-wheel independently driven (4WID) electric vehicle is one of the most promising topologies. Due to its over-actuated nature it ensures the necessary physical redundancy to introduce fault-tolerant control techniques. The disseminated x-by-wire technology provides the opportunity of controlling the steer of the vehicle, enhancing the degree of freedom of the system even more. For this reason, it will be considered a purely electric 4WID vehicle with steer-by-wire technology ensuring the control over the front wheels direction. At a physical level, it will be considered fault entry points in all the actuators (in-wheel motor and steering actuator) and the faults are going to be defined as loss-of-effectiveness of the actuators.

The platoon will be formed by n 4WID electric vehicles and all the constituent vehicles will be considered equipped with state of the art mobile communication. The communication link will be established through 5G mobile technology which promises faster, highly reliable, low-latency communication and provides a more suitable solution for vehicle communication. Communication delay or package loss will be considered in the resulting networked control system as a fault of the communication link. The relative distance of the vehicles is obtained through radar information and redundant information is accomplished by introducing a sensor in the front of the vehicle and a second one in the rear to measure the distance of the preceding vehicle as well.

3.2   Proposed Architecture

The proposed fault-tolerant platooning solution is presented in Fig. 2. The architecture is defined by two levels: the motion control and the high-level platooning control. Depending on the failure modes, the fault-tolerant control may decide which controller level will be used, and/or may change the control parameters. The fault-tolerant motion controller design will focus on actuator faults and will be responsible for quickly redistribute the control effort to the heathy actuators of the vehicle. This solution will prevent that actuator faults evolve to a hazardous situation. By promptly eliminating the fault we can detach the actuator fault of a single member from the platooning system, which allows an independent control design for the motion control and the string stability system.

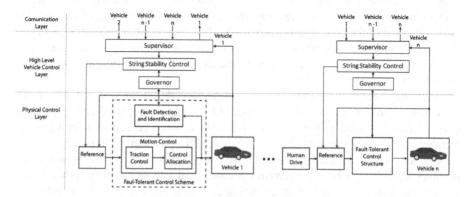

Fig. 2. Intelligent vehicle control system: diagram of the different control subsystems.

For the platooning system, a vehicle equipped with redundant distance sensor and car-to-car mobile communication is proposed. The communication link between the immediate neighbors of a given vehicle, although complex, may allow the implementation of a finite-horizon optimization technique based on a prediction horizon of the resulting system. To ensure the reliability of the information a sensor fusion algorithm, that penalizes the communication information in the presence of package loss or communication delay, is proposed. To ensure the safety of the platoon a command governor approach is also defined. This system may potentiate string stability of the system by dynamically change the standstill gap between vehicles, furthermore, other objectives can be achieved such us fault mitigations.

4   Discussion

The over-actuated nature of the 4WID topology provides an exciting opportunity to implement fault-tolerant control methodologies. The multiple degrees of freedom permit the development of an integrated motion control that tracks the vehicle yaw rate and longitudinal velocity by ensuring the total longitudinal force of the system and the external moment generated by the in-wheel actuators. These "virtual control variables" are then allocated to the available actuators. This control allocation algorithm ensures that the virtual control is correctly distributed to the actual physical actuators and the actuators faults are promptly addressed by assigning a higher weight to the faulty actuator. Nonetheless, the performance of the FTC is directly related to the responsiveness and accuracy of the Fault Detection and Identification (FDI) system.

The high-level controller should rely on the information provided by a supervisor algorithm (see Fig. 2) which has access to the sensor information of the vehicle and the information of the front and rear vehicle deliver by the communication link. The resulting networked control system is affected by communication delays, package lost and may even be subject to malicious attacks. All of these may be considered system faults that may ultimately evolve into a failure, dissolving this communication layer altogether. The proposed supervisor compares the measurement of the in-vehicle sensor with the information of the neighboring vehicles and assess the validity of the data. The reconfiguration of the controller is triggered by the supervisor that evaluates the degree of reliability of the communication layer. As the loss of package becomes more severe the supervisor assigns a greater weight to the in-vehicle sensor information to overcome this system uncertainty.

The string stability control will be achieved by designing a distributed model predictive controller. The solution is achieved by solving an optimal control problem over a given horizon. The main objective is to achieve coordination between the members of the platoon that are solving the MPC problem with locally relevant information, such as, state variables, costs and constraints of the system. The decomposition in n subsystems ensures a conceptual simplicity to the problem and reduces the computational burden in each agent of the system. In order to achieve a string stable system and simultaneously ensure the necessary physical and safe constraints a Command Governor approach will be explored to enforce these constraints by redefining the standstill gap [15].

5   Conclusion and Future Work

In the present work, the design of a Fault-tolerant platooning system is investigated. This study evaluates the main challenges in the automated highway system as well as the developments in motion control design that should be implemented to obtain a suitable solution to potentiate the string stability controller.

To obtain a fault-tolerant structure a supervisor algorithm that triggers the reconfiguration of the system in the presence of communication related faults or complete failure is proposed. A decentralized MPC solution is proposed for the string stable system combined with a command governor approach to ensure stability. In the future, the implementation of the supervisor algorithm in the platooning system and the string stable controller design will be investigated.

Acknowledgments. The authors would like to thank Dr. Ricardo Castro for his valuable comments and suggestions to improve the quality of the paper.

References

1. Shladover, S.E., Desoer, C.A., Hedrick, J.K., Tomizuka, M., Walrand, J., Bin Zhang, W., McMahon, D.H., Peng, H., Sheikholeslam, S., McKeown, N.: Automatic vehicle control developments in the PATH program. IEEE Trans. Veh. Technol. **40**(1 pt 1), 114–130 (1991)
2. Hedrick, J.K., Tomizuka, I.M., Varaiya, P.: Control issues in automated highway systems. IEEE Control Syst. **14**(6), 21–32 (1994)
3. Wong, H., So, K.K., Gao, X.: Bandwidth enhancement of a monopolar patch antenna with V-shaped slot for car-to-car and WLAN communications. IEEE Trans. Veh. Technol. **65**(3), 1130–1136 (2016)
4. Besselink, B., Turri, V., van de Hoef, S.H., Liang, K.Y., Alam, A., Mårtensson, J., Johansson, K.H.: Cyber–physical control of road freight transport. Proc. IEEE **104**(5), 1128–1141 (2016)
5. Zheng, Y., Li, S.E., Li, K., Borrelli, F., Hedrick, J.K.: Distributed model predictive control for heterogeneous vehicle platoons under unidirectional topologies. IEEE Trans. Control Syst. Technol. **25**, 899–910 (2016)
6. Ploeg, J., Shukla, D.P.D., van de Wouw, N., Nijmeijer, H.: Controller synthesis for string stability of vehicle platoons. IEEE Trans. Intell. Transp. Syst. **15**(2), 854–865 (2014)
7. Ploeg, J., Van De Wouw, N., Nijmeijer, H.: Lp string stability of cascaded systems: application to vehicle platooning. IEEE Trans. Control Syst. Technol. **22**(2), 786–793 (2014)
8. Kianfar, R., Falcone, P., Fredriksson, J.: A control matching model predictive control approach to string stable vehicle platooning. Control Eng. Pract. **45**(2010), 163–173 (2015)
9. Swaroop, D., Hedrick, J.K.: String stability of interconnected systems. IEEE Trans. Autom. Control **41**(3), 349–357 (1996)
10. Seiler, P., Sengupta, R.: An H∞ approach to networked control. IEEE Trans. Autom. Control **50**(3), 356–364 (2005)
11. Maschuw, J.P., Kessler, G.C., Abel, D.: LMI-based control of vehicle platoons for robust longitudinal guidance. IFAC Proc. vol. **17**(1 Part 1), 12111–12116 (2008)
12. Lu, X.-Y., Hedrick, J.K.: A panoramic view of fault management for longitudinal control of automated vehicle platooning. In: Dynamic Systems and Control, vol. 2002, pp. 723–730, November 2002
13. Lygeros, J., Godbole, D.N., Broucke, M.: A fault tolerant control architecture for automated highway systems. IEEE Trans. Control Syst. Technol. **8**(2), 205–219 (2000)
14. Axelsson, J.: Safety in vehicle platooning: a systematic literature review. IEEE Trans. Intell. Transp. Syst. **18**(5), 1033–1045 (2017)
15. De Castro, R., Brembeck, J.: A command governor approach for platooning applications. IEEE Intell. Veh. Symp. Proc. **1**, 978–984 (2017)

Simulation and Analysis

Systemic Model of Cardiac Simulation with Ventricular Assist Device for Medical Decision Support

Jônatas C. Dias[(⊠)], Jeferson C. Dias, Marcelo Barboza,
José R. Sousa Sobrinho, and Diolino J. Santos Filho

Escola Politécnica da Universidade de São Paulo, São Paulo, SP, Brazil
jxdias@ymail.com

Abstract. Biomedical Engineering uses computational simulation models that are increasingly refined to represent human physiological systems. These models allow changes and analysis *In Silico*, optimizing implementation in a real (human) system. This work explores the CVSim computational model of the cardiovascular system, developed and used by MIT and Harvard Medical School, since 1984. The purpose of this work is the prospect of a resilient and adaptive system that by obtaining a mass of data; associated to the hemodynamic behavior of the set: Heart and Ventricular Assist Device (VAD); through simulations, to predict the behavior of this system in an autonomous intelligent environment, which can support the decision making about possible adverse events that may occur. It is intended to consider the profile of the patient with heart disease and the exploration of data: *Big Data Analytics*.

Keywords: Cardiovascular · Simulation · Ventricular Assist Device (DAV)
In Silico · Context Rich · Big Data Analytics · Health 4.0

1 Introduction

The systems observed in nature tend to adapt in order to achieve the present dynamic equilibrium between the parts of these natural systems, creating an adaptive property, in many cases called homeostasis, which designates the process by which living organisms maintain is in steady state of the physiological system [1, 2]. Body temperature, for example: is maintained at its narrow levels, even with external thermal fluctuations.

Added to this concept, stress in biology as it is understood as the response, or the reaction, of the organism before the rupture of the normal balance of its functions [3, 4]. This reaction was called the General Adaptation Syndrome (SAG), with three distinct phases [3–5]: (1) Alarm or alert - occurs when there is an internal imbalance of the organism and the beginning of actions of coping with the stressing agent, normally triggered by the Sympathetic Autonomic Nervous System (SNAs); (2) Resistance - physiological and behavioral responses to neutralize the stressor; (3) Exhaustion - occurs in situations of prolonged internal imbalance, but the body continues to respond in a chronic way to this imbalance seeking restoration (homeostasis), causing an energy

© IFIP International Federation for Information Processing 2018
Published by Springer International Publishing AG 2018. All Rights Reserved
L. M. Camarinha-Matos et al. (Eds.): DoCEIS 2018, IFIP AICT 521, pp. 235–242, 2018.
https://doi.org/10.1007/978-3-319-78574-5_22

overload and exhaustion of the system [6–8]. It is perceveid then, in the situation of exhaustion, the need for an adjustment of the control variable of the homeostasis trigger threshold, the so-called "*Set Points*", perceived by Mrozovsky [9] when it proposes the use of the term reostease (*reos = change*) to extend the concept of homeostasis, emphasizing that appropriateness of the "*adjustment value*" is part of the way the adaptation mechanisms of natural systems operate, as observed in chronobiology, recurrent situations in pre-defined cycles that influence the balance of the organism. For example, in the hibernation cycle of some animals, the reduction of body temperature is not an imbalance of the system, otherwise the alert phase of homeostasis would be triggered, causing the system to deplete, instead, there is an adaptability of the biological system to this new condition.

There is still present in the natural systems a feature called resilience. The term resilience has several definitions, for example: "*The ability to recover and maintain adapted behavior after a damage.*" [10, 11]. "*The ability to maintain a stable equilibrium.*" [12–14]. These definitions, although adequate to this study, are linked to the area of psychology, related to the capacity of adaptation and overcoming adversities by the individual. In the exact sciences, the term integrates the studies on resistance of materials, used by the English Thomas Young in 1807, one of the first to use the term to define the capacity to return to the original state with the notion of elasticity [15]. After its discovery and practical applications in the field of physics, other areas of science borrowed the term, due to the possibility of extending its meanings beyond the resistance of the materials, being applied to explain characteristics of adaptability in diverse subjects.

The present article proposes the introduction of these concepts, described in the previous paragraphs, in a system that allows the medical team to deal with; especially with the variables of the engineering area, coming from the devices of ventricular assistance; with uncertainties in the aid of the medical diagnosis process, as well as in the time of response to the adversities of undesired health events of the monitored patient. The systemic proposal in this article is a prospection that combines the human potential (medical and medical team) with the potential of the computational systems, in this way the adaptability and the resilience of the system does not lie only in an artificial intelligence applied to a device, but to a whole which includes: (a) *In Vivo* that deals with a "*Patient/Device*" life cycle with the objective of creating a knowledge base with several records, to identify patterns and correlations of behavior that may alert the physician to risk conditions for a given patient in a predictive way; (b) *In Silico* using computational simulation models to represent human physiological systems, focusing on the "*Heart/DAV*" system; (c) *Analytics* with the objective of obtaining behavioral patterns, using the information obtained in *In Silico* and *In Vivo*, to provide decision support to the medical team.

2 Relation with Resilient Systems

The proposal of resilience observed in this article is directed to the concept of adaptation and overcoming (overcoming adversity or the stressor agent), in order to keep the "*Patient/Device*" system healthy despite adversities. This resilience is achieved with the proposal to unite the strong characteristics: (a) the human element that has creativity skills and problem solving ability through reasoning and judgment for clinical decisions; (b) with the information and communication technology, in obtaining information (not just data) in real time, allowing a rapid medical evaluation of the patient's state and decision making promoting the reestabilization of the system "*Patient/Device*" of adversities, resuming its state of equilibrium. This proposal is in line with the concepts of health 4.0 (here used to represent the integration of information technology, or medical technologies, to medical devices, medical equipment and e-health [16]) with the aim of early diagnosis [17].

3 Objective

The present article proposes to use the behaviors of the natural systems: the adaptivity (homeostasis and reostase) and the resilience; and implement these characteristics in the ventricular assist devices, since, at present, these devices do not have these characteristics. Careful implementation of these features requires a simulation model to systematically assess their viability. The main appeal to the use of simulation models is the fact that the real human system, under study, does not suffer any disturbance, leading to the development of an efficient project, before carrying out any physical change [18].

The use of simulation models is recommended for systems or environments with [19]: (a) dynamics (vary in time); (b) when they are interactive (presence of components that interact with each other, and this interaction affects the behavior of the whole system); (c) when they are complicated (numerous variables that interact in the system).

4 Method

The research has a theoretical delineation [20] and descriptive methodological, at the moment that it is sought to discover the existence of associations and correlations between variables [20] and proposes a method of experimentation using *In Silico* models, that is, computational simulation (see Fig. 1). Simulation is the process of designing a computational model of a real system and conducting experiments with this model in order to understand its behavior [21, 22].

Together with this method of approach, other technical procedures are necessary, such as: Data collection (observed in item 5 below) is followed quantitatively by means of a questionnaire (survey type), composed of previously established questions and documentary survey applied in the *In Vivo* phase (see Fig. 1). For the data treatment, the coding of the questions and a statistical analysis of the data are used, mainly to describe the variables in the *In Vivo* phase. Application of supervised predictive models and analysis of occurrence histories to determine dysgnostic and prognostic situations.

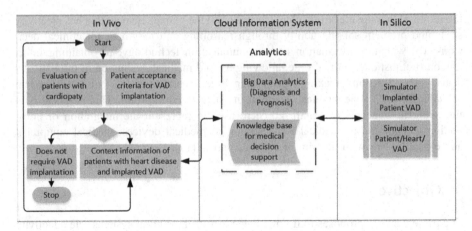

Fig. 1. Medical decision support system with adaptive control **Source:** Dias, J. C. 2017 (Author)

5 Data Collection and Handling

5.1 Current Data Collection and Processing

Traditionally, the information about the condition of the patient and the device is obtained through a pre-scheduled medical consultation, whose regularity or eventuality depends on each patient. At the time of the consultation the doctor performs the clinical examination and obtains information of the behavior of the device. However, this information compiles a set of data (a "*Data Log*"), which is basically a record in the order of occurrence of events. The doctor needs to interpret this data to take useful information to make decisions, for example: whether or not it is necessary to adjust the speed of the device.

Most of these devices do not have long term storage, causing the collected data to remain in a volatile condition (a data write overlay in the controller buffer) not moving to a persistence condition (a long term storage medium). Another issue observed in most of these devices is the absence of the trend assessment feature, or even cross-checking between these data to detect a patient's risk situation. There are, however, devices in which data is collected in the controller and sent to a remote system, such as a smartphone or notebook, for example, and directed to the attending physician. However, for all of these cases, the data do not undergo a process of data transformation into information, providing a diagnostic (descriptive) assessment adequate to the health professional, as well as providing a favorable condition of prognosis (predictive) medical. In general, the interpretation of these data is the responsibility of the physician and/or medical staff to perform evaluations from the events recorded in the "*Data Log*", or even alerts indicated by the device, unrelated to previous events or with the sequence of unwanted and chained events.

This current condition allows us to propose a new way of collecting and treating data from these ventricular assist devices, transforming data into useful information for decision making.

5.2 Proposed Data Collection and Treatment

A new data collection proposal will be supported by a collection cycle, Fig. 1 - *In Vivo*, intended for patients eligible for the implantation of a VAD, forming a Cloud Information System (phase 3), which can provide knowledge of patient profiles and their respective physiology, in response to behavioral variations, as well as medical interventions during this cycle [23].

This knowledge base will allow the identification of the predictability of failures (a failure, for this article, is an alert occurrence or a symptom unfavorable to the patient's health) in a standard way presented in phase 3 of this article. For each failure an action to resume the health of the patient will be administered by the physician, composing a base of causes with their respective effects and medical procedures. This is because the proposed method allows to simulate in computer exhaustively, the various faults and their possibility of occurrence. For each failure a procedure can be designed to avoid its occurrence or minimize its impact on the patient.

6 Proposal for Research Plan

The present proposal seeks to help the medical team to deal, especially with the variables of the engineering area coming from the devices of ventricular assistance. Providing a way to deal with this uncertainty by providing automated procedures, based on probabilistic reasoning, to aid in the diagnosis process. In order to attend to the expectation and to understand the behavior of the "*Heart/DAV*" system, this article proposes a research model divided in phases.

6.1 Proposal: PHASE 1 - *In Silico*

Biomedical engineering uses increasingly refined computer simulation models to represent human physiological systems. These models allow changes and *in silico* analysis, optimizing the implementation in a real system (human). This phase considers a CVSim computational simulation model of the cardiovascular system, developed and used by MIT and Harvard Medical School, since 1984. This model is intended to include an additional computational model that simulates the behavior of a continuous flow DAV. Forming a dichotomous system: "*Heart/VAD*", with the purpose of obtaining a mass of data associated to the hemodynamic behavior of this new system, through simulations. Figure 1 – *In Silico*.

6.2 Proposal: PHASE 2 – *In Vivo*

This phase is the center of the research and has two important challenges, which are: (1) the elaboration of a process containing a cycle of collection of information pertinent to the balance of performance of the VAD in the support of the blood pumping to the adequate tissue perfusion and (2) the identification of the variables that influence this balance. The information collected in all the states of the process will help in the identification of a set of influence variables in the behavior of the *"Patient/Device"* system, with the objective of identifying relationships and impact, Fig. 1 - *"In Vivo"*. The understanding of the behavior of these variables will provide the descriptive and predictive identification of this system.

6.3   Proposal: PHASE 3 – Information System/*Analytics*

Big Data Analytics technology will be responsible for obtaining behavior patterns allowing for diagnostic actions and medical prognosis, supporting the decision and minimizing the uncertainties so that the technological complexity involved is transparent to the physician. The systemic view should consider the use of variables that may provide a way of proceeding from diagnosis to prognosis.

A knowledge base based on simulations and a systematic that allows the learning in diverse situations, providing knowledge to overcome the uncertainties of an unstructured decision, guided by the various analyzes available by Big Data Anlytics that will provide diagnosis and medical prognosis.

Coupled to these phases (PHASE 1 - *In Silico*, PHASE 2 - *In Vivo*, PHASE 3 - *Analytics*): a decision support information system and a ventricular assist device control supported by artificial intelligence, complete a Support System to Medical Decision with Adaptive Control, Fig. 1. The adaptive term means that according to the patient's physiology and human anatomy in relation to their habits and way of life, the system calibrates itself, maintaining homeostasis and reostosis of the cardiovascular system, insofar as: (a) the patient makes demands distinct from the cardiac system (some physical requirement: movement, some exercise, etc.) and (b) to the extent that the human body undergoes changes, especially during aging. The decision and setting of *"Set Point"* can be superimposed on the automatic adjustment of the system by the doctor or medical staff, in this way the system adapts to different requirements of needs.

Information technology, based on the digitalization, are inspirations for the construction of this system. For example, Internet application of things (IoT), along with cloud computing (Cloud Cumputing), will allow the meeting of the various personalization characteristics in each VAD, with a *"Context Rich"* obtained for each patient. Allied to other information such as: patient physiology, implant technique to be performed; and the different types of devices; are directed to a common knowledge base, allowing the application of *Big Data Analytics* technology obtain correlations and standards for descriptive and predictive evaluations, providing better decision support parameters and monitoring of the "Patient/Device" health condition, that even the distance; can observe imminent real-time risk conditions.

7   Results and Discussion

It is observed that, in general, the monitoring systems that support the VADs, basically provide *"Data Log"* of the *"Patient/Device"* state and only at the moments of the query; when requested by the physician, or at times when the patient feels some symptomatic discomfort due to pathological reason; when the controller accompanying the patient unloads the "*Data Log*" for evaluation. This evaluation system not allow the physician to have a real-time view of events pertinent to the patient's condition.

A system with adaptativias characteristic and resilience; allied to information technology that makes it possible to extract information about behavior and trends; has the ability to verify situations and signal smarter alarms by allowing the crossing of variables under different conditions in real time.

Adaptivity and resilience are achieved through a joint work: man and machine, in such a way that the potentiality of each is orchestrated in equilibrium according to its potentialities. This balance is achieved by knowing the strengths and weaknesses of each of them, promoting health and the existence of a greater good; the life.

References

1. Cannon, W.B.: Organization for physiological homeostasis. Physiol. Rev. **9**(3), 399–431 (1929)
2. Oomen, P.J.A., Holland, M.A., Bouten, C.V.C., Kuhl, E., Loerakker, S.: Growth and remodeling play opposing roles during postnatal human heart valve development. Sci. Rep. 1–13 (2018)
3. Selye, H.: The general adaptation syndrome and the diseases of adaptation. J. Clin. Endocrinol. Metab. **6**(2), 117–230 (1946)
4. Cunanan, A.J., et al.: The general adaptation syndrome: a foundation for the concept of periodization. Sport. Med. 1–11 (2018)
5. Buckner, S.L., Mouser, J.G., Dankel, S.J., Jessee, M.B., Mattocks, K.T., Loenneke, J.P.: The General Adaptation Syndrome: potential misapplications to resistance exercise. J. Sci. Med. Sport **20**(11), 1015–1017 (2017)
6. Nelson, R.J.: An Introduction to Behavioral Endocrinology, 2nd edn. Sinauer Associates-Inc, Sunderland (2000)
7. Nansel, T.R., Thomas, D.M., Liu, A.: Efficacy of a behavioral intervention for pediatric type 1 diabetes across income. Am. J. Prev. Med. **49**(6), 930–934 (2015)
8. Webb, N.E., Little, B., Loupee-Wilson, S., Power, E.M.: Traumatic brain injury and neuro-endocrine disruption: medical and psychosocial rehabilitation. NeuroRehabilitation **34**(4), 625–636 (2014)
9. Mrosovski, N.: Rheostasis: The Physiology of Change, vol. 19, no. 2. Oxford University Press, New York, Oxford (1990)
10. Blum, R.W.: Risco e resiliência: sumário para desenvolvimento de um programa, vol. 16, no. 9, August 1977
11. Masoomi, H., van de Lindt, J.W.: Restoration and functionality assessment of a community subjected to tornado hazard. Struct. Infrastruct. Eng. **14**(3), 275–291 (2018)
12. Bonanno, G.A.: Loss, trauma, and human resilience: have we underestimated the human capacity to thrive after extremely aversive events? Am. Psychol. **59**(1), 20–28 (2004)
13. Chiang, Y.C., Ling, T.Y.: Exploring flood resilience thinking in the retail sector under climate change: a case study of an estuarine region of Taipei City. Sustainability **9**(9), 1650 (2017)
14. Nahayo, L., et al.: Extent of disaster courses delivery for the risk reduction in Rwanda. Int. J. Disaster Risk Reduct. **27**, 127–132 (2018)

15. Timoshenko, S.P.: History of Strength of Materials (Dover Civil and Mechanical Engineering). Dover Publications, New York (1953)
16. Aliança Brasileira da Indústria Inovadora em Saúde – ABIIS: Saúde 4.0 - Propostas para Impulsionar o Ciclo das Inovações em Dispositivos Médicos (DMAs) no Brasil. Aliança Brasileira da Indústria Inovadora em Saúde – ABIIS, São Paulo (2015)
17. Silva, E.B., Scoton, M.L.R.P.D., Dias, E.M., Pereira, S.L.: AUTOMAÇÃO & SOCIEDADE - Quarta Revolução Industrial, um olhar para o Brasil. Braspot, Rio de Janeiro (2018)
18. Stewart, R.: Simulation: the practice of model development and use (2004)
19. Pidd, M.: Computer Simulation in Management Science, Part I, 4th edn., p. 332. Wiley, New York (1998)
20. Mattar, S.L.S., Najib, F., Oliveira, B.; Motta, S.: Pesquisa de Marketing: metodologia, planejamento, execução e análise, 7th edn., Rio de Janeiro (2014)
21. Moza, A., et al.: Parametrization of an in-silico circulatory simulation by clinical datasets – towards prediction of ventricular function following assist device implantation. Biomed. Eng./Biomed. Tech. **62**(2), 123–130 (2017)
22. Centeno, M.A.: An introduction to simulation modeling. In: Winter Simulation Conference, pp. 15–22 (1996)
23. Dias, J.C., Dias, J.C., Filho, D.J.S.: Aplicação das tecnologias da indústria 4.0 em ambiente inteligente de tomada de decisão médica para pacientes com dav implantado. 14 Congr. da Soc. Lat. Am. Biomateriais, Orgãos Artif. e Eng. Tecidos - SLABO, pp. 74–81 (2017)

Simulation and Experiment on Electric Field Emissions Generated by Wireless Energy Transfer

E. N. Baikova[1,3], L. Romba[1,2], R. Melicio[4,5(✉)], and S. S. Valtchev[1,2]

[1] UNINOVA-CTS, Caparica, Portugal
[2] FCT of Universidade Nova de Lisboa, Lisbon, Portugal
[3] EST Setúbal, Instituto Politécnico de Setúbal, Setúbal, Portugal
[4] Departamento de Física, Escola de Ciências e Tecnologia, ICT,
Universidade de Évora, Évora, Portugal
ruimelicio@gmail.com
[5] IDMEC, Instituto Superior Técnico, Universidade de Lisboa, Lisbon, Portugal

Abstract. This paper presents a wireless energy transfer (WET) system operating at the frequency of tens of kHz. It treats the modeling and simulation of WET prototype and its comparison with experimental measuring results. The wireless energy transfer system model was created to simulate the electric field between the emitting and the receiving coils, applying the finite element method. The results from the simulation are compared to the measured values of the electric field emission from the wireless energy transfer equipment. In the recent years the interest in the WET technology, especially for the electric vehicles (EV) batteries charging, is rapidly growing. The WET systems pollute the environment by electromagnetic emissions. Due to the expanding use of this technology in industrial and consumer electronics products, the problems associated with the electromagnetic compatibility (EMC), and the adverse impact on the human health becomes highly important.

Keywords: Wireless energy transfer · Electric field · Modeling · Simulation
Prototype

1 Introduction

The attention to the electric transportation system is aimed to the improvement of the safety, including the safer battery charging, e.g. by contactless energy transfer [1–3]. In the last decade, the magnetic resonance technology is considered to be the most suitable for the contactless EV battery charging [2, 4, 5]. It does not require an accurately parking of the vehicle, being not so sensitive to the alignment [4, 5].

However, the WET systems represent the EMC problems. The electromagnetic field (EMF), through which the energy is transferred between the emitting coil and the receiving coil, is a high-frequency field and induces electromagnetic interferences (EMI) into the other electronic devices. This EMF has high intensity and so may influence the operation of the communication channel by which the emitter and the receiver exchange

© IFIP International Federation for Information Processing 2018
Published by Springer International Publishing AG 2018. All Rights Reserved
L. M. Camarinha-Matos et al. (Eds.): DoCEIS 2018, IFIP AICT 521, pp. 243–251, 2018.
https://doi.org/10.1007/978-3-319-78574-5_23

the information [4, 6]. Moreover, the electromagnetic field emitted by the WET system can induce high voltages and currents in the living organisms [4–8]. Thus, one of the fundamental concerns for the WET development is the conformity with EMF safety regulations and standards [9, 10].

The EMC issue of the WET system requires a detailed analysis of the EM processes in the living organisms exposed to EMF. A possibility to estimate the EMF impact on the living beings is by modeling and simulation. Recently, some papers were published, discussing the modeling and simulation of the electromagnetic process in WET systems [8, 11, 12].

In [8], a dosimetric numerical study is presented, assessing the exposure of the human in a proximity of a wireless EV battery charging prototype. The prototype is delivering a power of 560 W at operating frequency 85 kHz. The electric field induced in a homogeneous ellipsoid phantom and in an anatomical model of human body, exposed to a WET system prototype was evaluated.

In [11], the magnetic field from the WET system for realistic EV charging at a power level of 7 kW, is modeled and simulated using the Ansys Maxwell software. To validate the model, the simulated magnetic field is compared to the measured one. Finally, the electric field induced in a human body model by a leaked magnetic field from a WET system, is evaluated.

In [12], the scenarios of multiple WET systems are considered, working in parallel simultaneously. Using the Ansys Maxwell software and finite element method (FEM) the electric field and voltage, induced in the human body by the leakage EMF, are simulated and evaluated. The specific absorption rate in specific tissues of the human body, positioned at various orientations, are also simulated and assessed. The simulation results are compared to the basic margins defined in the international regulations, determined as boundaries for the human exposure to time-varying EMFs.

In the existing research papers on a WET system, more attention is dedicated to the magnetic field, which is logical having in mind that the transport of energy is made by the magnetic field. Few works deal with simulating and measuring of the electric field produced by WET systems. However, the coupling mechanisms of the EMF components, i.e. electric and magnetic fields, are different. Thus, for a more complete characterizing of the human exposure, both components must be simulated and measured [13].

In this paper, the modeling and simulation of the WET system operating at the frequency of 20 kHz, is presented. The results of electric field simulation are compared to the measurements of the electric field emissions radiated by a prototype of the WET system. The simulation and measurement results are compared to the international EMF safety standards and regulations.

2   Technological Innovation for Resilient Systems

The concept of system resilience is important and hyper-popular over the last few years [14]. It implies that the systems should be able to give safety and reliability, keeping the capability to act to sustain and rectify the infrastructure components without disturbing the service [15]. Some definitions of resilience comprise requirements on the recovery of the system, safety, including the safety of the humans, some parts of security and reliability of the system [16].

In the recent years, the market for EV is growing steadily. Most of EV battery supply today are unidirectional, i.e. they can only store electricity by charging the battery. Correspondingly, most of EV charging stations are designed for unidirectional electricity delivery to the vehicle [17]. With an increased infrastructure and market demand on the electric grid, the bidirectional, i.e. V2G-capable, EVs should be able to deliver energy from their batteries to the grid via bidirectional charging stations by using WET systems. The purpose of the smart grids is to reduce the energy costs, and in the same time, to reach a sustainable balance between the production and consumption, increasing the resiliency of the grid. Thus, using WET systems, EVs can improve the sustainability and resiliency of the power grids and the quality of electrical energy transferred to the customers [17].

On the other hand, as WET technology is becoming more extensively applied in high power applications, the human exposure to EMFs consequently increases [12]. Moreover, the electronic components of infrastructure may be impacted by the presence of oscillating EMFs associated with a WET system. This means that the WET system may become a source of hazard for human health and electronic equipment and, this way, reduce the infrastructure resilience. So, to attenuate the negative impact of the EM fields from wireless EV charging system and to increase the system resilience, the mitigation strategies should be employed. In [18, 19] some techniques for the reduction of the EMFs produced by wireless EV charging systems are proposed, applying various shielding methods, e.g. conductive, magnetic, active shield and reactive shield coil. By applying those shielding methods, the general resilience will be improved.

3   Modeling

The finite element method has been chosen for modeling of the WET system. A frequency-domain analysis was applied to evaluate the electric field distribution in the WET surrounding area.

The electromagnetic phenomena in the WET system are governed by the Maxwell equations, which are given below in their differential form [20]:

$$\nabla \times \mathbf{E} = -\frac{\partial \mathbf{B}}{\partial t} \tag{1}$$

$$\nabla \times \mathbf{H} = \mathbf{J} + \frac{\partial \mathbf{B}}{\partial t} \tag{2}$$

$$\nabla . \mathbf{D} = \rho \tag{3}$$

$$\nabla \cdot \mathbf{B} = 0 \tag{4}$$

The equation of continuity is given by:

$$\nabla \cdot \mathbf{J} = -\frac{\partial \rho}{\partial t} \tag{5}$$

where the vectors $\mathbf{E}$ is the electric field, $\mathbf{H}$ is the magnetic field, $\mathbf{D}$ is electric displacement field, $\mathbf{B}$ is magnetic flux density and $\mathbf{J}$ is conducting current density. The following group of constitutive equations connects the field vectors with the environment parameters:

$$\mathbf{B} = \mu_0\mu_r\mathbf{H} \quad \mathbf{D} = \varepsilon_0\varepsilon_r\mathbf{E} \quad \mathbf{J} = \sigma.\mathbf{E} \tag{6}$$

where μ_0 and ε_0 are magnetic permeability and electric permittivity of the free space, respectively, μ_r and ε_r are the relative permittivity and relative permeability of a medium, respectively, σ is the electric conductivity.

In the quasi-static approximation, the displacement currents defined by $\mathbf{D}$ are negligible, so the Maxwell Eqs. (1) and (2) in the frequency domain are given by:

$$\nabla \times \mathbf{H} = \mathbf{J} \tag{7}$$

$$\nabla \times \mathbf{E} = -j\omega\mathbf{B} \tag{8}$$

The magnetic flux density $\mathbf{B}$ in terms of the magnetic vector potential $\mathbf{A}$ is given by:

$$\mathbf{B} = \nabla \times \mathbf{A} \tag{9}$$

Combining the Eqs. (1) to (9) gives the Eq. (10), from where the magnetic vector potential $\mathbf{A}$ can be determined:

$$\nabla \times \left(\frac{1}{\mu} \cdot \nabla \times \mathbf{A}\right) + j\omega\sigma\mathbf{A} = \mathbf{J} \tag{10}$$

Next, the magnetic field distribution is obtained using (9). Lastly, the electric field is calculated in accordance with the constitutive Eq. (6) and the Ampere law (7).

4 Proposed Design and Simulation

The proposed design of the WET system model including coils, ferrite cores, aluminum profiles and surrounding (air) domain, has been simulated using the numerical software. Due to the symmetry of the WET system, 1/4 of the full structure, which is shown in Fig. 1, was modeled and simulated. The model used for the electromagnetic simulation with size in mm shown in Fig. 1.

The WET charger system consists of ferrite cores for the emitter and receiver coils made of copper with the section of 1 mm^2 and of aluminum profile. The number of turns in the emitter coil is $N_e = 36$ and in the receiver coil is $N_r = 4$. The simulation was performed by applying to the WET system the first harmonic of the rectangular voltage waveform, neglecting the higher order harmonics.

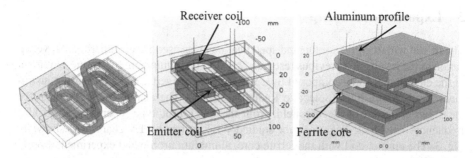

Fig. 1. Modelling set of ferrite cores, aluminum profiles and inductive coils.

The magnitudes of the electric field from 262 W WET system were simulated for the air gap between the emitter and the receiver coils of 20 mm. The electric field distribution for 20 kHz frequency range with contours lines, which represent the magnitude of the electric field, is shown in Fig. 2. The bar on the right is illustrating the values of E-field (V/m) in certain parts of the model.

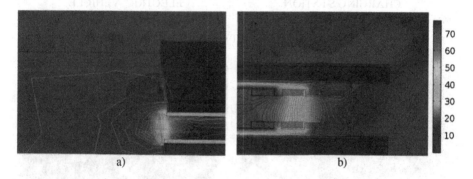

Fig. 2. WET system E-field generated at the air gap of: (a) side view; (b) cross-section view.

According to the International Commission on Non-Ionizing Radiation Protection (ICNIRP) Guideline, the reference level for general public exposure to time-varying electric field for the frequency range of 3 kHz to 10 MHz is 83 V/m [9]. As it is shown in Fig. 2, the electric field intensity is below the exposure limits defined by ICNIRP. However, in the regions limited by the ferrite cores and aluminum profiles the value of electric field reaches 75 V/m, i.e. is close to this limit. These results also confirm that the strength of the electric field decreases rapidly with the distance from the WET system and reaches 5.72 V/m at the distance of 200 mm.

From the Fig. 2 it can be observed that the ferrite core and aluminum profile are effectively confining and guiding the EMF.

5   Experimental Setup

The model with the given setup and specifications was used for an experimental verification of the results of a previous research [21, 22]. The equipment for the measurement of the electric field was composed by a WET power converter and a measuring instrument Narda SRM-3000 (Selective Radiation Meter). The analysis is focused on the measured emission of electric field, from the WET prototype, operating at 20 kHz frequency. The block diagram of the studied WET system for EV charging is shown in Fig. 3. The emitter coil with the ferrite core, aluminum profile and experimental setup of the WET prototype is shown in Fig. 4.

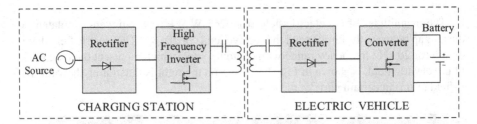

Fig. 3. Block diagram of the WET system for EV battery charging

Fig. 4. 3D vision of emitter coil and experimental setup of the WET prototype.

Applying the Fourier series expansion, the square wave AC input which supplies the emitter circuit can be transformed into a sum of odd harmonics. So, to estimate the electromagnetic interference impact from the WET prototype with output power value of 262 W, the RMS values of harmonics generated by the WET system with an air gap of 20 mm were measured.

The used measuring equipment was able to evaluate frequencies starting from 100 kHz. In this way, the amplitudes of the harmonics were measured from the 5th harmonic, whose frequency corresponds to 100 kHz. The minimum squares method was used to obtain the 1st and the 3rd harmonics amplitudes. According to this method, the experimental data were attuned with a coefficient of determination $R^2 = 0.9999$. The high value of the coefficient of determination shows that the trend line is quite exactly consistent with the data. The amplitudes of the 25 odd harmonics, including the 1st and the 3rd ones calculated by minimum squares method, is shown in Fig. 5.

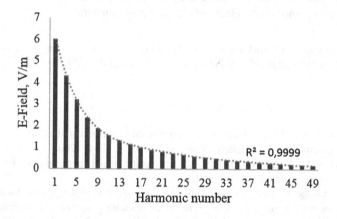

Fig. 5. Odd harmonics amplitudes generated by the WET prototype.

According to ICNIRP Guidelines [9] for situations with simultaneous exposures to fields of multiple frequencies, these exposures have a cumulative effect. In this case, the following criteria [9] must be applied:

$$\sum_{j=1\,Hz}^{10\,MHz} E_j / E_{R,j} \leq 1 \tag{11}$$

where E_j is the electric field strength at frequency j, $E_{R,j}$ is the reference level of electric field strength at frequency j according to [9].

The total level of electric field radiated by WET prototype operating at 20 kHz frequency is $E_j = 29.71$ V/m and $E_{R,j} = 83$ V/m. Applying (11) the criteria value is 0.36, i.e. below 1. Thus, this result confirmed that the level of the electric field generated by WET prototype is below the exposure limits defined by ICNIRP [9].

By comparing the simulated and measured results of the electric field 1st harmonic, it can be concluded that the simulation results are in agreement with the measured data: the simulated electric field −5.62 V/m and the measured electric field −6.03 V/m. The

WET system considered in this research, in fact, is not a linear system. Thus, the discrepancy between simulated and measured results can be explained by the linearity assumptions taken in the computational simulation.

6 Conclusions

The WET system described here is modeled and simulated at the frequency of 20 kHz. The electric field density was simulated at an air gap of 20 mm between the emitter and the receiver, and then measurements were executed at the same physical and geometrical conditions. The simulated electric field values are compared to the measured ones. The comparison shows an error of roughly 6.7% between the simulation and the reality. The present study is useful for a more complete knowledge about the EMI, not only about the magnetic field component produced by the WET system, but also to take a deeper view into the influence of the electric field on the human health.

Acknowledgement. This work was funded in part by the Center of Technology and Systems of Uninova and the Portuguese FCT-PEST program UID/EEA/00066/2013.

References

1. Boys, J.T., Covic, G.A., Green, A.W.: Stability and control of inductively coupled power transfer systems. IEE Proc. Electr. Power Appl. **147**(1), 37–43 (2002)
2. Li, S., Mi, C.C.: Wireless power transfer for electric vehicle applications. IEEE J. Emerg. Sel. Top. Power Electron. **3**(1), 4–17 (2015)
3. Imura, T., Okabe, H., Hori, Y.: Basic experimental study on helical antennas of wireless power transfer for electric vehicles by using magnetic resonant couplings. In: Proceedings of the IEEE Vehicle Power and Propulsion Conference, Dearborn, USA, pp. 936–940 (2009)
4. Kim, S., Park, H.-H., Kim, J., Kim, K., Ahn, S.: Design and analysis of a resonant reactive shield for a wireless power electric vehicle. IEEE Trans. Microw. Theory Tech. **62**(4), 1057–1066 (2014)
5. Gao, Y., Farley, K.B., Ginart, A., Tse, Z.T.H.: Safety and efficiency of the wireless charging of electric vehicles. Proc. Inst. Mech. Eng. Part D J. Automobile Eng. **230**(9), 1196–1207 (2015)
6. Baikova, E.N., Valtchev, S.S., Melicio, R., Krusteva, A., Fernão Pires, V.: Study of the electromagnetic interference generated by wireless power transfer systems. Int. Rev. Electr. Eng. **11**(5), 526–534 (2016)
7. Christ, A., Douglas, M.G., Roman, J.M., Cooper, E.B., Sample, A.P., Waters, B.H., Smith, J.R., Kuster, N.: Evaluation of wireless resonant power transfer systems with human electromagnetic exposure limits. IEEE Trans. Electromagn. Compat. **55**(2), 265–274 (2013)
8. Pinto, R., Lopresto, V., Genovese, A.: Human exposure to wireless power transfer systems: a numerical dosimetric study. In: Proceedings of the 11th European Conference on Antennas and Propagation (EUCAP), Paris, France, pp. 1–3 (2017)
9. International Commission on Non-Ionizing Radiation Protection: Guidelines for limiting exposure to time-varying electric and magnetic fields for low frequencies (1 Hz–100 kHz), Health Phys. **99**, 818–836 (2010)

10. IEEE Standard for Safety Levels with Respect to Human Exposure to Radio Frequency Electromagnetic Fields, 3 kHz to 300 GHz, IEEE Std. C95.1 (2005)
11. Laakso, I., Hirata, A.: Evaluation of the induced electric field and compliance procedure for a wireless power transfer system in an electrical vehicle. Phys. Med. Biol. **58**, 7583–7593 (2013)
12. Wen, F., Huang, X.: Human exposure to electromagnetic fields from parallel wireless power transfer systems. Int. J. Environ. Res. Public Health **14**(2), 1–15 (2017)
13. International Agency for Research on Cancer (IARC), Non-ionizing radiation, part 2: radiofrequency electromagnetic fields, IARC Monograph, vol. 102, Lyon, France (2013)
14. Woods, D.: Four concepts for resilience and the implications for the future of resilience engineering. Reliab. Eng. Syst. Saf. **141**, 5–9 (2015)
15. Sridhar, S., Hahn, A., Govindarasu, M.: Cyber–physical system security for the electric power grid. Proc. IEEE **100**(1), 210–224 (2012)
16. Sheard, S.: A framework for system resilience discussions. In: INCOSE International Symposium, vol. 18(1), pp. 1243–1257 (2008)
17. Monteiro, V., Gonçalves, H., Ferreira, J.C., Afonso, J.L.: Batteries charging systems for electric and plug-in hybrid electric vehicles. In: New Advances in Vehicular Technology and Automotive Engineering, 1st ed., pp. 149–168 (2012)
18. Ahn, S., Pak, J., Song, T., Lee, H., Byun, J.-G., Kang, D., Choi, C.-S., Kim, E., Ryu, J., Kim, M., Cha, Y., Chun, Y., Rim, C.-T., Yim, J.-H., Cho, D.-H., Kim, J.: Low frequency electromagnetic field reduction techniques for the on-line electric vehicle (OLEV). In: IEEE International Symposium on Electromagnetic Compatibility, Fort Lauderdale, Florida, USA (2010)
19. Kim, M., Kim, S., Chun, Y., Park, S., Ahn, S.: Low frequency electromagnetic compatibility of wirelessly powered electric vehicles. In: International Symposium on Electromagnetic Compatibility, Tokyo (2014)
20. Šolín, P.: Partial Differential Equations and the Finite Element Method, pp. 269–291. Wiley, New York (2006)
21. Baikova, E.N., Valtchev, S.S., Melício, R., Pires, V.M.: Electromagnetic interference impact of wireless power transfer system on data wireless channel. In: Camarinha-Matos, L.M., Falcão, A.J., Vafaei, N., Najdi, S. (eds.) DoCEIS 2016. IAICT, vol. 470, pp. 293–301. Springer, Cham (2016). https://doi.org/10.1007/978-3-319-31165-4_29
22. Baikova, E.N., Valtchev, S.S., Melicio, R., Fernão Pires, V.: Wireless power transfer impact on data channel. In: IEEE International Symposium on Power Electronics, Electrical Drives, Automation and Motion, Capri, Italy, pp. 582–587 (2016)

Check for
updates

Simulation and Analysis of Surface
Plasmon Resonance Based Sensor

Paulo Lourenço[1,2(✉)], Manuela Vieira[1,3], and Alessandro Fantoni[1,3]

[1] ISEL - Instituto Superior de Engenharia de Lisboa, Instituto Politécnico de Lisboa,
Rua Conselheiro Emídio Navarro, 1, 1959-007 Lisbon, Portugal
[2] Departamento de Engenharia Eletrotécnica, Faculdade de Ciências e Tecnologia, FCT,
Universidade Nova de Lisboa, Campus da Caparica, 2829-516 Caparica, Portugal
pj.lourenco@campus.fct.unl.pt
[3] CTS-UNINOVA, Departamento de Engenharia Eletrotécnica,
Faculdade de Ciências e Tecnologia, FCT, Universidade Nova de Lisboa,
Campus da Caparica, 2829-516 Caparica, Portugal

Abstract. In this paper, we will be presenting the results obtained through Finite-Difference Time Domain simulations on a photonic sensing architecture. This device consists on a dielectric/metal/dielectric sensing structure. Under adequate conditions, when electromagnetic energy strikes the different dielectrics interface, these devices develop surface plasmon resonances which are extremely sensitive to refractive index variations, thus being able to be used as sensing structures. Considering their minute dimensions, monolithic integration is attainable and by incorporating cost-effective materials in their manufacture, devices' mass production may be efficient and information and communication technological systems' resiliency will be greatly facilitated. Next, this architecture is analysed under amplitude and refractive index sensitivity perspectives, its performance is analysed and considerations about its use as a sensing device are contemplated. Finally, conclusions of our work are presented and future development directions are described.

Keywords: Photonics · FDTD simulations · Surface plasmon resonance
Fano interference

1 Introduction

Surface Plasmon Resonance (SPR) occurs when a polarized Electromagnetic (EM) field strikes a metallic surface at the separation interface between the metal and an insulator. It is characterized by the interface's conduction electrons resonant oscillation which is also known as plasmon waves. These electron charge density waves result from the energy absorption at the resonant wavelength subtracted from the spectrum of the incident EM field. Hence, the energy amplitude of reflected spectrum is reduced at the resonant wavelength. Since this wave is generated at the boundary between the metallic surface and the external medium, this phenomenon is very sensitive to alterations in the surrounding environment, namely the refractive index, and may be used in sensing

© IFIP International Federation for Information Processing 2018
Published by Springer International Publishing AG 2018. All Rights Reserved
L. M. Camarinha-Matos et al. (Eds.): DoCEIS 2018, IFIP AICT 521, pp. 252–261, 2018.
https://doi.org/10.1007/978-3-319-78574-5_24

structures. This detection principle has been widely explored in many real time and label free sensor applications in areas such as gas detection or biological/chemical sensing, amongst many other.

Contemporaneous sensor applications rely essentially on noble metals, namely gold (Au) and silver (Ag) and these materials are not adequate to mass production and large dissemination. This brings us to this paper's main driving research purpose:

– Is it possible to utilize cost-effective plasmonic materials instead of gold or silver in SPR based sensing devices and in such dimensions to comply with the lab-on-chip concept [1]?

The remaining of this paper is organised as follows: - Resilient systems association is presented in the next section, where some examples of how photonics has contributed to system's resiliency is pointed out. Section 3 is where an SPR based sensor is described and simulations results are discussed to determine device's performance and, finally, conclusions are drawn in the last section.

2   Resilient Systems Association

Information and communication technology (ICT) systems have been playing an important and increasing role for several years, as far as reliable control of automated tasks is concerned, in many sectors of contemporaneous society. Even in traditional sectors of our economy, technological proliferation is a fact and the trend seems to indicate an increasing penetration in every aspect of our lives for several years to come.

As this situation evolves so does the complexity of cyber threats, as well as their ever-increasing performers' abilities. Hence, it is no longer realistic to assume that ICT systems are able to withstand all possible attacks and that their infrastructures may not be compromised. Instead, contemporaneous and future systems should be engineered under the assumption that these architectures have been, or might be, under an effective threat and, despite of that, they are still able to carry on normal operation [2].

To accomplish this, these systems must rely on resiliency with its inherently associated cost. The challenge is the development of efficient and reliable resilient systems in a cost-effective manner. For this purpose, it is our belief that the inclusion of photonic devices in contemporaneous and future ICT systems will provide the necessary means to achieve previously mentioned requirements on efficiency, reliability and cost effectiveness. Following presented examples and, for that matter, most cases whenever the requirement of increasing a system's resiliency is present, share a common denominator: - Flexibility on a pliable system of this nature comes always with redundancy of elements. For the first example, a multiplexed reserve wavelength has been added and, on the second, an extra optical fibre path was required. The inclusion of extra components comes with an associated cost increase, hence our proposal for an efficient and cost-effective SPR based sensor that would present itself as an economical alternative whenever such redundancy is required.

The proposed device, and to the best of our knowledge, is the only one using this configuration, Aluminium (Al) as the active metal to develop a Surface Plasmon Wave (SPW) and a-Si:H for the propagating waveguide. This structure is easily manufactured, given its low complexity, and cost-effective, thus being more adequate for large scale production, were it based on noble metals such as gold or silver. These devices would be suitable, after customization for whichever specific task, to be included on a resilient network of sensors, e.g. monitoring combustion gases in forests to prevent wildfire episodes. Maybe in a different network architecture but with similar functionality as the case of the second example mentioned in the following subsections.

We will be next referring to some architectures where photonic devices have been included, e.g. from a hybrid metropolitan area packet switched network patented method up to a resilient long-distance remote sensor system, to provide the reader with some examples where systems' resiliency has been reinforced through the inclusion of photonic elements.

2.1 Hybrid Packet Switched Network

The patent this subsection refers to presents an alternative architecture to traditional Metropolitan Area Networks. These networks are usually based on Synchronous Optical Network (SONET) technology which, due to signalling time requirements, are not able to consistently assure faster than 50 ms restoration time in the event of a connection failure. There have been other attempts that did overcome this restoration time limit, e.g. the Resilient Packet Ring (RPR) or the Multiprotocol Label Switching (MPLS) with Fast Reroute, but due to their complexity they are not cost effective.

This method has been designated Photonic Resiliency and Integrated Switching Mechanism (PRISM) and is depicted in Fig. 1 on a ring configuration. PRISM is able to provide faster than 50 ms protection switching with off-the-shelf switches/routers by using photonic switches (SP1 to SP6) and dual ($\lambda_p \rightarrow$ primary wavelength and $\lambda_r \rightarrow$ reserve wavelength) wavelength division multiplexing (WDM),

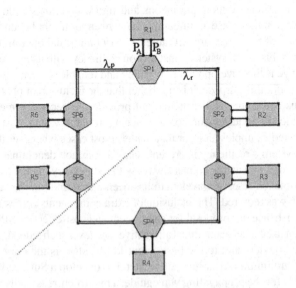

Fig. 1. PRISM architecture on a ring configuration [3].

without requiring either Medium Access Control (MAC) layer modification or non-standard signalling between switching nodes [3].

2.2 Long Distance Resilient Single Mode Fibre Based Sensor

When it is required to remotely monitor structures of vital importance such as seismic activity at the bottom of the ocean, mechanical stress on dams or bridges, human safety at risk due to abnormal temperature rising, etc, this can be accomplished through the use of Fibre Bragg Grating (FBG) based sensors. These structures usually rely on the EM energy amplification provided by either an Erbium-Doped Fibre Amplifier (EDFA) or a Raman amplifier.

It is not unusual to find these devices operating without any protection scheme and with the FBGs placed serially within a standard Single Mode Fibre (SMF). Resiliency is non-existent this way for in the event of a lost connection, the whole system is not able to operate anymore. Plus, serially disposed FBGs turn the necessary lasing wavelengths amplitude equalisation into an extremely difficult task to perform.

Hence, in order to overcome previous limitations, Fernandez-Vallejo et al., developed an experimental setup [4] where resiliency is included and the FBGs are in a parallel configuration at the end of the sensing network, allowing control for better power stability in all lasing wavelengths through the variable attenuators, as shown in Fig. 2. Here, resiliency is achieved through the inclusion of a secondary fibre path and a photonic switch which commutates paths if disruption is detected on the operating one.

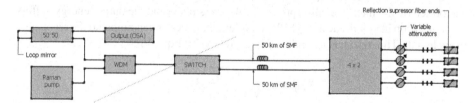

Fig. 2. Multiwavelength Raman fibre laser remote sensing [4].

3   Surface Plasmon Resonance Based Sensor

Sensors based on SPR phenomenon are highly sensitive to variations of the surrounding environment refractive index and have been thoroughly investigated by the research community, resulting in several commercial applications contemporaneously in use for chemical and biomolecular detection because of their label free features, high sensitivity and real-time monitoring capabilities. Nevertheless, there is still a trending demand for cost effective and monolithic integrated devices able to fulfil the lab-on-chip concept [1] which would enable mass production and wide implementation in many technological areas.

Nowadays SPR based sensors rely mainly on noble metals such as gold (Au) or silver (Ag). The more often used active metal in these devices is Au for it has better chemical stability and exhibits a higher shift of the resonance peak wavelength to variations of the sensed analyte, although its resonance curve possesses a wide spectral width which worsens its performance specially when operating on higher energies. Sensors utilizing Ag as the active metal benefit from a narrower spectral width but are easily oxidized due to its poor chemical stability and it is not a cost-effective element. Moreover, both Au and Ag cannot be deposited directly over silicon (Si) and a transitional material is required, which increases the complexity of the manufacturing process.

Al is also a plasmonic metal, it is the most abundant metal on Earth's crust and, for this reason, it is ~425 times less expensive than Ag and ~25000 times cheaper than Au. It benefits from a narrow resonance curve spectral width, even narrower than Ag, which enables operation at higher energies and can be deposited directly over Si, although is also affected by poor chemical stability for it can be easily oxidized. Nevertheless, through a surface passivation process that creates a robust protective layer [5], corrosion and material degradation caused by oxidizing agents usually employed in biosensing tests can be prevented and chemical stability is achieved. Another way to provide the necessary chemical stability is also used in many devices and is achieved through a bimetallic alloy, consisting on a thin layer of Chromium (Cr) atop the Al one.

3.1 Theoretical Background

The optical phenomenon of SPR is characterized by a propagating transversal magnetic (TM) wave supported by a metal-dielectric interface. When the TM polarized wave is incident on this interface in a way that its propagation constant matches the SPW supported by the interface, absorption of light takes place and the output energy suffers a sudden drop in amplitude at the SPW wavelength, which is denominated the resonance wavelength. This resonance condition is expressed by Eq. 1 [6]:

$$\beta_{SPW} = \kappa_0 \sqrt{\frac{\varepsilon_m \varepsilon_d}{\varepsilon_m + \varepsilon_d}}. \tag{1}$$

Where β_{SPW} is the SPW propagation constant, κ_0 is the operating wavenumber and ε_m and ε_d are the dielectric permittivity for metal and insulator, respectively. As can be inferred by the above equation, this matching condition is highly sensitive to slight changes on the refractive index of the surrounding environment.

Performance of SPR based sensors is mainly evaluated by their sensitivity, detection accuracy (DA) and resolution. When working with sensors on spectral interrogation [7], the refractive index of the sensing layer ($n_{sensing}$) is assumed as the independent variable and to which the resonance wavelength ($\lambda_{resonance}$) depends on, i.e. a $\delta n_{sensing}$ variation on the sensing layer refractive index will correspond a shift $\delta\lambda_{resonance}$ on resonance wavelength. Thus, sensitivity (S_n) of a spectral interrogation sensor can be expressed as Eq. 2:

$$S_n = \frac{\delta\lambda_{resonance}}{\delta n_{sensing}} \left[nm/RIU\right]. \tag{2}$$

These devices' detection accuracy (DA) depends on how accurate and precisely they are able to detect $\lambda_{resonance}$ and, consequently, their sensing layers $n_{sensing}$. The accuracy in detecting $\lambda_{resonance}$ further depends on the spectral width of the SPR dip. A narrower spectral width leads to higher detection accuracy. Therefore, expressing the spectral width of the SPR response curve referenced to an arbitrary level of transmitted power as $\delta\lambda_{FWHM}$, detection accuracy of the sensor can be defined as the reciprocal of $\delta\lambda_{FWHM}$. Therefore, for spectral interrogation devices, detection-accuracy is thus defined as in Eq. 3:

$$DA = \frac{1}{\delta\lambda_{FWHM}}. \tag{3}$$

being $\delta\lambda_{FWHM}$ the SPR resonance curve's Full Width Half Maximum (FWHM).

Resolution (Δn) is the minimum variation in the refractive index that can be detected by an SPR based sensor. This parameter also depends on the spectral resolution of the measuring device (spectrometer of some sort: - external spectrum analyser or wavelength monitoring diode, or a fully integrated, lab-on-chip concept [1], wavelength detector), $\delta\lambda_{DR}$. Hence, assuming a shift of $\delta\lambda_{resonance}$ on the resonance wavelength corresponding to a $\delta n_{sensing}$ variation of the sensing layer refractive index, resolution might be defined as in Eq. 4:

$$\Delta n = \frac{\delta n_{sensing}}{\delta\lambda_{resonance}}\delta\lambda_{DR}. \tag{4}$$

Fano resonances can be usually observed in light scattering, transmission and reflection spectra of resonant optical systems and can be described as the asymmetric peaks in a spectral response generated by the interaction between a discrete and a continuum of states at the same energy level. Under Coupled-Mode Theory (CMT) and considering ω_{upper} and ω_{lower} as the modes' frequencies, a_{upper} and a_{lower} as their amplitudes, γ as the damping rate which includes both ohmic and radiative losses, and Ω as the coupling parameter, the interaction between the resonant modes can be explained by Eqs. 5 and 6, where s^+ and s^- as the EM flux density either entering and exiting the resonator, respectively, c as the background reflection coefficient without exciting the resonant mode, and $\alpha_{upper/lower}$ as the coupling coefficient between the external flux and the first resonant mode.

$$\begin{cases} \dfrac{da_{upper}}{dt} = i\omega_{upper}a_{upper} + \gamma a_{upper} + \Omega a_{lower} + i\alpha_{upper}s^+ \\ \dfrac{da_{lower}}{dt} = i\omega_{lower}a_{lower} + \gamma a_{lower} + \Omega a_{upper} + i\alpha_{lower}s^+ \end{cases}. \tag{5}$$

$$s^- = cs^+ + i\alpha_{upper}a_{upper} + i\alpha_{lower}a_{lower} \tag{6}$$

Then, the reflection coefficient r is given by Eq. 7 that represents an asymmetric line shaped Fano resonance.

$$r = \left| \frac{s^-}{s^+} \right|^2 = \left| \frac{\alpha_{upper}^2 \left(i\omega_{lower} - i\omega + \gamma \right)}{\left(i\omega_{lower} - i\omega + \gamma \right) \left(i\omega_{upper} - i\omega + \gamma \right) + \Omega^2} \right|^2. \tag{7}$$

3.2 Numerical Simulations of an SPR/Fano Waveguide Sensor

The proposed SPR based sensor is of great development simplicity and is depicted in Fig. 3. It consists on an input waveguide, the sensing area formed by the central waveguide with a thin metal layer placed on top and an output waveguide. These are a-Si:H waveguides and their simulation has been conducted considering the operating wavelength $\lambda = 1.55\,\mu m$ and, as far as refractive index and extinction coefficient are concerned, a dispersive model [8] for good quality hydrogenated amorphous silicon (approximately $2.5\,dBcm^{-1}$ attenuation at the operating wavelength). Input and output waveguides are 235 nm wide to assure single mode operation and the width of the central waveguide was programmatically searched by OptiFDTD [9], being 240 nm the optimized found value. An identical procedure was conducted to search for the optimal width and length of the metal thin layer on top of the central waveguide and the encountered values were 18 nm and 500 nm for its width and length, respectively. The optimization consisted on a Visual Basic script (a feature of the simulation tool) that iterates through a given range of the selected dimension, while simulating EM propagation through the device and collecting the reflected spectrum. Then, collected data is analysed and the iteration with the best modulation depth is selected.

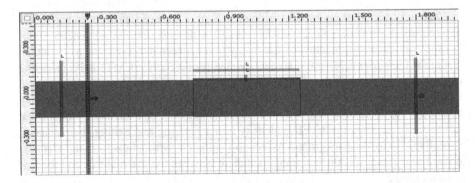

Fig. 3. SPR based sensor workspace layout.

A comparison platform for this device when based on metallic elements such as Ag or Au and when Al based, would require an approach that considered optimal conditions for each of these three elements. Preliminary simulation results showed that each ideal configuration is unique. Consequently, simulation parameters would be different in each case, namely the metallic structure dimensions. Hence, establishing such a benchmark will be considered in future developments of our work.

To assure that our simulations reflect as much as possible physical materials behaviour, Lorentz-Drude dispersive model from OptiFDTD [9] library has been used for the

Al thin layer and, for the aqueous medium that is the sensor's surrounding environment, its extinction coefficient [10] at the operating wavelength of our simulations has also been accounted for.

The input field consisted on the TM fundamental mode, a Gaussian modulated continuous wave (GMCW), with 1 A/m amplitude and at the operating wavelength of 1.55 μm. Four observation lines were placed to monitor, from left to right and bottom up, reflection, the SPR propagating wave, the power level penetrating the analyte and transmittance. There was also an observation point placed near the end of the output waveguide to monitor the dynamic convergence of the propagating EM pulse. Results obtained for the reflectance considering previously mentioned optimal conditions are shown in Fig. 4.

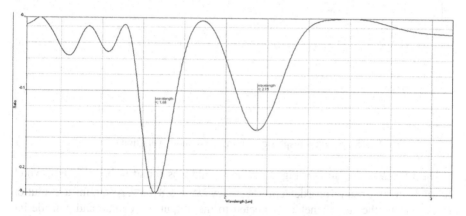

Fig. 4. Normalized reflection wavelength impulse response in the range 1.033 to 3.1 μm.

As can be observed, two main dip features are present in this figure. The first one at the approximate wavelength 1.66 μm and the second one at the approximate wavelength 2.15 μm. Considering Eq. 1 and water and Al permittivity values at operating wavelength, the SPR should be verified at the approximate wavelength 2.052 μm, which is not far from the second observed dip feature. Analytical and simulated resonant wavelengths might be considered as in good agreement because the former data has been collected from a different source [10]. However, there is a narrower dip feature observed at a shorter wavelength, which offers the prospect of a better performance sensing capability.

This resonance might be explained by an interference phenomenon known as Fano resonance [11]. In this particular case, there are two independent modes propagating with similar propagation constants along the lower and upper surfaces of the Al thin layer and their nearfield interaction due to close proximity originates the interference condition.

3.2.1 Results Analysis

Still referring to Fig. 4, our proposed SPR/Fano based sensor exhibits good modulation depth. According to Eq. 2 and through the observation of Fig. 5, which represents the optimal spectral response of our proposed Fano based device when the refractive index of surrounding environment is $n = 1.325$ and $n = 1.425$, the resonance wavelength was red shifted by ~24 nm, which results in a sensitivity $S_n = $ ~240 nm/RIU.

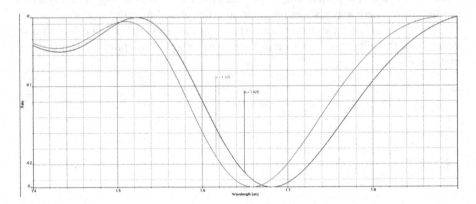

Fig. 5. Spectral response to a refractive index variation of 0.1.

The FWHM calculated in Fig. 5 for $n = 1.325$ was found to be ~172 nm, which according to Eq. 3 results on a detection accuracy $DA = 5.8\,\mu m^{-1}$. Assuming a resolution of 0.02 nm for the spectrometer connected to the output waveguide and considering Eq. 4, this sensor's resolution would be $\Delta n = 8.3333e^{-5}$.

4   Conclusions and Future Work

The proposed SPR/Fano based sensing device, developed with cost-effective materials such as Al for the active metal and a-Si:H for the waveguide supporting elements, is able to detect refractive index variations of the surrounding environment at the near infrared operating wavelengths. Other operating wavelength ranges are available through customization. This sensor architecture has been modelled with dispersive materials, losses included, to reflect as much as possible physical reality. Its performance is not exceptional, particularly when compared to prism based devices (e.g. Kretschmann configuration), but it is sufficient for most eventual implementations. Device integration of one or many of these structures is easily attainable given development simplicity and involved minute dimensions, thus fully complying with the lab-on-chip concept [1].

Future work will consist on this device's experimental implementation (through Plasma Enhanced Chemical Vapour Deposition facilities at ISEL) to validate the obtained results, setting up a performance benchmark between Al and noble metals based devices and will also consider research on different SPR/Fano architectures in order to improve either performance and chemical stability against oxidizing agents, and trying to preserve simplicity at the same time to facilitate monolithic integration.

Acknowledgments. This work was supported by FCT (CTS multi annual funding) through the PIDDAC Program funds (UID/EEA/00066/2013) and by the IPL IDI&CA 2016 EXOWAVE project. A special word of recognition to professors Miguel Fernandes and Yuri Vygranenko, whom have contributed with their invaluable expertise and will be of key importance for the experimental setup.

References

1. Haeberle, S., Mark, D., Von Stetten, F., Zengerle, R.: Microfluidic platforms for lab-on-a-chip applications. Microsyst. Nanotechnol. **9783642182**(9), 853–895 (2012)
2. Bodeau, D.J., Graubart, R.D., Laderman, E.R.: Cyber resiliency engineering overview of the architectural assessment process. Procedia Comput. Sci. **28**, 838–847 (2014)
3. Koley, B.: Method and apparatus for photonic resiliency of a packet switched network. US 2008/0050117 A1 (2008)
4. Fernandez-Vallejo, M., et al.: Resilient long-distance sensor system using a multiwavelength Raman laser. Meas. Sci. Technol. **21**(9) (2010)
5. Canalejas-Tejero, V., Herranz, S., Bellingham, A., Moreno-Bondi, M.C., Barrios, C.A.: Passivated aluminum nanohole arrays for label-free biosensing applications. ACS Appl. Mater. Interfaces **6**(2), 1005–1010 (2014)
6. Maier, S.A.: Plasmonics: Fundamentals and Applications. Springer, New York (2007). https://doi.org/10.1007/0-387-37825-1
7. Cennamo, N., Massarotti, D., Conte, L., Zeni, L.: Low cost sensors based on SPR in a plastic optical fiber for biosensor implementation. Sensors (Basel) **11**(12), 11752–11760 (2011)
8. Fantoni, A., Lourenco, P., Vieira, M.: A model for the refractive index of amorphous silico for FDTD simulation of photonics waveguides. In: Proceedings of the International Conference on Numerical Simulation of Optoelectronic Devices, NUSOD, no. L, pp. 167–168 (2017)
9. Optiwave Photonic Software (2017). https://optiwave.com. Accessed 20 Apr 2017
10. Refractiveindex.info database (2017). http://refractiveindex.info/. Accessed 20 Apr 2017
11. Zafar, R., Salim, M.: Enhanced phase sensitivity in plasmonic refractive index sensor based on slow light. IEEE Photonics Technol. Lett. **28**(20), 2187–2190 (2016)

Monitoring Systems

Monitoring of Actual Thermal Condition
of High Voltage Overhead Lines

Balint G. Halász[✉] ⓘ, Bálint Németh, Levente Rácz, Dávid Szabó, and Gábor Göcsei

Budapest University of Technology and Economics,
18 Egry József street, Budapest 1111, Hungary
halasz.balint@vet.bme.hu

Abstract. The main issue of the article is to investigate and improve the usability of existing line rating calculation methods (Cigré, IEEE) in DLR systems, based on years of measured and forecasted weather and conductor data, including the differences of various calculation models (white-box model). In addition to the use on conventional physical model, the application of soft-computation methods is considered (black-box model). The aim is to create a neural network capable of recognizing patterns based on the data of previous years and the actual current values of the wires. In this way, it is not only possible to fine-tune, but also accelerate the applied calculation of maximum load capacity.

Keywords: Dynamic line rating · DLR · Grid resilience · Transmission line
Soft computing · Neural network · Black box model

1 Introduction

The increasing demand of transmission lines capacity is more and more important in the interest of consumers' qualitative supply. Dynamic Line Rating is a cost-efficient technology to maximize the network-capable power.

According to traditional approach a static rating was defined for transmission lines capacity, where the worst case influencing parameters are taken into consideration. During normal operation circumstances the power carried on the transmission line cannot exceed this static value. Contrarily, the dynamic line rating takes into consideration the changing of environmental parameters and the line rating is adapted to these parameters. If ambient conditions are more favorable than worst case used for static rating, the conductor could carry more current. Therefore, the defined dynamic line rating is higher than static rating most of the time and the operational safety can be increased rest of the time [1–3].

In our work, more than 2 years of archive weather and SCADA data were used to make conclusions about the thermal behavior of the bare overhead line, and the usability of forecasted weather parameters. The used data contained the actual weather data provided by national weather service, measured local weather parameters by equipped weather stations along the examined 220 kV double circuited line and also forecasted data from the national weather service.

© IFIP International Federation for Information Processing 2018
Published by Springer International Publishing AG 2018. All Rights Reserved
L. M. Camarinha-Matos et al. (Eds.): DoCEIS 2018, IFIP AICT 521, pp. 265–273, 2018.
https://doi.org/10.1007/978-3-319-78574-5_25

2   Relationship to Resilient Systems

Due to the continuous growth of the consumer demand, the integration of renewable energy sources and other related changes in the market issues, number of problems and challenges with the operation and utilization of the existing network have been identified. The need for a higher level of transmission capacity for the transmission network is one of the major challenges in the electricity network. A possible way to improve the transmission capability of the existing overhead power lines in electric power network is the use of dynamic line rating (DLR) system. On the one hand this grid management solution enables the optimization of transmission path capacity thus increasing the resilience of the transmission grid to a new level. Furthermore, the globally changing weather condition by climate change could mean increased mechanical load through the eventually occurring ice loads on the surface of conductors [4]. Because of these phenomena, the elements of high voltage overhead lines have to suffer increased loads, which can cause several damages, such as in the last years.

The well-monitored mechanical and electric conditions of power lines in which DLR system is working, the risk of high degree of icing due to weather conditions or the thermal overload of power line could be reduced, by the realization of alerts for the transmission system operator, who could make steps before an eventual breakdown of the transmission line. In addition, during normal operation circumstances the power carried forward over the transmission line is not allowed to exceed static line rating value, while an additional amount of capacity is available in a large percentage of time. Contrarily, the Dynamic Line Rating takes into consideration the changing of environmental parameters – actual and forecasted weather conditions – and the line rating is adapted to these parameters. To increase the resilience of the transmission line several DLR methods have been investigated.

3   Standardized Calculation Methods of Ampacity

The conductor temperature is influenced by the heating effects as the Joule and magnetic heating effect, solar heating, corona heating and the cooling effects which are natural and forced convection, radiative heat loss, evaporative cooling (Fig. 1).

Basically, the maximum ampacity of a conductor is usually limited by the maximal allowable temperature. The limit comes from the allowed maximal temperature of the conductor – which can be withstood by the conductor without significant degradation of elements –, and the minimal height of the conductor above the ground, which depends on the conductor sag, therefore the determination of transmission capability is in close relation with the calculation of the thermal behavior of the conductor [5].

For the calculation of the temperature of a current-carrying conductor, standardized methods exist, which contain methodologies to quantify the heating and cooling effects of each impact. Two internationally used standards are the following:

– CIGRÉ Guide for Thermal Rating Calculations of Overhead Lines [6]
– IEEE Standard for Calculating the Current - Temperature of Bare Overhead Conductors [7]

These proceedings contain the description of calculating models of the line using heat balance equation of the overhead line conductor. The models contain the same heating and cooling effects, but use different physical approaches such as the specification of the wind caused convective cooling effect. The Cigré model requires more input parameters about the conductor than the IEEE model, which likely results in more realistic values.

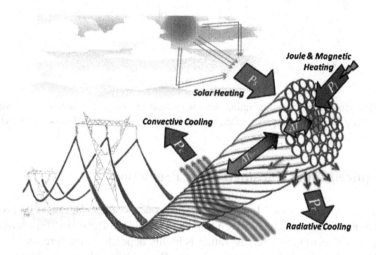

Fig. 1. The heating and cooling effects on the conductor [6]

3.1 Precipitation Cooling Method

The heat loss of the conductor due to precipitation is not contained by IEEE and Cigré guides, while there can be a significant heat loss effect on the conductor's surface due to these phenomena, for example in case of windy and rainy circumstances, when the increased electrical energy generated by wind plants load the OHLs.

There is an available method, which can be used for the quantification the cooling effect of both raining and snowing on bare stranded conductors [8]. It contains several steps, such as the calculation of the airborne liquid water content from raining or snowing precipitation rate (in different ways), the liquid mass flux density impinging onto the conductor and the determination of boiling point of liquid water (Clausius-Clapeyron equation); the result of the calculation is the heat loss with dimension of Wm^{-1}, therefore it fits to the IEEE and Cigré models also. The validity of the model is verified by sensor measurements [8].

A surplus ampacity can be achieved by the implementation of the calculation of heat loss due to different type of precipitations as the Fig. 2 shows, which is one result of our case study based on the properties of an existing 220 kV overhead line (OHL).

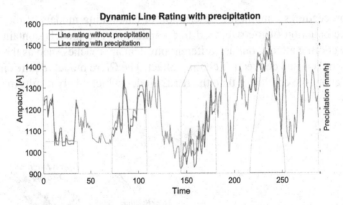

Fig. 2. Precipitation effect on line rating: yellow curve is the amount of rain, red is the ampacity without raining, blue is the ampacity of the line taking into consideration the cooling effect of rain (Color figure online)

4   Required Input Parameters for Prediction

During actual weather-based line rating calculations, weather data must be provided. Predetermined line rating can be given also by available forecasted weather parameters. In this case the accuracy of the line rating is highly depends on the forecasted weather data, which will be more inaccurate over time. Hence, it is only worthwhile to specify the predetermined line rating only a few hours ahead [9].

Weather station along the line provides high resolution data in real-time. These instruments generally provide data in every 5–10 min. Contrarily, the weather services applied in Europe provide forecasted weather data with only 1–3 h time resolution. Its accuracy is also lower, because the measured weather parameter can change a lot in this 1–3 h period, especially in case of wind speed.

4.1   Weather Forecast Processing Method

The different weather services give forecasted data with resolution of hours, which results in steps in the value of each weather parameters in function of the time, which means also steps in ampacity. Based on our investigation, if these step-like functions are taken into account during the determination of prevailing ampacity, the inaccuracy is relatively high compared to the ampacity calculated with the use of real weather parameters. To decrease the magnitude of deviation of the forecasted weather data from the real predictable weather circumstances, the data process has to contain an additional important operation. Between two weather forecast values, the measured parameter can be estimated by interpolation; therefore, the line rating can be calculated with finer time resolution and in a more accurate way.

The results of our research showed that the interpolation method can be used well in case of slowly changing weather parameters like solar radiation or ambient temperature as Fig. 3 shows, but the available resolution gained by interpolation is lower in case of fast-changing wind. This proceeding accuracy also depends on the forecast accuracy itself.

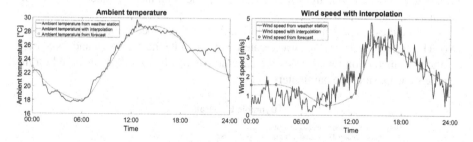

Fig. 3. Ambient temperature and wind speed with interpolation

5   Extended White Box Model

As it is presented in the previous paragraphs, three different procedures were developed to use the calculations for rating prediction:

- Cooling effect of the precipitation
- Considering wind effect in different OHL sections
- Improved forecasted weather data processing

As it is shown in Fig. 4 the extended white box model provides a more realistic way to determine the ampacity of an OHL. The extra load capacity is provided by precipitation modelling is available if weather conditions are appropriate, therefore the transmission line exploitation is increased. At the same time the operational safety of the grid is also developed by considering wind effect in different OHL sections. During this method, the line rating determination is based on the worst-case scenario, thus the conductor temperature shall nowhere exceed the maximal conductor temperature

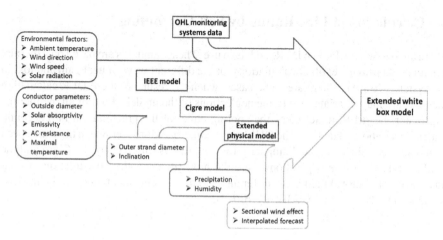

Fig. 4.   Scheme of extended white box model

specified by the manufacturer. The qualitative and quantitative additional information of weather variables between discrete forecast values is gained by improved forecast weather data processing method, therefore the expected conductor temperature is available in a more accurate way.

Based on these experiences gained during the weather process and model extension a MATLAB graphical interface was created. It can handle the measured and forecasted weather parameters on which basis the calculations can be executed for based on the specific transmission line parameters as Fig. 5.

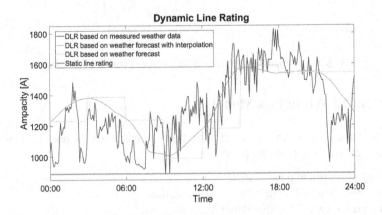

Fig. 5. Calculations based on extended white box model

Overhead line monitoring systems can be used for model validation; accordingly, the extended white box model error can be minimized. During operation all the external parameters changing, like terrain around the transmission line, absorptivity and emissivity factors, etc. To follow these changes with the model, sensors are required, therefore the model can be dynamically adjusted to these mutations.

6   Calculation of Line Rating by Soft Computing

There are two widely known DLR models in use. These standards are based on a thermal equation and provide explicit calculations for the different ways of heat gain and loss in the system. However, there are some cases, when the results of these DLR calculation methods are inappropriate, due to the neglections of the models. First of all, neither the Cigré, nor the IEEE model takes the cooling effect of the precipitation into account, which is an important factor in the OHL's heat equation. Moreover, when the wind speed is above 5 m/s, the calculated ampacity for the OHL is different by the Cigré and the IEEE standards. These neglections were the motivation for the investigation of non-analytic computational methods – different from the current calculations of the international standards (e.g. CIGRÉ, IEEE) [6, 7, 10].

6.1   Applicability of Soft Computing

There are several types of soft computing methods, such as neural networks, fuzzy systems, genetic algorithms, etc. For this model, there is a need for a capability of processing big database quickly. Moreover, the ability of learning and optimization is also not negligible. Based on these facts, the use of neural networks may be the appropriate method for the DLR model [11, 12].

6.2   The Structure of the Neural Network Model

Basically, the neural network requires 4 inputs and generates 1 output via activation functions: the inputs include the environmental parameters and the output is the temperature of the OHL. The main advantage of this method is that sensors attached on the OHL could provide real-time conductor temperature for the learning method of the network. With the application of sensors all of the environmental parameters can be taken into account for the calculation of DLR. Furthermore, it is possible to validate the efficiency of the network with measured parameter.

There are two main steps after the learning process in the line rating calculation in the investigated model. At first, the trained network can calculate the temperature of the OHL from environmental parameters and the SCADA current determined with static rating. After this, it is possible to estimate the ampacity of the OHL, in the following step.

In Fig. 6 the result of the training, validation and testing data of the network can be seen. According to the diagrams, the black dots fit well the solid line which means a linear relationship between the targets and outputs. This also can be seen from the R value, which is the indicator of the relationship. If R is close to 1, the relationship is strong, so in our case the learning process of the network was successful.

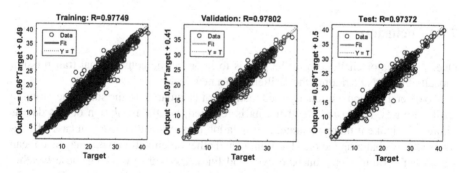

Fig. 6. Learning process of the neural network

The second part of the determination of DLR is the calculation of ampacity. Knowing the conductor temperature by the use of trained neural network and forecasted weather input data, the difference is given between this temperature and the maximum permissible temperature of the OHL. This difference is the unexploited potential on the OHL, which allows carrying higher current.

6.3   Results of the Simulations

Several simulations were carried out to determine both the temperature and the DLR of the inspected OHL. By the use of real-time temperature of the OHL via the installed sensor, in the learning method, it is possible to calculate the OHL temperature with an error lower than 7% as it can be seen on first diagram in Fig. 7. It means, that in most of the cases the error was about 1 °C. This error can be reduced with the fine-tune of the neural network. According to the second diagram in Fig. 7 the calculated DLR value approaches the result of the Cigré model, but never exceeds that in our study. This means that the investigated neural network model shifts the DLR to the safety and security in this case.

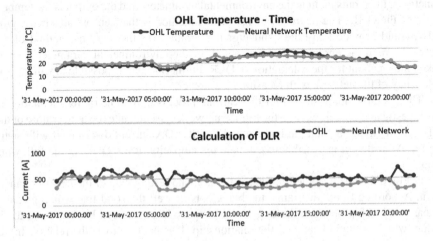

Fig. 7. Calculations based on extended black box model

7   Conclusion

One of the major challenges in the grid is to improve the conventional transmission capacity methods upgrading the resilience to a new level.

To extend the conventionally used methods of calculation of the static line rating to DLR purposes, forecasted weather data is required, but the resolution of this data is limited. To make it more accurate, interpolation methods can be used, in case of relatively slowly changing factors. The raining and other precipitation types may also mean significant cooling factor on bare overhead lines, therefore by taking these weather phenomena into account, the physical calculation could be more realistic. During the research the wind speed was concluded to be the most influencing factor, thus it is necessary to segregate the OHL into smaller sections. Based on these studies a new extended white box model has been investigated.

Another way to determine the DLR of the OHL could be possible via the use of black box model. In this model the application of neural networks could be appropriate to determine the conductor temperature. This model requires sensors installed on the wire to validate the calculations that contains all the environmental parameters. From this line temperature the DLR could be calculated using Joule–heat. However, for more accurate results further development is necessary.

In general, the precise knowledge of the prevailing thermal behavior of the conductor can make the operation of the grid more reliable, even in case of extreme weather circumstances.

Acknowledgments. This work has been being developed in the High Voltage Laboratory of Budapest University of Technology and Economics within the boundaries of BEST PATHS project, which is an international project. BEST PATHS aims the development of innovative technologies to increase the flexibility and the transmission capacity of the existing panEuropean transmission network. BEST PATHS stands for "BEyond State-of-the-art Technologies for rePowering Ac corridors and multi-Terminal HVDC Systems". It is co-funded by the European Commission under the Seventh Framework Program for Research, Technological Development and Demonstration.

 SUPPORTED BY THE ÚNKP-17-2-I AND ÚNKP-17-2-II NEW NATIONAL EXCELLENCE PROGRAM OF THE MINISTRY OF HUMAN CAPACITIES

References

1. McCall, J.C., Goodwin, T.: Dynamic line rating as a means to enhance transmission grid resilience. In: 2015 Grid of the Future Symposium, Cigré (2015)
2. McCall, J.C., Servatius, B.: Enhanced economic and operational advantages of next generation dynamic line rating systems. In: 2016 Grid of the Future Symposium, Cigré (2016)
3. Seppa, T.O.: Reliability and real time transmission line ratings, 18 June 2007. http://www.nexans.com/US/2008/reliability_whitepaper.pdf. Accessed 23 June 2017
4. Péter, Z.: Modelling and simulation of the ice melting process on a current carrying conductor. In: UNUVERSITÉ DU QUÉBEC (2006). http://constellation.uqac.ca/492/. Accessed 4 Sept 2017
5. Morgan, V.T.: The thermal rating of overhead-line conductors. Electr. Power Syst. Res. **5**, 119–139 (1982)
6. Cigré 601 WG B2.43 – Guide for thermal rating calculations of overhead lines (2014)
7. IEEE Standard for Calculating the Current Temperature of Bare Overhead Conductors, Transmission and distribution committee of the IEEE Power Engineering Society IEEE Std 738-2006 (2006)
8. Pytlak, P., Musilek, P., Lozowski, E., Toth, J.: Modelling precipitation cooling of overhead conductors. Electr. Power Syst. Res. **81**(12), 2147–2157 (2011)
9. Hall, J.F., Deb, A.K.: Prediction of overhead line ampacity by stochastic and deterministic models. IEEE Trans. Power Deliv. **3**, 789–800 (1988)
10. Holbert, K.E., Heydt, G.T.: Prospects for dynamic transmission circuit ratings. In: The 2001 IEEE International Symposium on Circuits and Systems, Sydney, ISCAS 2001, May 2001
11. Altrichter, M., Horváth, G.: Neural networks (in Hungarian) (2016). http://www.tankonyvtar.hu/hu/tartalom/tamop425/0026_neuralis_4_4/adatok.html. Accessed 17 Nov 2017
12. Borgulya, I.: Neural Networks and Fuzzy Systems (in Hungarian). Dialóg Campus Kiadó, Budapest (1998)

Resilient Energy Harvesting System for Independent Monitoring Nodes

Alberto Gutiérrez-Martínez[(✉)] and Enrique Romero-Cadaval[(✉)]

Power Electrical and Electronic Systems R+D Group,
School of Industrial Engineering,
University of Extremadura, Avda. de Elvas, s/n, 06006 Badajoz, Spain
agutierrez@peandes.es, eromero@unex.es

Abstract. A study of multisource energy harvesting is presented to propose the design of a multisource energy harvesting system. This system is resilient, robust, modularly connects most of available energy sources, and has a low power consumption. In this design, maximum power point is used only in the source converter, which reaches this point clamping the maximum power point to the output voltage of the source with the battery voltage.

Keywords: Resilience · Energy harvesting · Multisource
Battery management · Maximum power point · Converter

1 Introduction

Many researchers around the world are currently striving to incorporate the Internet of Things (IoT) into daily life. Devices that are unable to supply their own energy or must operate wirelessly are significantly more difficult to connect to the grid than their wired and/or self-powered counterparts. Energy harvesting has become a natural way to continuously supply energy to small, stationary hardware platforms, allowing monitor nodes to achieve perpetual operation [1]. Energy harvesting can be considered on both the macro- and micro-scale. Macro-scale energy harvesting is used to harness energy in the range of kilowatts to megawatts, usually used to feed the grid. Conversely, micro-scale energy harvesting refers to scavenging small amounts of energy, in the range of nanowatts to milliwatts [2]. Micro-scale energy harvesting (which will be referred to just as "energy harvesting" from here on out) can rely on many environmental factors, like sunlight, wind, vibration, etc., to provide sufficient and continuous energy to power nodes [3]. After sourcing the energy, the next step is storing it. Most solutions in the literature rely on conventional rechargeable batteries; because the current technology of batteries allows a limited number of recharge cycles, this method can reduce the lifespan of a given device [4]. Once the battery has reached its finite lifespan, it must be replaced. If this device is placed in remote location, battery replacement could become an expensive and difficult task [5]. For some devices, like wireless nodes, the energy consumption required to operate as a network router can be considerable [6]. Multisource energy harvesting can overcome this obstacle.

© IFIP International Federation for Information Processing 2018
Published by Springer International Publishing AG 2018. All Rights Reserved
L. M. Camarinha-Matos et al. (Eds.): DoCEIS 2018, IFIP AICT 521, pp. 274–281, 2018.
https://doi.org/10.1007/978-3-319-78574-5_26

The aim of this document is to propose a design for a multisource energy harvesting system. The following sections include a description of the relationship between multisource energy harvesting systems and resilient systems of the IoT, the classification of a typical energy harvesting multisource system, and the design of a reliable, robust, and modular Multisource Energy Harvesting System.

2 Relationship to Resilient Systems

Resilience for complex systems is defined as the rate at which the system returns to normal conditions after an extreme event [7]. The capability of a system to absorb and adapt to external shocks is an important factor in its resilience [8]. This proposed design attempts to improve the resilience of the IoT's network and its power supply.

Wireless systems have greatly improved resilience in networks. Conventional wired network cables are subject to natural disasters, human events, etc., and take longer to install than wireless systems. Additionally, the medium of wireless transmission is not nearly as susceptible to damage as cables, and can be reset in the event of a disaster or malfunction. Wireless systems also hold a strong advantage when it comes to potential reconfiguration, as there is little physical change necessary.

Wireless systems are also superior in terms of power supply stability, as they are not connected to the grid, and thus protected from grid faults. Energy harvesting techniques allow a device to achieve autonomy via a perpetual energy, making the device more resilient. As a single source can be irregular and unpredictable [9], a multisource energy harvester should improve the system, and make it more independent from environmental conditions.

Once the system no longer depends on the electrical grid for energy, and does not need wires to connect to the Internet, it can be placed anywhere. Adding more devices close by makes it possible to create a resilient IoT network.

In Fig. 1, the resilience of the network is shown; in the case of a fault in a device, Fig. 1(b), the network is able to keep working because a fault is not able to disrupt the path of the information.

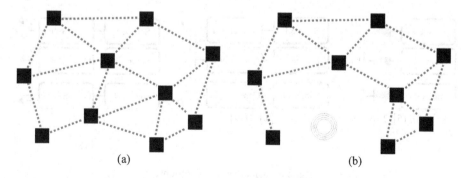

(a)   (b)

Fig. 1. Resilient wireless network.

3 Classification of Multisource

The first generation of a multisource energy harvesting system was shown in 2003, where solar and vibration energies were used to feed a wireless sensor node. In this system, a capacitor is charged directly from the source, and a voltage regulator is chosen to step-down the voltage to 1.2 V [10]. In 2006, the AmbiMax Platform was presented; it used modular buck converter topologies with MPPT to charge superca-pacitors, where all loads are directly connected to a battery [11]. A similar system is presented in MPWiNodeX using water flow, sun, and wind as sources. In this case, the MPPT was implemented by an MCU, consuming a large amount of power [1]. A more modular system was presented in 2011, in which energy from each source is regulated by an LDO regulator, power sources can work together in parallel, and each module is optimized in its operation [2]. Another modular multisource energy harvesting system connects each energy harvesting system to a supercapacitor through a diode [9]. A lithium polymer battery is connected in parallel with the supercapacitor, for when the solar energy is not enough to feed the load. At the end, a buck-boost converter adapts the voltage to the wireless network. A general scheme of multisource topologies is shown in [12] that are based on connecting the energy harvesting blocks to the load or energy combiner through diodes or converters. Moreover, the system presented con-nects the converter to the energy combiner, which connects the source to the load or the battery. In a commercial energy harvesting system with battery management [13–16], the capacitor is connected to the battery through a switch. The battery feeds the load when the energy source in unable to, and charges when the sourced energy is greater than the device's consumption. Common systems are possible to sort by dividing the system in source converter or energy management and load DC-DC converter.

The most typical source converters are diodes or DC/DC converters for energy management, as seen in Fig. 2(a) and (b). Typically, this energy charges a capacitor or a battery. The advantage of one converter for each source is that it is possible to use each source without depending of the voltage in that capacitor or battery. A less typical way to manage the energy is connecting each source converter individually to the battery or the load, like in Fig. 2(c).

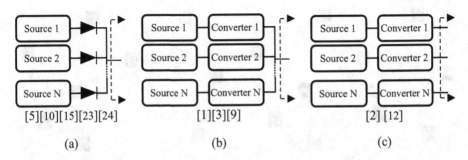

Fig. 2. Source converter options.

Typical options for energy management are shown in Fig. 3. Although there are different ways to store and transform the energy, a storage location – even if minimal – is always necessary. In Fig. 3(a) the energy source is connected to a capacitor, and needs a converter to adapt the voltage to the load. Figure 3(b) is similar to Fig. 3(a) but the converter is not necessary if battery voltage is equal to load voltage. Figure 3(c) represents commercial systems, in which the source charges a capacitor and, when there is enough energy, the battery.

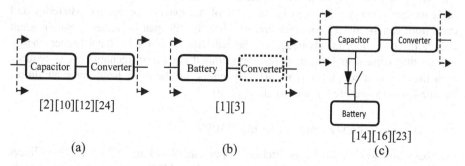

Fig. 3. Power management options. (The converter is a control unit to disconnect the load)

4   Losses in Energy Harvesting

One of the big problems with energy harvesting is the loss of energy. Because micro-scale energy harvesting operates in the range of microwatts to milliwatts [2], small losses can be a large problem. This section explores the potential reasons for these losses.

4.1   Threshold Voltage

A problem with the conventional scheme is that much of the ambient power is wasted. This is because if the ambient power is not high enough to both drive the system and recharge the battery at the same time, then the battery must drive the system. Any power from the ambient source must be discarded in this case [17]. Similarly, energy is wasted when the source voltage is below the voltage threshold, which can be solved by using a DC/DC elevator to reduce the voltage threshold and use the energy, though it is minimal.

4.2   Control Circuits Power Consumption

Power consumption of the control circuits can be a big problem due to the low amount of energy generated by the system; in some cases, the control circuit could waste most of the collected energy [9].

4.3   Conversion Losses

Efficiency is an important consideration when choosing a voltage regulator for low power applications. The efficiency of linear regulator depends on the ratio of the input to output voltages. DC/DC converter are around 90% efficient [10].

4.4   Storage

All energy harvesting systems need some accumulation of energy due to their low level of scavenged energy. This can be in form of a capacitor or battery. Batteries and capacitors can be used in similar circuits, but the lifespan of batteries suffer when deeply discharged. The advantage of the batteries is that they have higher energy density than capacitors [10]; however, their limited number of recharge cycles reduces their lifespan to a couple of years [4]. Capacitors, on the other hand, have an almost infinite lifespan and are simpler to charge [10].

4.5   Maximum Power Point Tracking (MPPT)

One basic method for MPPT is fractional open-circuit voltage, which makes a linear relationship between the open-circuit voltage and its MPP voltage, while taking into account the irradiance and temperature. Still, this is based on observations and must be empirically determined for each type of panel [18]. It momentarily opens the output of the panel [19], or uses a pilot cell [20]. To get a true MPPT, the circuit complexity and the power consumption will increase [18]. Open-circuit voltage MPP is used for piezoelectric generators when the MPP voltage is the half of the open circuit voltage [22].

Another approach to the MPP is to clamp the output terminals of the solar panel to a rechargeable battery to force the solar panel to operate at a point of a determined voltage. This has the advantage of minimizing energy losses in the transfer mechanism, but depends on a careful choice of battery and solar panel in order to ensure that the operating point of the system remains close to the maximum point power (MPP) [21].

Figure 4 shows the experiment using an evaluation board based on [16, 21], with MPPT using the fractional open-circuit voltage method and clamping the output terminals of the photovoltaic panel to the battery. Moreover, shows the behaviour with piezoelectric using the same board for the possible mix photovoltaic and piezoelectric generation. In (a) a rechargeable lithium-ion battery of 3,7 V is used, and the open-circuit voltage of the panel is 6,5 V. In (b) is used a piezoelectric generator charging a 2.2 mF capacitor for 100 s is used.

The result Fig. 4(a) show that in this experiment we obtain better results through clamping of the MPP because the total energy in 40 ms is 220 µJ with MPPT and 320 µJ with clamping the output. In accordance with this, our design implements clamping the source to the MPP. Also in Fig. 4(b) we obtain better results clamping the output with the capacitor instead of using a MPP, harvesting a total energy of 40 mJ with the clamping and 16.7 mJ with the MPP.

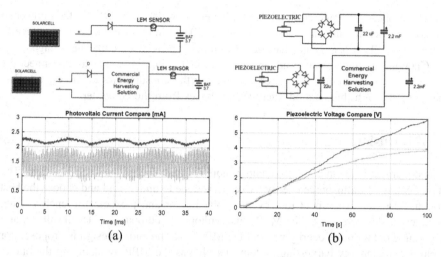

Fig. 4. Energy Generated with MPPT algorithm or without. (a) Photovoltaic comparison with MPPT (red) and clamping the output (blue) measuring the input current [mA] in a 3,7 V battery (b) Piezoelectric generation comparison of MPPT (red) and clamping the output (blue) measuring the voltage [V] in a capacitor of 2,2 mF. (Color figure online)

5 Proposed Multisource System

The final design for the multisource system is show in Fig. 5. In this design, each source is connected to a boost converter, and the voltage is clamped with the battery voltage. The battery is directly connected to the harvester, and the diode of the boost converter should behave like a short circuit if the source voltage is higher than battery voltage.

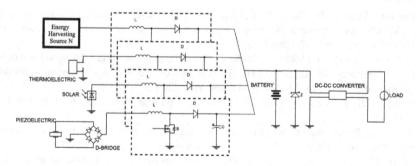

Fig. 5. Design proposed.

MPPT is only necessary for voltages below that of the battery voltage, although, for values above the battery voltage, MPP is produced by clamping this voltage. It is necessary to choose sources with the MPP close to the battery voltage.

The proposed storage component is a battery in order to constantly feed power to the node, which requires a great deal of energy. Also, in order to clamp the voltage to obtain the MPP, a battery is necessary, assuming it provides a constant voltage.

Clamping voltage has two advantages. The first being the ability to disconnect the control logic of the boost converter, eliminating the consumption of the circuit. Furthermore, through this method, there are no losses from the converter.

6   Conclusion

This document is about ways to obtain a more resilient energy harvesting system. The classification of multisource energy harvester is done to understand and choose the best option. Also, an experiment to scavenge solar indoor and piezoelectric energy using commercial device is done with MPPT and without, for comparison. The result is that it is possible obtain more energy without the MPPT. At the end, a design is proposed for a multisource energy harvesting system that obtains the MPP by clamping the output voltage of the source.

References

1. Morais, R., Matos, S.G., Fernandes, M.A.: Sun, wind and water flow as energy supply for small stationary data acquisition platforms. Comput. Electron. Agric. 64(2), 120–132 (2008)
2. Colomer-Farrarons, J., Miribel-Català, P., Saiz-Vela, A., Samitier, J.: A multiharvested self-powered system in a low-voltage low-power. IEEE Trans. Power Electron. 58(9), 4250–4263 (2011)
3. Park, C., Chou, P.H.: AmbiMax: autonomous energy harvesting platform for multi-supply wireless sensor nodes. Center for Embedded Computer Systems (2006)
4. Corbus, D., Baring-Gould, I., Drouilhet, S., Gervorgian, V., Jimenez, T., Newcomb, C., Flowers, L.: Small wind turbine testing and applications development. In: Windpower 1999 (1999)
5. Liu, J.-Q., Fang, H.-B., Zheng-Yi, X., Mao, X.-H., Shen, X.-C., Chen, D., Liao, H., Cai, B.-C.: A MEMS-based piezoelectric power generator array for vibration energy harvesting. Microelectron. J. 39, 802–806 (2008)
6. Corke, P., Valencia, P., Sikka, P., Wark, T., Overs, L: Long-duration solar-powered wireless sensor networks. In: EmNets 2007: Proceedings of the 4th Workshop on Embedded Networked Sensors, pp. 33–37. ACM, New York (2007)
7. Holling, C.S.: Resilience and stability of ecological systems. Annu. Rev. Ecol. Syst. 4, 1–23 (1973)
8. Arab, A., Eskandarpour, R., Khodaei, A.: Improving power grid resilience through predictive outage estimation. In: North American Power Symposium (2017)
9. Li, H., Zhang, G., Ma, R., You, Z.: Design and experimental evaluation on an advanced multisource energy harvesting system for wireless sensor nodes. Sci. World J. 2014, 13 (2014)
10. Roundy, S., Otis, B.P., Chee, Y.-H., Rabaey, J.M., Wright, P.A.: 1.9 GHz RF transmit beacon using environmentally scavenged energy. Optimization 4(2), 4 (2003)

11. Park, C., Chou, P.H.: AmbiMax: autonomous energy harvesting platform for multi-supply wireless sensor nodes. In: Proceedings of the 3rd Annual IEEE Communications Society on Sensor and Ad hoc Communications and Networks (Secon 2006), pp. 168–177 (2006)
12. Umaz, R., Wang, L.: An Energy combiner design for multiple microbial energy harvesting sources. In: Great Lakes Symposium on VLSI, pp. 443–446 (2017)
13. Texas Instruments: bq25504: Ultra Low-Power Boost Converter With Battery Management for Energy Harvester Applications (2011)
14. Texas Instruments: bq25570: Nano Power Boost Charger and Buck Converter for Energy Harvester Powered Applications (2013)
15. Linear Technology: LTC3331: Nanopower Buck-Boost DC/DC with Energy Harvesting Battery Charger (2014)
16. STMicroelectronics: SPV1050: Ultralow power energy harvester and battery charger
17. Park, C., Chou, P.H.: Power utility maximization for multiplesupply systems by a loadmatching switch. In: Low Power Electronics and Design ISLPED '04 (2004)
18. López-Lapeña, O., Penella, M.T., Gasulla, M.: A new MPPT method for low-power solar energy harvesting member. IEEE Trans. Power Electron. **57**(9), 3129–3138 (2010)
19. Simjee, F.I., Chou, P.H.: Efficient charging of supercapacitors for extended lifetime of wireless sensor nodes. IEEE Trans. Power Electron. **23**(3), 1526–1536 (2008)
20. Dondi, D., Bertacchini, A., Brunelli, D., Larcher, L., Benini, L.: Modeling and optimization of a solar energy harvester system for self-powered wireless sensor networks. IEEE Trans. Ind. Electron. **55**(7), 2759–2766 (2008)
21. Raghunathan, V., Kansal, A., Hsu, J., Friedman, J., Srivastava, M.: Design considerations for solar energy harvesting wireless embedded systems. In: Proceedings of the 4th International Symposium on Information Processing in Sensor Networks, p. 64. IEEE Press, Piscataway (2005)
22. Kawai, N., Kushino, Y., Koizumi, H.: MPPT controlled piezoelectric energy harvesting circuit using synchronized switch harvesting on inductor. In: Industrial Electronics Society, IECON 2015 - 41st Annual Conference of the IEEE (2015)
23. Hande, A., Polk, T., Walker, W., Bhatia, D., Jonsson, E.: Indoor solar energy harvesting for sensor network router nodes. Microprocess. Microsyst. **31**(6), 420–432 (2007)
24. Vankecke, C., Assouere, L., Wang, A., Durand-Estebe, P., Caignet, F., Dilhac, J.-M., Bafleur, M.: Multisource and battery-free energy harvesting architecture for aeronautics applications. IEEE Trans. Power Electron. **30**(6), 3215–3227 (2015)

Development and Testing of Remotely Operated Vehicle for Inspection of Offshore Renewable Devices

Romano Capocci(✉), Edin Omerdic, Gerard Dooly,
and Daniel Toal

Centre for Robotics and Intelligent Systems (CRIS),
University of Limerick, Limerick, Ireland
{Romano.Capocci,Edin.Omerdic,Gerard.Dooly,
Daniel.Toal}@ul.ie

Abstract. This paper presents novel aspects of a Remotely Operated Vehicle (ROV) designed to inspect offshore renewable energy devices. The relationship of some design aspects of the ROV to resilient systems is discussed, focusing on the navigation system, control system and novel reconfigurable propulsion system. The design and development of these aspects of the ROV are presented with initial test results illustrated. Finally, conclusions and future work suggestions are put forward.

Keywords: Remotely Operated Vehicle · Observation class ROV
Inspection class ROV · Underwater robotics · Inertial navigation system (INS)
Offshore renewables

1 Introduction

As our insatiable demand for energy continues to escalate and our knowledge of how carbon emissions contribute to climate change increase it is becoming more important to use sources of renewable energy technologies. Marine renewable energy technologies are seen as part of the energy mix now and into the future, allowing governments to further diversify their energy generation technology uses. Currently offshore wind is generating electricity throughout the world whereas wave and tidal energy technologies require further research and development (R&D), investment and validation in order for them to become more financially viable and reduce the levelised cost of generating electricity (LCOE).

One of the main areas in which LCOE can be reduced is in the operational and maintenance (O&M) costs. Offshore operations can be significantly high. Wave and tidal energy converters will be situated in high energy sites to enable them to extract maximum energy from their environment. This furthers complication for O&M as access may be limited and weather windows small, further increasing costs.

To aid in the reduction of these O&M costs an observation class Remotely Operated Vehicle (ROV) has been developed. This ROV can be utilised during the deployment, installation and operational phases of these marine renewable energy

© IFIP International Federation for Information Processing 2018
Published by Springer International Publishing AG 2018. All Rights Reserved
L. M. Camarinha-Matos et al. (Eds.): DoCEIS 2018, IFIP AICT 521, pp. 282–289, 2018.
https://doi.org/10.1007/978-3-319-78574-5_27

installations. During the operational phase the ROV will allow for periodic inspection and post storm inspection of converters, moorings and foundations, reducing the need for commercial divers to be employed in this difficult and potentially dangerous environment. The ROV has been designed so that a two man team can deploy, operate and recover it from a small vessel, reducing the need for a large ROV support vessel.

To ensure that the ROV can operate within these energetic environments it has been designed to accommodate a number of novel aspects, including a reconfigurable multi-layer thruster propulsion system, integration of state-of-the-art navigation suite (fibre optic gyro INS with GNSS, DVL and USBL aiding sensors), and fault-tolerant control system with multi-modal user interface.

2   Relationship to Resilient Systems

In the context of the offshore oil & gas and marine renewable energy industries, it is extremely important to integrate resilience into the design of subsea vehicles. Any loss in communications, power or thruster usage may mean that the ROV and/or the support vessel can be damaged or lost. This can be expensive and can cause a threat to the persons on the support vessel. There is also the possibility of the vehicle causing damage to the installations in which they are carrying out inspections on.

The designed ROV is a complex system consisting of a number of subsystems – electrical, telemetry, propulsion, navigation and control, mechanical frame and buoyancy. A number of features have been designed into the ROV to allow it to become a more resilient system, including reconfigurable propulsion system, state-of-the-art navigation system and fault-tolerant control system.

The reconfigurable propulsion system offers the possibility to utilise two different types of thrusters; one configuration offers $8 \times$ high-thrust output thrusters (4 vertical, 4 vectored horizontal) and the other configuration offers the possibility to use 12 thrusters producing a high level of redundancy, with each configuration offering advantages to operate in difficult conditions, increasing the resilience of the ROV.

The navigation system comprises of state-of-the-art fibre optic gyro based inertial navigation system (INS) integrated with aiding sensors: Global Navigation Satellite System (GNSS) antenna and module, and a Doppler Velocity Log (DVL), providing extremely accurate navigational data. To maximize processing speeds and manoeuvrability the most advanced control allocation algorithms are developed in the bottom hardware layer of control system, with active thrusters fault diagnosis & accommodation features built in. The thruster fault-tolerant control system uses a hybrid approach for control allocation, based on integration of the pseudoinverse and the fixed-point iteration method [7]. This solution minimises a control energy cost function, the most suitable criteria for underwater applications. With the control system the pilot has access to various modes of operation – Full Pilot Navigation mode, Semi-Autonomous mode, control in absolute frame or relative to reference frame attached to support vessel, which can be extremely useful in difficult conditions as the ROV can react to local disturbances automatically allowing the pilot to concentrate on close-up inspection of the subsea structures, and Full Autonomous mode. This increases resilience as it allows the ROV to be operated in difficult conditions.

3 Background

Over the last twelve years, researchers in the Mobile & Marine Robotics Research Centre (MMRRC) at the University of Limerick have been engaged in science collaborative and engineering led seabed survey projects, including technical - design, integration and offshore support, and survey operations carrying out detailed survey projects acquiring high-resolution bathymetric, sidescan and video imagery/maps. The team has further developed real-time Virtual Underwater Laboratory (VUL), and real-time high-resolution sidescan sonar simulators, for use in laboratory testing, training and offshore operations support. Experience gained through years of development and field trials has been used as a driving wheel to develop OceanRINGS - a suite of smart technologies for subsea operations, designed to be integrated with any ROV – ship combination. The OceanRINGS is Internet/Ethernet-enabled ROV control system, based on robust control algorithms, deterministic network-oriented hardware and flexible, highly adaptive 3-layer software architecture, [7]. System validation and technology demonstration has been performed over the last eight years through a series of test trials with different support vessels and UL-made smart ROV LATIS. Operations included subsea cable inspection/survey, wave energy farm cable to shore routing, shipwreck survey, ROV-ship synchronisation and oil spill/HNS incident response.

Researchers at UL are developing the next generation of ROV control system (OceanRINGS$^+$). The highly adaptive 3-layer software architecture of OceanRINGS$^+$ includes fault-tolerant control allocation algorithms at the bottom layer, transparent interface between an ROV and supporting platforms (surface platforms, surface/subsea garages and/or supporting vessels) in the middle layer and assistive tools for mission execution/monitoring/supervision in the top layer. Software modules have been developed for advanced control modes such as auto compensation of ocean currents based on ROV absolute motion, robust speed/course controller with independent heading control, semi and full auto pilot capabilities, auto-tuning procedure for low-level controllers, ROV high precision dynamic position & motion control in absolute earth-fixed frame, or relative to target or support platform/vessel. The overall system provides installation of remote displays on the ship's bridge during different research cruises, providing real-time 2D and 3D view of the ship and ROV. Remote displays present all mission critical information to the ship bridge in a simple and intuitive way, yielding improved situation awareness and better understanding of mission progress. ROV dynamic positioning (DP) relative to support vessel in surface and subsea operations has been successfully tested in challenging weather conditions (sea state 3+), [7].

4 Design and Development of ROV

The ROV has been designed to be large enough to ensure that all essential telemetry, propulsion, video, power, and navigational equipment can be mounted. The weight has been kept to a minimum to ensure that a two-man team can deploy, operate and recover it from a small vessel. The dimensions of the vehicle are 1 m long, 600 mm high and 650 mm wide.

4.1   Propulsion System Design

Reconfigurable propulsion system includes two thruster configurations (Table 1):

- **Configuration 1**: Eight VideoRay M5 thrusters configured in two layers: Horizontal Layer with four thrusters and Vertical Layer with four thrusters.
- **Configuration 2**: Twelve Blue Robotics T200 thrusters configured in three layers: Horizontal Layers L0 and L1 with four thrusters each and Vertical Layer with four thrusters.

Table 1. Thruster configurations.

Configuration 1: 8 x M5 Thrusters	Configuration 2: 12 x T200 Thrusters
Horizontal Plane: HT1, HT2, HT3, HT4 Vertical Plane: VT1, VT2, VT3, VT4	Horizontal Plane (L0): HT1, HT2, HT3, HT4 Horizontal Plane (L1): HT1, HT2, HT3, HT4 Vertical Plane: VT1, VT2, VT3, VT4

Experimental data of bollard pull tests have been acquired at a test tank at University of Limerick. The results are shown in Fig. 1. VideoRay M5 thrusters were operated with 48VDC and controlled using RS485 interface. The Blue Robotics T200 thrusters were operated with 16VDC and controlled by FPGA-generated PWM control signals (NI myRIO). As indicated in Fig. 1, the M5 thrusters consume considerably more power than the T200s, with maximum consumption at 540 W and 230 W respectively. Similarly, the M5s produce a maximum forward thrust of 9.1 kg compared to 3.4 kg forward thrust produced by the T200. The horizontal thrusters are orientated in a vectored configuration, allowing for more accurate surge, sway and heading control.

An essential component within the reconfigurable thruster system is the control system, which will be discussed in more detail in the Control System Design section.

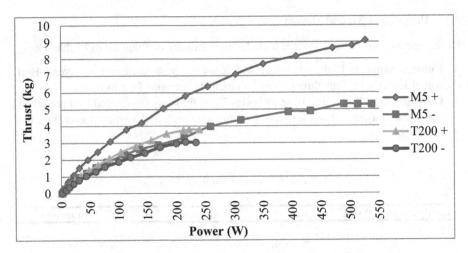

Fig. 1. Thruster characteristics' comparison test results.

4.2   Navigation System Design

Practically all observation class ROVs have a basic sensor suite consisting of a depth sensor and a heading sensor. Although depth sensors give accurate depth information the magnetometers used for heading information can have slow output rates with delays of up to 2 s observed [1, 2]. These magnetometers are also subject to dynamic interferences from thrusters and external interference from ferrous structures. These sensors' accuracy can be further enhanced by using an inertial navigation system (INS), the most common type based on micro-electro-mechanical sensor (MEMS) technology. However these sensors can also be highly prone to drift [3].

To ensure that the ROV's control system is resilient it is dependent on the feedback of an accurate navigational sensor suite. This is achieved by combining a number of different navigational sensor technologies - a GNSS antenna and module, a fibre optic gyro-based INS and a Doppler Velocity Log (DVL). It is not common for observation class ROVs to be equipped with a navigational sensor suite that can provide such high accuracy. This gives an advantage to this ROV in applications where the observation of dynamic offshore renewable device targets is required. In order for the ROV to move in and around these targets it is essential to have accurate navigation feedback to prevent damage to the systems.

The GNSS module utilised is a u-blox NEO-7P. This module can provide sub metre accuracy by coupling a precise point positioning (PPP) algorithm with the satellite-based augmentation system (SBAS). The NEO-7P module provides access to the GPS or GLONASS satellite networks. The module also provides a timepulse, synchronised Coordinated Universal Time (UTC), which is used to synchronise the full navigation system.

The INS integrated into the ROV is a state-of-the-art Rovins Nano by iXBlue. It is an extremely accurate dead reckoning navigation sensor based on fibre optic gyroscope technology. It can provide positional accuracy of 0.5% distance travelled and station keeping of <1 m.hr-1, both when coupled with a DVL. It can also provide roll and pitch dynamic accuracy of 0.1°, heading dynamic accuracy of 0.5°, and a resolution of 0.001° [4].

The DVL employed in the ROV is a Nortek 1 MHz system. The high operating frequency of the DVL's transducers provides long term accuracy of $\pm 0.2\%/\pm 1$ mm.s^{-1} and resolution of 0.01 mm.s^{-1} [5]. It has a maximum and minimum altitude of 50 m and 0.2 m respectively.

The INS receives navigational data and timestamp from the GNSS system and the DVL, integrates this data with its own, runs the data through an internal Kalman filter algorithm and outputs accurate navigational data to the ROV's control system.

4.3   Control System Design

To ensure that the ROV can be controlled in difficult conditions the high power, high thrust and accurate navigation system needs to be coupled with a strong control system. The ROV is equipped with a full real-time embedded control system, which performs all necessary data processing and synthesis online, aboard the vehicle in real-time.

The control system is split into two tasks – regulation task and actuator selection task. The regulation selection task, which is controlled by the Topside Mission Builder, the Arbitration module and the Synthesis module, is utilised to determine the total control efforts of the ROV. The actuator selection task, which is controlled by the Control Allocation module, maps the total control efforts of the ROV onto individual actuator settings. Further reading and information can be found in [6]. In short, the control system combines normalised force and moment command outputs from the virtual joystick (vector τ_{VJ}) and the low level controllers (vector τ_{LLC}) to produce vectors $\tau_{d\ HT}\ \tau_{d\ VT}$, which represent the total normalised control efforts to be exerted by the thrusters. The Control Allocation module converts the τ_d forces and moments to individual outputs to the individual actuators. In OceanRINGS+ control architecture the virtual control space is separated into horizontal (HT) and vertical (VT) subspaces. Virtual control input for HT subspace consists of normalised surge (τ_X) and sway (τ_Y) force and yaw moment (τ_N) and heave force (τ_Z) and roll (τ_K) and pitch (τ_M) moments for VT subspace.

5   Initial Testing of ROV and Results

Initial trials of the ROV have taken place in a wave and current flume tank, with further field trials being conducted in a freshwater lake.

During initial trials the ROV's low-level controllers have been first tuned and then tested in various operating conditions. Tests have also been conducted to gather the vehicle's initial top speed, which has resulted in 1.25 m.s^{-1}.

Each DOF has a dedicated Low Level Controller, as shown in Table 2. Performance of LLC in real world environment (freshwater lake) is shown in Fig. 2. Graphs in upper (lower) half show time responses of LLC in HT (VT) subspace. The first row of graphs show desired and actual time responses of surge speed, sway speed and yaw rotation, while the second row shows LLC outputs (normalised surge and sway forces and yaw moment). The third row displays desired and actual time responses of heave (depth), roll and pitch rotation. Finally, the fourth row shows LLC outputs (normalised heave force and roll & pitch moments). Despite the high level of coupling due to effects of environment and nature of control problem, each LLC has produced excellent performance in tracking set points, yielding rapid responses with negligible overshoots.

Table 2. Low level controllers: inputs and outputs.

Horizontal subspace (HT)			Vertical subspace (VT)		
DOF	Input	Output	DOF	Input	Output
Surge	u_d (m/s)	τ_X	Heave	Z_d/A_d (m)	τ_Z
Sway	v_d (m/s)	τ_Y	Roll	R_d (deg)	τ_K
Yaw	Y_d (deg)	τ_N	Pitch	P_d (deg)	τ_M

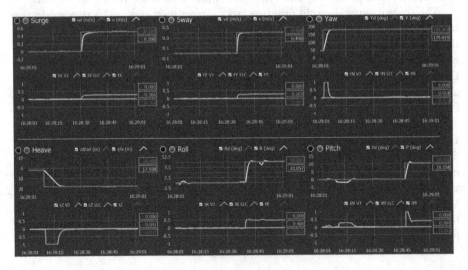

Fig. 2. Low level controllers' output

6 Discussion

Upon completion of the ROV design it can be noted that the vehicle's combined navigation and control system allow for extremely accurate control of the ROV through the water column. After initial tests proved that the ROV could travel at speeds of 1.25 m.s^{-1} further improvements could be made to the design to ensure that the top forward speed of the vehicle could be increased. One suggestion would be to house all the power and telemetry in one oil-filled enclosure, reducing drag and reducing the weight significantly. This weight decrease would have a knock-on effect on the size of the buoyancy block, further reducing the system's weight and drag.

Another realisation occurred during the initial testing phase when two thrusters developed faults. These faults were not detected and the control of the ROV dramatically decreased. An automatic thruster fault diagnosis system and control accommodation would be a strong solution for this issue in the future.

7 Conclusions and Future Work

The ROV has been designed to allow for observation of offshore renewable energy devices. The novel aspects of the vehicle have been discussed in relation to resilient systems. The design of the novel features of the ROV has been discussed, with initial test results described. Some design suggestions are put forward to further enhance the characteristics of the ROV and allow it to become a more resilient system.

Work is currently underway in the design of a thruster fault diagnosis and isolation system, using a combination of software and hardware. Further work needs to be undertaken to develop the software so that thruster faults can be automatically diagnosed and accommodated for.

References

1. Bandala, M., Salgado, T., Chávez, R.: Multi-rate sensor fusion for underwater heading estimation. Ind. Robot Int. J. **41**, 347–350 (2014)
2. Dukan, F., Ludvigsen, M., Sorensen, A.J.: Dynamic positioning system for a small size ROV with experimental results. In: OCEANS 2011 IEEE, Spain, pp. 1–10. IEEE (2011)
3. De Agostino, M., Manzino, A.M., Piras, M.: Performances comparison of different MEMS-based IMUs. In: IEEE/ION Position, Location and Navigation Symposium, pp. 187–201. IEEE (2010)
4. iXBlue Rovins Nano. https://www.ixblue.com/products/rovins-nano. Accessed 16 Nov 2017
5. Nortek: Nortek and DVLs. http://www.nortek-as.com/en/products/dvl. Accessed 16 Nov 2016
6. Omerdić, E., Toal, D., Nolan, S., Ahmad, H.: Smart ROV$_{LATIS}$: control architecture. In: UKACC International Conference on CONTROL 2010, Institution of Engineering and Technology, pp. 787–792 (2010)
7. Omerdić, E., Toal, D., Dooly, G.: OceanRINGS: smart technologies for subsea operations. In: Gal, O. (ed.) Advanced in Marine Robotics. Lambert Academic Publishing, Saarbrücken (2013)

Energy Distribution Systems

Some Significant Problems of Lightning Protection in Flexible Energy Systems

Zoltán Tóth(✉) ⓘ, István Kiss, and Bálint Németh

Department of Electric Power Engineering, Budapest University
of Technology and Economics, Egry József utca 18, Budapest 1111, Hungary
toth.zoltan@vet.bme.hu

Abstract. The lightning protection of different components of flexible systems are more and more important because the produced energy quotient increases by the time. Proportionally, the risk and the number of the lightning strikes and the probability of the fault are increasing, too. Nowadays the renewable power plants take a determinative part in the energy sector and therefore, it is necessary to calculate with the loss after a lightning strike or because other overvoltage caused fault.

Lightning protection of overhead lines is a widely examined topic, too. Standard and enhanced calculation methods are used to estimate risk arising from lightning stroke. However, there are some special cases, like parallel lines, crossing lines, or lines running into a substation, when the exact geometry of the arrangement has to be taken into consideration.

Among these arrangements, there are some interesting examples, where large wind turbines are situated near a transmission line or a substation. Lightning protection of large-scale wind turbines is itself a very difficult topic, but the effect of their proximity to the lines on the risk of lightning stroke is also an interesting one.

Keywords: Lightning protection · Renewable energy plan · Transmission line
Photo voltaic · Wind turbine · Probability Modulated Attractive Space

1 Introduction

Shielding effect of nearby objects on the number of strokes in case of critical sections of a transmission line is an important question. Such a critical section can be the one near a substation or a tower equipped by telecommunication devices, etc. There are different methods applied for the calculation of the number of strokes as it was overviewed in [1]. The method described in that paper was selected to use it in the examination of the following problem.

Until nowadays high number of wind turbines were installed all over the world. Some of them were installed quite close to a transmission line. Regarding that the wind turbines are much higher than the transmission lines they have a significant effect on the expected number of strokes. The question is, how this effect can be calculated and how the results can be taken into consideration.

There is another interesting situation with lot of questions that it is necessary to install lightning protection system for the installed solar farms or not [1]. The protection of the

© IFIP International Federation for Information Processing 2018
Published by Springer International Publishing AG 2018. All Rights Reserved
L. M. Camarinha-Matos et al. (Eds.): DoCEIS 2018, IFIP AICT 521, pp. 293–299, 2018.
https://doi.org/10.1007/978-3-319-78574-5_28

inverters is clear because that is one of the main cost. It is possible to be 1/3 of the whole costs. In this paper there will be shown a calculation for a small solar power system, which could be equivalent as a part to a bigger one.

2   Relationship to Resilient Systems

The lightning protection and the design of a lightning protection system (LPS) is an important part of the transmission systems, too like the LPS of a building (e.g. family house, hospital, skyscapers, etc.).

Nowadays with the dynamic system regulation (like DLR – Dynamic Line Rating), for which is a possible solution for the transmission of the produced energy from renewable sources, it is necessary to consider other aspects during the designing and planning. This method could not work well if there is no high efficiently lightning protection. Therefor it is relevant to deal in this aspect with the protection of transmission lines and wind turbines.

3   About the PMAS Method

The PMAS method (Probability Modulated Attractive Space) is a mode to determinate the expected frequency of lightning strokes attached to a certain object. The basis of this method is similar to the electro-geometric model (EGM), [2].

From the point of the application of PMAS it is very important to determinate the boundary (50% regression surface) of the attractive space. According to the experience, this boundary differs from the one generated according to pure geometric method (points that have the same distance from the object and the ground/air termination). In additional to, the boundary surface is polarity-dependent.

In case of high objects, a certain number of upward lightning strokes appear. (Their number increases with increasing height.) When the upward leader is initiated by the electric field of the space charge in the thundercloud, striking point cannot be defined in such a way as in case of downward lightning. However such examination which results in the "attachment point" of lightning in the previous case is still interesting.

The striking point is that point, where the endpoint of the downward leader is at the moment, when the return stroke is starting. The distance, between the striking point and the point of strike, is the striking distance, which is dependent on the lightning current. There are different methods to describe this dependence [3, 4].

$$N_F = N_G \int_V \frac{dP}{dr} dV \, b \tag{1}$$

$$N_{Fa} = N_G \int_{V_a} \frac{dP}{dr} dV \tag{2}$$

In the Eqs. (1) and (2) the P is the probability, the r is the striking distance, V is the volume, N_G is the number of the strikes for a km^2 and a year.

In reality, the attractive space is an abstract term. Although it can be considered, that from a given striking point the lightning attachment will happen to the closest object, it happens only with a given probability. (Let us denote it by b.) So, in the reality, the attractive space of an object gives that points where the probability is higher that the stroke will be attached there than to other point. Therefore, the interface are determined by the points, where the probability of the stroke is b = 0.5. (Equation 1 and for the attractive space only the Eq. 2.) To each point of the attractive space it is possible to give a dP/dr value:

$$\frac{dP}{dr} = \frac{kp}{\sqrt{2\pi}\,r} \exp\left(-\frac{1}{2}k^2 p^2 \left(\ln\frac{r}{r_m}\right)^2\right)$$
(3)

Where

k   is the parameter which is dependent to polarity of the lightning,
p   is a value between 1.2 and 2,
r_m is the median value of the striking distance
r   is the striking distance.

The attractive space for positive and negative polarity is different.

For a PMAS simulation it is necessary to know the dependence of these parameters on different influencing factors, like environment, geography, climate and the arrangement of the objects.

4   Formulation of the Problem for Wind Turbines Near Transmission Line

In this paper, simulations was made to determinate the equivalent or collection area, A_{eq} (see (1)) depending on the relative position of the wind turbine and the transmission line as it is illustrated in Fig. 1. In this arrangement, it is possible to see calculations has been made for the following cases. First, when the bottom of the wind turbine and a tower have the same x-coordinate. In the second case, the wind turbine is in the same distance from two towers.

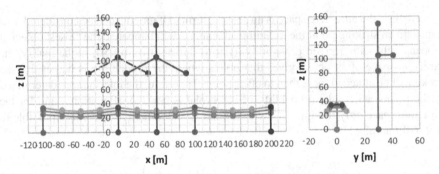

Fig. 1. The arrangement in two positions (position one: brown; position two: red). (Color figure online)

The wind turbine has the following parameters:

Table 1. The parameters of the wind turbine.

Type	Vestas V80 HU
Height of the nacelle	105 m
Length of blades	45 m

The parameters of the transmission line are the following (Table 2):

Table 2. The parameters of the transmission line.

Height of overhead wires	35 m
Height of phase wires	(31, 31, 26) m
Position of phase wires from centre	(5, 5, 7) m
Sag	4 m
Distance between the towers	100 m

Regarding that most of the calculation methods result in the so-called equivalent area (or collection area), that gives the number of stroke according to (Eq. 4), our goal is to calculate this parameter. In (4) N represents the number of stroke/year to the object, N_g denotes the annual ground flash density, and A_{eq} is the equivalent area.

$$N = N_g A_{eq} \tag{4}$$

As a reference, A'_{eq} was calculated according to the equation from [4] for the previously mentioned transmission line. According to this $A'_{eq} = 7000 \, \text{m}^2$ (by Eq. 5 and Fig. 2.)

$$A'_{eq} = 2 B_a l \tag{5}$$

where B_a is the half-width of an equivalent area (A_{eq}) (Table 3).

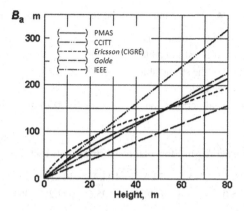

Fig. 2. Comparison of the width of equivalent area, obtained from several calculation methods [6].

Table 3. Calculated $A_{eq}[m^2]$ for the overhead lines.

A_{eq}/I	1 kA	2 kA	5 kA	10 kA	20 kA	35 kA	50 kA	100 kA
Shielding wire	64 453.3	63 234.4	60 282.6	53 658.4	43 238.8	20 946.3	10 004.9	1 201.0
Phases	1 581.6	1 063.6	912.4	0.0	0.0	0.0	0.0	0.0
Total	66 034.9	64 298.0	61 195.0	53 658.4	43 238.8	20 946.3	10 004.9	1 201.0

For the further calculations distance between the wind turbine and the centre line of the transmission line is denoted by "y".

5   Study Case: The Necessity the Lightning Protection of a Solar Panel String Table

In this study case there will be shown a calculation for the efficiency of a lightning protection system for a smaller solar farm. The parameters of the solar panel string table are the following (Table 4).

After the calculation, the results are in the Tables 5 and 6.

The air-termination reduces the risk of direct lightning stroke significantly. However, the comparison of the cost of selected air-termination system to the expected cost of damage in the unprotected case is very necessary [7, 8]. It is especially useful to see how the resultant risk is changing with the lightning current.

6   Summarizing the Results

The calculated values by the PMAS method and by [5] have different result. The value of $A'_{eq} = 7000 \, m^2$ and the calculated for the overhead lines without turbines give the result which is shown in the Table 1.

Table 4. The parameters of the solar panel string table

Width (x) of the string table	17 m
Depth (y) of the string table	3.39 m
Height (z) of the string table	2.62 m
Angle of the table	40°
High of the air termination	0.75 m

Table 5. Calculated $A_{eq}[m^2]$ for solar panel string table (without air termination)

I	1 kA	2 kA	5 kA	10 kA	20 kA	35 kA	50 kA	100 kA
$A_{eq}[m^2]$	1255.0	1247.3	1196.0	1105.6	927.2	491.7	254.7	35.7

Table 6. Calculated $A_{eq}[m^2]$ for solar panel string table (with air termination)

$A_{eq}[m^2]/I$	1 kA	2 kA	5 kA	10 kA	20 kA	35 kA	50 kA	100 kA
Panels	218.3	217.0	201.7	182.8	151.4	78.2	39.3	5.5
Air termination	1329.9	1319.8	1272.5	1177.5	992.2	533.6	278.6	38.6

The calculated values for our arrangement give the expected values. The wind turbine has a risk reduction effect from the point of the transmission line. Therefore the striking frequency (for the case of power lines) decreases [9]. The measure of this effect depends on the relative position of the wind turbine and the transmission line. From the point of PMAS it can be considered that the attractive space of the transmission line is reduced by the one of the wind turbine. This can be interesting when a critical section of a line is situated near the wind turbine. Although the phase wires are protected by the grounded wires (thus the reduction effect of their equivalent area is not significant), the severity of the back-flashovers and the secondary effects are depending on the number strokes to the grounded overhead wires. In case of a solar power plant with a well-selected and well-designed lightning protection system, that is possible to reduce the probability of the risk of a fault caused by a lightning strike. But it is important to know the high of this risk and be more reliable calculation for the expected stroke.

7   Conclusion

In the paper was shown, the wind turbine has a risk reduction effect from the point of the transmission line. Therefore the striking frequency decreases. The measure of this effect depends on the relative position of the wind turbine and the transmission line. From the point of PMAS it can be considered that the attractive space of the transmission line is reduced by the one of the wind turbine.

The solar power should be in the middle point of the lightning protection, too. With a lightning protection system, that is possible to reduce the probability of the risk of a fault caused by a lightning strike. But it is important to know the high of this risk and be more reliable calculation for the expected stroke. There should be other researches for another important situation when there will take into account the environment, too.

References

1. Rousseau, A., et al.: Direct Lightning Protection Risk Assessment on PV Systems, APL (2017)
2. Kiss, I., et al.: Improved method for the evaluation of shielding effect of objects near medium voltage transmission lines. In: International Conference on Lightning Protection, ICLP 2014, Shanghai, China (2014)
3. Horváth, T.: Computation of Lightning Protection. Wiley, New York (1990)
4. Horváth, T.: Interception efficiency of lightning air termination systems constructed with rolling sphere method. In: 28th International Conference on Lightning Protection, ICLP 2006, Kanazawa, Japan, 18–22 September 2006
5. Dellera, L., et al.: Lightning stroke simulation by means of the leader progression model, parts I and II. IEEE Trans. Power Deliv. **5**, 2009–2029 (1990)
6. Horváth, T.: Efficiency of lightning protection shielding of overhead lines and substations. In: 8th International Symposium on High Voltage Engineering, Yokohama, Paper 73.03, vol. 4, pp. 241–244 (1993)
7. Horváth, T.: A new method for design of the air termination system of high voltage overhead stations. In: 24th ICLP, Birmingham, UK (1998)
8. Horváth, T.: Gleichwertige Fläche und relative Einschlagsgefahr als charakteristische Ausdrücke des Schutzeffektes von Blitzableitern, vol. 11. Internationale Blitzschutzkonferenz, München (1971)
9. Tóth, Z., Kiss, I.: Evaluation of striking frequency in case of wind turbines with multiple receptors. In: International Symposium of Winter Lightning, ISWL 2017, Joetsu, Japan (2017)

Effect of Enhancing Distribution Grid Resilience on Low Voltage Cable Ageing

Gergely Márk Csányi$^{(\boxtimes)}$ [ID], Zoltán Ádám Tamus, and Péter Kordás

Department of Electric Power Engineering, Budapest University
of Technology and Economics, Egry József Str. 18, Budapest, 1111, Hungary
{csanyi.gergely, tamus.adam}@vet.bme.hu, kordas.peter@gmail.com

Abstract. Enhancing system resiliency is getting more and more attention due to the climatic changes, growing number of earthquakes, floods and terrorist attacks. One pillar of system resiliency is distributed generation. As a result of distributed generation reverse power flow can occur causing the aggregate load in certain cases to surpass the rated capacity of the distribution cables without being detected. In this paper the effect of short term thermal ageing is investigated modeling the short term overloads. The samples were exposed to 4 cycle of 3 h thermal ageing at 110 °C and 125 °C and the changes of the insulation were followed by a mechanical (Shore D) and an electrical (loss factor) measurement method. The results suggest that the short term ageing processes have a measureable effect of the insulations and the calculated central frequencies are good indicators of the duration of the thermal ageing and the central loss factor values are decent indicators of the hardness of the insulation.

Keywords: Distributed generation · Low voltage cables · System resiliency
Cable ageing · PVC cable · Smart systems · Loss factor · Tan δ

1 Introduction

All across Europe the low voltage distribution cables are valuable assets. Hence, the electricity provider companies aim to prevent outages, therefore assessing the condition of the cables and optimizing replacement strategies are the main goals for them [1]. Low voltage distribution cables are now facing new challenges due to environmental stresses (e.g. desertification) and novel appliances connected to the grid (e.g. solar panels, electric vehicles etc.). As a result of these, the aggregated load can cause undetected overloads by the fuses in just parts of a cable section since the consumers are not equally distributed nor in power consumption neither in time. Hence, the load can surpass the cables' rated capacity elevating its temperature that decreases the expected lifetime [2, 3]. Previous study have shown that even a short term overload at 125 and 140 °C cause changes in the condition of a LV cable insulation. However, in that paper the effect was measured after two ageing period and only on the cable's jacket insulation [4]. In this paper same type cable specimens are investigated at two lower temperatures and after 4 ageing periods. Loss factor measurements were carried out on the jacket and on the core insulations as well to answer the following research questions: how does the short

© IFIP International Federation for Information Processing 2018
Published by Springer International Publishing AG 2018. All Rights Reserved
L. M. Camarinha-Matos et al. (Eds.): DoCEIS 2018, IFIP AICT 521, pp. 300–307, 2018.
https://doi.org/10.1007/978-3-319-78574-5_29

term thermal stresses affect the PVC insulated LV cables and is there any good marker to follow this kind of degradation?

2   Relationship to System Resiliency

The distribution system resilience means the ability of the system to withstand rare and serious events and recover from them. The enhancement of the distribution grid is lying on 4 main pillars, namely the smartening of the grid, hardening the grid against rising environmental impacts (e.g. the effect of flooding, extreme heat or high winds), distributing the generation and building resilience on demand [5]. These needs and directions in the electric power industry make the different assets of the grid face new challenges.

The low voltage (LV) distribution cables are now or in the near future have to withstand extra stresses due to the high variation of load and reverse power flow caused by distributed generation, electric vehicles etc. The connection of the novel appliances can result the aggregate load in exceeding the rated capacity of the currently operating cable systems which elevates the operating temperature that decreases the lifetime of the cable insulations. However, the resilient distribution systems rely on microgrids with distributed generation, the existing distribution cable network is used for power-flowing [6, 7]. Therefore, in case of extreme events, the distribution cables play key role and their essential function can be fulfilled if the real condition and environmental parameters of distribution cables are known. Nowadays, by means of new technologies, such as Smart Sensors and Embedded Intelligence, the exact temperatures of the cables can be followed more precisely and if the correlation between the temperature stress and the insulation lifetime is known, better predictions can be made, a reliable asset management system can be developed [4], which improves the distribution grid resiliency.

3   Experimental

3.1   Samples and Thermal Ageing

During investigation six NYCWY 0.6/1 kV $4 \times 10/10$ mm^2 low voltage cables were used and half-meter-long samples were prepared from it. The build-up of the cable can be seen in Fig. 1. The parts from inside to outside in order: four copper conductors (1), PVC core insulations (2), filling material (3), copper wire and tape screens (4) and PVC jacket (5). The core insulations are colored to blue, black, grey and brown. Maximum operating temperature is 70 °C while the conductors can heat up to 160 °C during short circuiting, no information was given by the manufacturer regarding thermal ageing.

5. 4. 3.   2. 1.

Fig. 1. The construction of the cable

When measuring the jacket an inner electrode was created by connecting wire and tape screens of the cables to all of the core conductors while an outer electrode was created by wrapping an aluminum foil around each cable jacket.

The samples were introduced to 4 cycles of accelerated thermal ageing. The cable specimens were grouped to two and were aged for 3 h in each cycle at 110 °C and 125 °C, respectively.

Arrhenius equation was used to calculate the equivalent ageing durations from 125 °C to 110 °C, the activation energy was 80 kJ/mol [8, 9]. Table 1 shows the calculated equivalent ageing times.

Table 1. Equivalent ageing duration at different temperatures [hour]

80 kJ/mol	3 h @ 125 °C	6 h @ 125 °C	9 h @ 125 °C	12 h @ 125 °C
110 °C	7.7	15.5	23.2	30.9

3.2 Measurement

The investigation included both mechanical and electrical measurements. The mechanical measurements were carried out by a Shore D hardness tester and the polarization processes were investigated by tan δ measurement [9]. On the cable jackets the hardness was measured 10 times. However, since the standards (ASTM etc.) require at least 4 mm thickness of the material and the thickness of the jacket insulation is around 2 mm, the measured hardness values are only for comparison purposes. By a Wayne-Kerr 6430a impedance analyzer tan δ values were measured at 5 V at 15 different frequencies in the 20 Hz…500 kHz range. During the jacket measurements an inner electrode was formed by connecting all cores and wire and tape screens and this was connected to 1000 V while by wrapping an aluminum foil to the cable jacket an outer electrode was created and was grounded. During the measurement the temperature was 23.1 ± 1.0 °C. Each electric measurement started by examining proper connections.

3.3 Results

Shore D measurement
Shore D hardness is a dimensionless value between 0 (soft) and 100 (hard). Figure 2 plots the measured hardness values against the ageing time at 110 °C The equivalent ageing times from Table 1 was to transform the 125 °C values to 110 °C values.

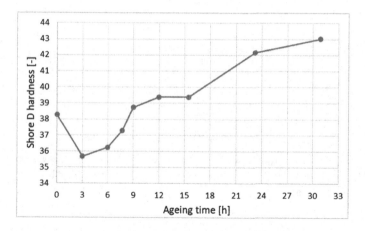

Fig. 2. Effect of thermal ageing on Shore D hardness of the cable jacket insulation

It can be stated that the jacket insulation initially softened by ageing, nevertheless by time the hardness exceeded the initial values. This initial softening effect is surprising since when the plasticized PVC samples are exposed to thermal stress the dominant process is usually the loss of plasticizer that decreases the conductivity of the insulation while increasing the hardness of PVC [10]. The opposite of this that is experienced here. Due to the short ageing time dehydrochlorination can be neglected. Hence, these results suggest that the plasticizer initially contained by the inner core insulations and the filling material migrates to the cable jacket insulation, over dominating the vaporization of the plasticizer from the jacket.

Tan δ measurement
Figure 3 shows the loss factor on the jacket and the blue core insulations after ageing at 110 °C and 125 °C, respectively. The different colored core insulations did not differ significantly hence here the blue core is plotted.

The results show that the jacket and core insulation have their peak values at different frequencies and the former have higher peak value. The jacket and core insulations react similarly to thermal ageing: the peaks of the tan δ values on the inspected frequency range shifted left, meaning the slower polarization processes became more significant, however this change is slower in the jacket. The other similarity is that in all cases the curves shifted upwards and then followed a descending order by ageing. It is important to note that the 50 Hz loss factor increased significantly after ageing in all cores, while on the jacket a clear trend cannot be identified.

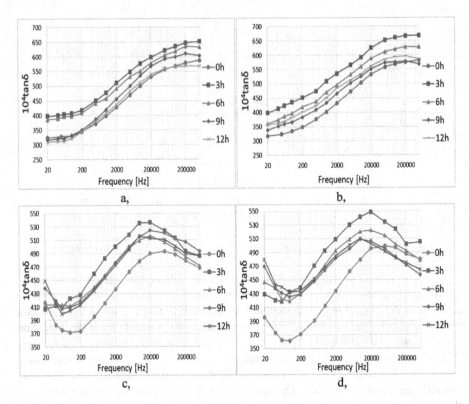

Fig. 3. Jacket ageing 10^4 tan δ at 110 °C (a,), and 10^4 tan δ at 125 °C (b,) Blue core ageing 10^4 tan δ at 110 °C (c,), and 10^4 tan δ at 125 °C (d,) (Color figure online)

4 Discussion of the Results

To examine the movement of the polarization spectrum two quantities were calculated namely the central frequencies and central tan δ values. These values were calculated by summing the multiplication of the logarithm of the frequencies by the measured tan δ values and dividing this sum by the sum of the measured tan δ values in case of central frequencies and divided by the sum of the logarithm of the frequencies in case of central tan δ values. These values were then used as an exponent to a base 10. Figures 4 and 5 show the central frequencies and Fig. 6 shows the central tan δ values to each core and jacket and plotted against the ageing time at 110 °C.

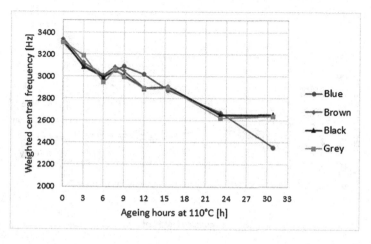

Fig. 4. Changing of central frequencies of the cable core insulations by ageing at 110 °C (Color figure online)

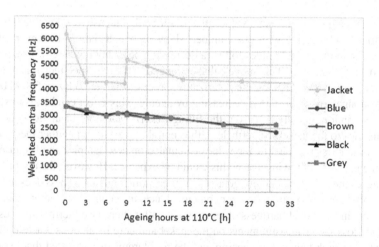

Fig. 5. Changing of central frequencies of the cable jacket and core insulations by ageing at 110 °C (Color figure online)

The curves at Fig. 4 show that by ageing the weighted central frequencies of the core insulations decrease by ageing, however a hump can be observed after 9–12 h of ageing. Hence, the central frequency value is a good marker of thermal ageing in case of short term stresses. This descent trend proves the assumption made regarding the polarization spectrum shifting towards the slower time constants. This result also suggest that the main process taking place as a result of thermal ageing is the loss of plasticizer content of the material since the less dipolar molecules the lower intensity of polarization processes taking place. The central frequencies on the jacket show similar trend but with significantly higher values and bigger humps after 9 and 12 h of ageing. Figure 2 also have these humps but in the opposite direction because the loss of plasticizer means harder insulations and vice versa.

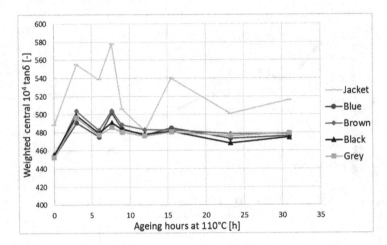

Fig. 6. Changing of central 10^4 tan δ of the cable jacket and core insulations by ageing at 110 °C (Color figure online)

The changing of central tan δ values, that describe the intensity of polarizations in the 20 Hz to 500 kHz range with one number, show higher deviation. However, after an ascending trend the central values follow a decreasing order both in case of jacket insulations and core insulations. This suggest that compared to the new specimen due to thermal ageing until about 8 h of ageing the polarization processes in the 20 Hz to 500 kHz range became stronger and then started decreasing by ageing. This effect can be also explained by the assumption, namely that from the filling material and core insulations the plasticizer migrates to the jacket. Nevertheless, the central loss factor on neither the jacket nor the core insulations went below the values calculated on new samples and because of the increasing hardness of the insulation this would be expected. However, the distribution of plasticizer in the jacket is not homogeneous i.e. measuring higher hardness does not necessarily imply lower all depth plasticizer content. Hence, the central loss factor values seem to be decent indicators of hardness of an insulation after short term thermal stresses.

All the loss factor measurements on the jacket and core insulations react similarly to short term but high temperature ageing and the measurements suggested that even these short thermal stresses can initiate degradation processes. These results can be used to determine whether a new cable have already suffered a relatively short term thermal overload, giving the possibility for the electricity providers to handle this problem and save precious years of lifetime. Although the results and the indicators seem to be promising to follow the degradation of LV PVC cables, further investigations are needed to determinate threshold values.

5 Conclusion

The enhancement of the distribution grid is becoming more and more important every day due to the numerous new challenges that a distribution network has to face.

In this paper the effect of the short term overloads were examined on low voltage PVC insulated cables' jacket and core insulations. These affect the lifetime of the insulations. The samples were exposed to periodic thermal stress at two different temperatures. After each ageing cycle Shore D hardness measurements were carried out on the jacket of the insulation and loss factor measurements on the jacket and all the core insulations. The calculated central frequency values proved to be good indicators to follow the duration of the ageing process and the central loss factor values shew reasonable relation to the change of the hardness of the insulation. The results suggested the loss of the plasticizer from the insulations, however at the beginning of the ageing the measurement results suggested the migration of plasticizer from the cores and filling material to the jacket insulation.

All in all, the investigation suggested that two calculated quantities can be used to follow the duration of ageing: the central frequency and the central loss factor values.

References

1. Kruizinga, B., Wouters, P.A.A.F., Steennis, E.F.: Fault development on water ingress in damaged underground low voltage cables with plastic insulation. In: 2015 IEEE Electrical Insulation Conference (EIC), pp. 309–312, June 2015
2. Höning, N., Jong, E.D., Bloemhof, G., Poutré, H.L.: Thermal behaviour of low voltage cables in smart grid-related environments. In: IEEE PES Innovative Smart Grid Technologies, Europe, pp. 1–6, October 2014
3. Trichakis, P., Taylor, P.C., Lyons, P.F., Hair, R.: Predicting the technical impacts of high levels of small-scale embedded generators on low-voltage networks. IET Renew. Power Gener. 2(4), 249–262 (2008)
4. Csányi, G.M., Tamus, Z.Á., Varga, Á.: Impact of distributed generation on the thermal ageing of low voltage distribution cables. In: Camarinha-Matos, L.M., Parreira-Rocha, M., Ramezani, J. (eds.) DoCEIS 2017. IAICT, vol. 499, pp. 251–258. Springer, Cham (2017). https://doi.org/10.1007/978-3-319-56077-9_24
5. Hosseini, S., Barker, K., Ramirez-Marquez, J.E.: A review of definitions and measures of system resilience. Reliab. Eng. Syst. Saf. 145, 47–61 (2016)
6. Chen, C., Wang, J., Qiu, F., Zhao, D.: Resilient distribution system by microgrids formation after natural disasters. IEEE Trans. Smart Grid 7(2), 958–966 (2016)
7. Chen, C., Wang, J., Ton, D.: Modernizing distribution system restoration to achieve grid resiliency against extreme weather events: an integrated solution. Proc. IEEE 105(7), 1267–1288 (2017)
8. Ekelund, M., Edin, H., Gedde, U.: Long-term performance of poly(vinyl chloride) cables. part 1: Mechanical and electrical performances. Polym. Degrad. Stab. 92(4), 617–629 (2007)
9. Tamus, Z.Á., Németh, E.: Condition assessment of PVC insulated low voltage cables by voltage response method. In: International Conference on Condition Monitoring and Diagnosis, CMD 2010, pp. 721–724 (2010)
10. Nagy, A., Tamus, Z.Á.: Effect of dioctyl phthalate (DOP) plasticizing agent on the dielectric properties of PVC insulation. In: 2016 Conference on Diagnostics in Electrical Engineering (Diagnostika), pp. 1–4, September 2016

Suppression of Conducted Disturbances During the Partial Discharge Monitoring of Industrial Cable Systems

Richárd Cselkó$^{(\boxtimes)}$ ⓘ and István Kiss

Group of High Voltage Technology and Equipment,
Budapest University of Technology and Economics,
18 Egry József u., Budapest 1111, Hungary
cselko.richard@vet.bme.hu

Abstract. Modern industry is seeking to create more reliable, fail-safe and flexible systems. It is shown in the article that diagnostic measurements on the equipment of the physical infrastructure is one of the enabling technologies to reach these goals. The paper deals with a technological question of a diagnostic method, noise suppression during partial discharge measurement. A summary is given of noise mitigation techniques and a novel method applicable to cables is discussed, making use of two high frequency current transformers. The method is evaluated by simulation of an idealized model and by measurement. Based on the latter, the issues to be solved in the future are identified and presented.

Keywords: Diagnostics · Energy systems · Reliability · Noise suppression
Smart sensors

1 Introduction: Partial Discharge Measurement as a Diagnostic Tool

Diagnostic measurements and online monitoring of high voltage power equipment have already proved their usefulness in increasing the resiliency of energy supply and optimizing maintenance and replacement actions. There are several drivers to extend these technologies to low voltage systems, e.g. control and measurement cables in industrial systems [1]. One of the most important ones is the need to prove the reliability of safety equipment in case of lifetime extension of nuclear power plants. One of the possible methods is partial discharge (PD) measurement, detecting the local defects in insulations. This is particularly important, as these may lead to actual failure, when performance of the cables was of utmost importance: during a loss-of-coolant accident. The goal of this research is to improve the applicability of PD testing of low voltage industrial cables.

The most important issue of PD testing is noise suppression, especially in case of on-site measurements on industrial systems. The measurement of PDs consists in catching tiny current impulses emerging from a defect within an insulation system. In the conventional method the charge of these current impulses are taken into account [2].

© IFIP International Federation for Information Processing 2018
Published by Springer International Publishing AG 2018. All Rights Reserved
L. M. Camarinha-Matos et al. (Eds.): DoCEIS 2018, IFIP AICT 521, pp. 308–316, 2018.
https://doi.org/10.1007/978-3-319-78574-5_30

The sensitivity of the measurement goes down to the picocoulomb range in case of field tests and two orders lower in case of special factory testing; this sensitivity makes the measurement susceptible to noise. The frequency range of the conventional measuring devices is contaminated by several noise sources, like power electronics and switching phenomena. Noise can couple into the measurement system basically from the voltage source (conducted disturbances) and directly into the test specimen. Accordingly, noise mitigation is a basic need during PD measurements.

2   Relationship to Resilient Systems: The Importance of the Physical Condition of Assets in the Creation of More Resilient Systems

The resiliency of any system is fundamentally determined by two factors: the architecture of the system and the properties of the individual components; this work concentrates on the latter. In general, the ability of a component or more precisely, a population of components to withstand a stress factor can be handled statistically and represented by a distribution function. As the components age and their condition deteriorates, the distribution function shifts to lower stress values, as it is shown in Fig. 1. Accordingly, any deteriorated component of a system poses an increased risk of failure, while components in good condition may also withstand stresses well above the planned nominal value.

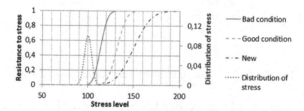

Fig. 1. Susceptibility of an element or population to failure during its lifetime (after [3])

Furthermore, this implies that some components, if necessary, might be overstressed, increasing the flexibility of system [4]. Dynamic line rating is already applied on transmission power lines, however, its extension to transformers and the medium voltage network might be an option to meet further demands, e.g. the fast charging of electric cars [5].

These have been recognized directly or indirectly in almost all smart grid definitions. The creation of the smart grid includes the improvement of the supply quality and reliability, as well as the integration of intermittent energy sources (such as wind energy) in the system, with the requirement of increased flexibility. These can be achieved only if the appropriate diagnostic and monitoring techniques are available and integrated in the system [6].

3 Noise Suppression Methods Applied for PD Measurements

Noise suppression can be achieved by physical methods and signal post-processing. The main advantage of the physical methods is that the noise is suppressed before digitization. Some methods are applicable only in laboratory (e.g. shielding), while others may be used in field testing, as well.

Bridge or balanced circuits are applied mostly for laboratory measurements [7], as they require the test object(s) to be separated from ground, as it can be seen in Fig. 2. The principle of its operation is that in the two arms of the bridge, if balanced correctly, the same impulse current is flowing through for an external signal (i.e. coming from the H.V. generator) and generating zero differential voltage at the detector. On the contrary, the PD impulses originated in the test object flow in opposite direction in the arms, generating a measurable voltage at the detector.

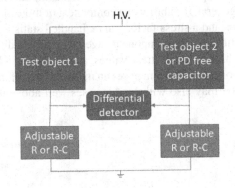

Fig. 2. Balanced circuit for partial discharge measurement

One exception where field application has been achieved is the so-called balanced permanent coupler, which makes use of the construction of large electric machines [8, 9], where the test objects are a pair of symmetrical splits of the phase winding.

Another method is based on the measurement of the signals with two sensors installed physically apart from each other, as shown in Fig. 3. The time of arrival of the signals are determined, and the signals that reach sensor 2 first are rejected as noise.

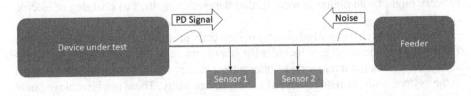

Fig. 3. Time-based discrimination of external noise (after [10])

Increasing the frequency range of PD measurement up to the UHF range on several types of equipment has some advantages, one of them being better noise rejection. In a measurement arrangement developed for power transformers, where the conventional measurement is combined with a UHF sensor, external noise is rejected via gating. The transformer tank acts as a Faraday-cage, and the bushing as a low-pass filter. Accordingly, noise in the UHF range cannot penetrate the tank. Noise suppression is achieved by enabling the conventional signal only if the UHF sensor detects PD. With this method, the noise suppression of the UHF method is exploited, while the well-known evaluation methods and standardized limits of the conventional method are preserved (Fig. 4).

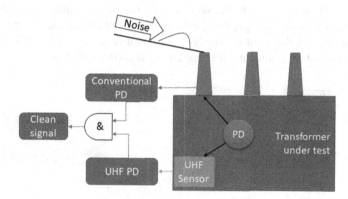

Fig. 4. Suppression of external noise by gating (after [11])

Gating of the signal is also applied in the opposite way [10]. In this case, an external antenna detects the radiated noise, and if it exceeds a limit, it blocks the signal before the PD measuring device. The disadvantage of both methods is that noise and PD can occur at the same time or very close to each other, allowing noise to penetrate in the first, while blocking useful signals in the second case.

Besides the physical methods, several means have been developed for processing the already acquired signals that allow noise suppression. It is well beyond the scope of this paper to give even an enumeration of the available methods; a comprehensive review with short description of the techniques can be found in [10], only some examples are given here. One method with high reputation is the phase-resolved PD pattern method (PRPD) that investigates the connection between the PD phenomena and the phase angle of the AC test voltage. Another method applied with high success is the time-frequency mapping method that plots the individual discharges in a coordinate system based on their length and frequency content. The resulting plot is then clustered by means of fuzzy logic, and then the various clusters shown in the PRPD plot [12]. Some of the available methods are able to find the useful signal even if the noise level is higher. However, if noise is by orders higher than the signal, there is a definite need for physical noise suppression to allow the acquisition of a usable signal.

4   Novel Solution for Suppressing External Noise for Distributed Elements

The idea of a novel method to cancel conducted disturbances in case of the measurement of distributed elements has been presented in [13]. The evaluation of the method is presented here by simulations and measurements, assessing the practical applicability and enumerating the issues resulting from the non-ideal realization.

The method makes use of the architecture of the balanced circuits and the distributed element behavior of the test specimen, exactly that it can be modeled with its characteristic impedance at the input terminals. The regular arrangement is shown in Fig. 5, while the modified arrangement with two sensors is shown in Fig. 6.

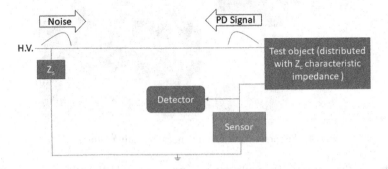

Fig. 5. Regular measurement arrangement with one sensor

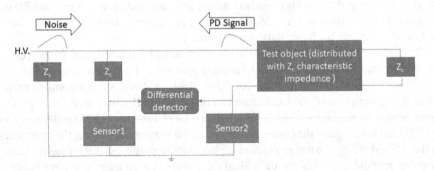

Fig. 6. Theoretical arrangement of the novel method for suppression of external noises in case the test object behaves as a distributed elements

An extra arm of the measurement is added that has an impedance equal to the characteristic impedance of the cable. A sensor picks up the signal from this sensor, while another is attached to the ground lead of the cable that is carrying the same current as the phase conductor but in the opposite direction.

5   Evaluation of the Two-HFCT Method

The theoretical arrangement in Fig. 6 has been implemented in circuit analysis software. The model and the results are shown in Figs. 7 and 8. The impulse originating inside the cable causes voltage drop with different sign in the two sensors, yielding a larger differential signal, while external noise causes cancelling signals.

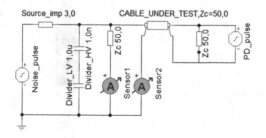

Fig. 7. Idealized simulation model

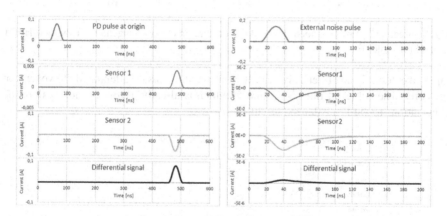

Fig. 8. Simulation of the circuit in Fig. 6 in ideal conditions

In practice, the arrangement has been realized by the application of high frequency current transformers (HFCT), which are regularly applied for PD diagnostics, especially in case of cables [14]. Their bandwidth matches very well with the bandwidth of the PD signals reaching the terminals [15]. It is also advantageous that their lower cutoff frequency is higher than in case of the conventional PD detectors.

During the realization of the measurement, the most important difference is that the characteristic impedance cannot be directly connected to the test voltage at the near and the far end of the cable, because the test voltage must be blocked by capacitors.

The capacitor should be dimensioned so that it acts as a high impedance for the power frequency test voltage, while a small impedance for the high frequency PD impulses. Though it is possible to achieve, it complicates the case. Figure 10 shows measurement results taken from the arrangement shown in Fig. 9b. A PD calibrating impulse has been

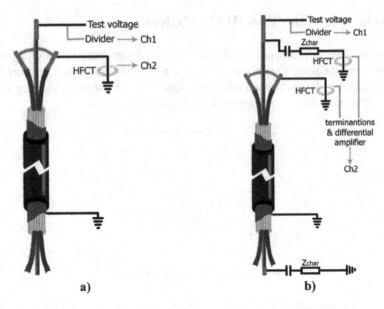

Fig. 9. Regular PD measurement with HFCT (a) and practical arrangement of the two-HFCT method for cables (b)

injected externally and internally. It is obvious that the noise is attenuated, however, not perfectly. The achieved ratio is around 5, which is already useful, but lower than the expectations towards a balanced circuit.

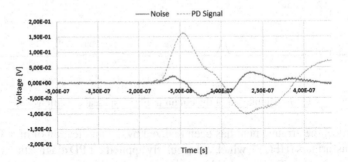

Fig. 10. Output of the realized circuit, when the same impulse is injected externally at the test voltage (Noise) and internally into the cable (PD Signal)

The main reason of the low efficiency of the noise cancellation can be followed in Fig. 11a. The output of the two sensors is similar, but some amplitude and phase error occur, which cause the differential signal to be different from zero. In Fig. 11b the amplification of the PD signal can be followed.

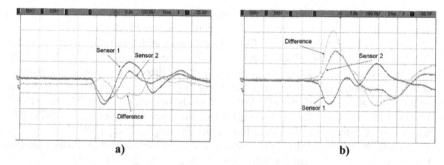

Fig. 11. Impulses measured by the HFCTs in case of external noise (a) and partial discharge originating in the cable (b)

6   Conclusions and Future Work

The paper has shown that there is a strong relationship between the resiliency of a system and the condition of the physical assets it is composed of. One of the methods to improve the condition is diagnostic tests, e.g. PD measurement.

A novel technique for the suppression of external conducted noises during PD measurement has been presented and evaluated. The method can be categorized in the physical noise elimination techniques. It makes use of the distributed element behavior of the test object and the ease of the application of high frequency current transformers. Its advantages are the suppression of noise before digitization and the detection of PD signals occurring simultaneously with noise impulses.

The simulations have shown the theoretical applicability of the method. The realized circuit has to differ from the theoretical arrangement, which causes the noise suppression to be imperfect. One path of the future work has to be the corrections to the impedances so that the response of the two sensors can be adjusted to be similar. This is expected to increase the noise suppression to some orders, which will allow the measurement of useful signals much weaker than the noise level. If this is achieved, the applicability of electronic voltage generators, that ease field testing, will be tested.

Further work in general will concentrate on the noise coupled directly into the cable that must be filtered with other methods.

References

1. Tamus, Z.A., Berta, I.: Application of voltage response measurement on low voltage cables. In: 2009 IEEE Electrical Insulation Conference, Montreal, QC, pp. 444–447 (2009)
2. High-voltage test techniques - Partial discharge measurements, IEC 60270:2000 Standard
3. Ford, G., et al.: Asset Management of Transmission Systems and Associated CIGRÉ Activities, CIGRÉ Working Group C1.1, Paris (2006)
4. Csányi, G.M., Tamus, Z.Á., Varga, Á.: Impact of distributed generation on the thermal ageing of low voltage distribution cables. In: Camarinha-Matos, L.M., Parreira-Rocha, M., Ramezani, J. (eds.) DoCEIS 2017. IAICT, vol. 499, pp. 251–258. Springer, Cham (2017). https://doi.org/10.1007/978-3-319-56077-9_24

5. Catterson, V.M., et al.: The impact of smart grid technology on dielectrics and electrical insulation. IEEE Trans. Dielectr. Electr. Insul. **22**(6), 3505–3512 (2015)

6. Morshuis, P.H.F., Bernstein, B.S.: IEEE DEIS and smart grid: how to fit in. In: 2010 IEEE PES Innovative Smart Grid Technologies Conference Europe, Gothenburg, pp. 1–3 (2010)

7. Horii, K.: Development of tuning type partial discharge detector with noise discrimination and computer aided automation system. In: Proceedings of Second International Conference on Properties and Applications of Dielectric Materials, Beijing, vol. 2, pp. 624–627 (1988)

8. Kurtz, M., Stone, G.C.: Partial discharge testing of generator insulation. In: 1978 IEEE International Conference on Electrical Insulation, Philadelphia, PA, pp. 73–77 (1978)

9. McDermid, W.: How useful are diagnostic tests on rotating machine insulation? In: 19th Electrical Electronics Insulation Conference, Chicago, IL, pp. 209–211 (1989)

10. Stone, G.C.: Partial discharge diagnostics and electrical equipment insulation condition assessment. IEEE Trans. Dielectr. Electr. Insul. **12**(5), 891–904 (2005)

11. Kraetge, A., Hoek, S., Koch, M., Koltunowicz, W.: Robust measurement, monitoring and analysis of partial discharges in transformers and other HV apparatus. IEEE Trans. Dielectr. Electr. Insul. **20**(6), 2043–2051 (2013)

12. Montanari, G.C., Cavallini, A.: Partial discharge diagnostics: from apparatus monitoring to smart grid assessment. IEEE Electr. Insul. Mag. **29**(3), 8–17 (2013)

13. Cselkó, R., García, M.M., Berta, I.: Partial discharge characteristics of NYCY low-voltage cables. In: 2013 IEEE Electrical Insulation Conference, Ottawa, ON, pp. 138–141 (2013)

14. Gillie, R., Nesbitt, A., Ramirez-Iniguez, R., Stewart, B.G., Kerr, G.: Statistical analysis of simultaneous partial discharge measurements from IEC60270, HFCT and HFCT EMI methods. In: 2015 IEEE Electrical Insulation Conference (EIC), Seattle, WA, pp. 454–457 (2015)

15. Guo, J.J., Boggs, S.A.: High frequency signal propagation in solid dielectric tape shielded power cables. IEEE Trans. Power Deliv. **26**(3), 1793–1802 (2011)

Author Index

Printed in the United States
By Bookmasters